Radio Propagation in the Urban Scenario

For a listing of recent titles in the
Artech House Antennas and Propagation Library,
turn to the back of this book.

Radio Propagation in the Urban Scenario

Giorgio Franceschetti
Antonio Iodice
Daniele Riccio

ARTECH
HOUSE

BOSTON | LONDON
artechhouse.com

Library of Congress Cataloging-in-Publication Data
A catalog record for this book is available from the U.S. Library of Congress

British Library Cataloguing in Publication Data
A catalog record for this book is available from the British Library.

ISBN-13: 978-1-63081-856-2

Cover design by Mark Bergeron

© **2023 Artech House**
685 Canton St.
Norwood, MA

10 9 8 7 6 5 4 3 2 1

*In memory of my wife, Giuliana
(Giorgio)*

*To my parents, Franco and Lucia
(Antonio)*

*To my children,
Camilla and Lorenzo
(Daniele)*

Contents

Preface

This book is aimed at providing telecommunication engineers with the theoretical and practical tools to plan radio coverage in cellular networks, to design radio links, to predict connectivity in wireless networks, and to ensure that the systems they are designing comply with regulations pertaining to exposure of the general public to electromagnetic fields. The following audiences will find the book useful:

- Academic and industry researchers, public institution members, and practicing engineers involved in wireless network planning, with a limited knowledge of electromagnetic propagation in complex environments. They can use the book to acquire the basic concepts governing wireless system design and operations, and to better understand how the wireless channel influences overall system performance. In addition, they can use it as a sort of handbook on available propagation models.
- Students in a master's-level course, planning to work in the wireless communication industry. They can use this book as a textbook on radio propagation in urban scenarios.
- Engineers and technicians in public administration, responsible for ensuring compliance with regulations on the exposure of the general public to electromagnetic fields. They can use it to gain a better

knowledge of wireless systems and to understand the rationale underlying international and national regulations on citizens' exposure to electromagnetic fields.

The reader is assumed to have basic knowledge of mathematics (calculus, algebra, vectors, and matrixes) and physics. A basic knowledge of electromagnetic fields and signal processing is also useful, but it is not mandatory, because Chapter 2 offers a thorough discussion of the fundamentals of electromagnetic propagation.

The book is organized as follows. Chapters 1–3 cover the basic concepts of electromagnetic wave propagation and radiation, as well as the asymptotic techniques employed to study propagation in a complex environment, when obstacles are large with respect to wavelength.

Next, Chapter 4 presents basic models for propagation in free space and over a flat or spherical Earth. In addition, Chapter 5 illustrates some important phenomena involved in wireless propagation in complex environments: multipath, fading, and delay spread. These phenomena characterize the wireless channel in urban areas.

The remainder of the book is more application-oriented: Chapters 6 and 7 discuss available models and tools to predict radio-wave propagation in urban areas, providing the illustrative example of a ray-tracing solver. Subsequently, Chapters 8 and 9 cover the new propagation scenarios arising with 5G telecommunication systems and introduce international and national regulations on the exposure of the general public to electromagnetic fields.

Finally, Chapter 10 summarizes the primary concepts and tools discussed throughout the book with concluding remarks and future perspectives.

Each chapter is complemented by examples of practical applications of the presented concepts.

1

Introduction

Modern society is strongly dependent on wired and wireless digital communications. As we write this book, all the world is engaged in the development and deployment of the latest-generation 5G *new radio* wireless systems. Accordingly, engineers all over the world are facing the challenges and the design problems posed by the necessity of wirelessly connecting a huge number of devices, most of which operate in complex urban scenarios. This book presents models and tools to face such challenges properly. As a matter of fact, it is aimed at providing the reader with the basic concepts governing wireless system design and operations and to illustrate how the wireless channel influences overall telecommunication system performance.

Although the fundamental mathematical tool to study and characterize wireless propagation—namely, the Maxwell's system of equations—dates back to more than 150 years ago, the fast evolution of wireless telecommunications, summarized in Section 1.1, calls for more and more refined methods to master propagation phenomena in complex scenarios like urban areas.

1.1 Historical Notes

The existence of electromagnetic waves was theoretically hypothesized in 1865 by James Clerk Maxwell in his paper "A Dynamical Theory of the

Electromagnetic Field" [1], where he also correctly suggested that light is an electromagnetic wave with very fast oscillations. Experimental demonstration of the existence of electromagnetic waves at radio frequency (radio waves) was provided about 20 years later by Heinrich Hertz [2]. The enormous potential applications of radio waves to telecommunications were then envisioned and exploited by Guglielmo Marconi, who, in the years from 1895 to 1905, conducted a series of experiments on the transmission of radio waves over increasing distances, thus paving the way to wireless telegraphy and to radio broadcasting. Since then, also thanks to the invention of vacuum tube amplifiers (1910s), efficient oscillators (1910s to 1940s), and solid-state amplifiers (1960s), wireless communications have developed at a very rapid pace: In Europe and in the United States, the first commercial radio broadcasting services started their operations in the 1920s; television broadcasting companies started their transmissions in the 1940s in the United States and in the 1950s in Europe; satellite communications began their operations in the 1960s; the first mobile phone systems were introduced in the 1980s; digital communications became ubiquitous around the turn of the millennium; and the beginning of this century saw the development of interconnected wired and wireless data transmission networks, forming the current World Wide Web. As a result, the entire region of the electromagnetic spectrum, from very low frequencies to the millimeter-wave band, is currently employed for wireless communications.

1.2   Electromagnetic Spectrum

Electromagnetic waves are characterized by their frequency (i.e., the number of oscillations per unit of time). In the International System (SI), frequency is measured in cycles per second (i.e., in hertz). As illustrated in Chapter 2, frequency f is strictly related to wavelength λ (i.e., the spatial period of the wave, measured in meters). In fact, for electromagnetic waves propagating in vacuum, $\lambda = c/f$, where c is the velocity of light in vacuum. Accordingly, wavelength (in vacuum) can be used in the place of frequency to characterize an electromagnetic wave.

The entire electromagnetic spectrum of frequencies can be subdivided into a few broad regions (Figure 1.1): waves with frequency lower than 300 GHz (i.e., a wavelength longer than 1 mm) are called radio waves; the region from 300 GHz to about 400 THz[1] (i.e., from 1 mm to about 0.7 μm) corresponds to

---

1. 1 kHz = 10^3 Hz, 1 MHz = 10^6 Hz, 1 GHz = 10^9 Hz, 1 THz = 10^{12} Hz.

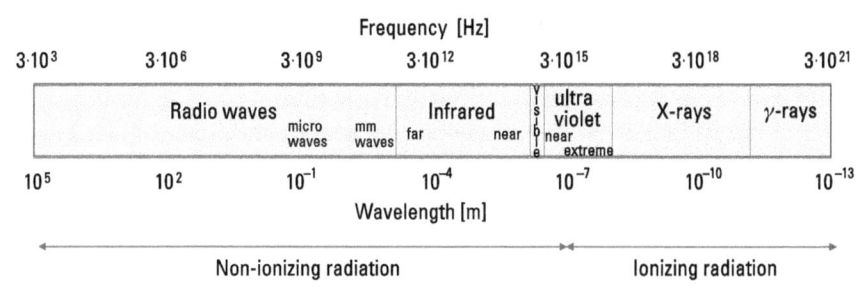

Figure 1.1 Electromagnetic spectrum.

infrared radiation; *visible light* is made of waves of wavelengths from about 0.7 μm (red) to about 0.4 μm (violet); the *ultraviolet* region includes wavelengths shorter than about 0.4 μm and longer than 10 nm; *X-rays* have wavelengths shorter than 10 nm and longer than about 10 pm; and finally *gamma rays* have wavelengths shorter than about[2] 10 pm. For reference, we recall that atomic radii of most elements range from 50 to 100 pm. Radiation at wavelengths shorter than about 100 nm (extreme ultraviolet, X-rays, and gamma rays) is able to ionize atoms and is therefore called *ionizing radiation*, whereas radiation at longer wavelengths (near-ultraviolet, visible-light, infrared radiation, and radio waves) is called *nonionizing radiation*. Ionizing radiation has been determined to be carcinogenic to humans, as the energy carried from the photon at such high frequencies is able to break molecular and atomic bonds[3]; therefore, it is not used for communication purposes. In the telecommunication field, visible light and infrared radiation are currently employed almost exclusively for wired communications via optical-fiber cables (with some exceptions reported in [3]), whereas radio frequencies are used for both wired and wireless communications. Therefore, this book focuses on radio-wave propagation, namely, on waves at frequencies from 3 kHz to 300 GHz. The upper-frequency part of this region is further subdivided into the bands of *microwaves* [from 1 to 30 GHz (i.e., wavelengths from 1 to 30 cm)] and *millimeter waves* [from 30 to 300 GHz (i.e., wavelengths from 1 mm to 1 cm)]; in addition, the entire

---

2. 1 mm = 10^{-3} m, 1 μm = 10^{-6} m, 1nm =10^{-9} m, 1 pm = 10^{-12} m.

3. Each photon (the quantum of the electromagnetic field) carries an energy $E_p = hf$, where h is the Plank's constant. For ionizing radiation, the frequency f is high enough to make this energy sufficient to tear an electron from an atom. Note that, apart from here, throughout this book we consider only the classical (i.e., nonquantum) theory of electromagnetic fields, because it is sufficient in practice for wireless system design.

radio frequency spectrum is conventionally subdivided in several bands, as listed in Table 1.1 [4]. A finer subdivision of the microwave band is sometimes employed, especially in radar and satellite communication applications, as reported in Table 1.2 [5]. It is finally noted that the band from 300 GHz to 3 THz (the terahertz band) is sometimes included in the region of radio waves rather than in the infrared one [6].

Table 1.1
Radio Frequency Bands and Corresponding Wireless Communication Applications

Band	Abbreviation	Frequency Range	Wavelength Range	Wireless Communication Applications
Very low frequency	VLF	3–30 kHz	100–10 km	Time signals, communication with submarines
Low frequency	LF	30–300 kHz	10–1 km	Time signals, AM longwave broadcasting, RFID, amateur radio
Medium frequency	MF	0.3–3 MHz	1–0.1 km	AM medium-wave broadcasting, amateur radio, wireless telegraphy
High frequency	HF	3–30 MHz	100–10 m	Shortwave broadcasts, amateur radio, over-the-horizon aviation communications, RFID
Very high frequency	VHF	30–300 MHz	10–1 m	FM broadcasts, television broadcasts, aviation and mobile communications, amateur radio
Ultra-high frequency	UHF	0.3–3 GHz	1–0.1 m	Television broadcasts, mobile phones, wireless LAN, mobile communications, satellite communications
Super-high frequency	SHF	3–30 GHz	10–1 cm	Wireless LAN, satellite communications, satellite television broadcasting, 5G mobile phones
Extremely high frequency	EHF	30–300 GHz	10–1 mm	High-frequency microwave radio relay, wireless LAN 802.11ad

AM = amplitude modulation, FM = frequency modulation, LAN = local area network, RFID = radio frequency identification.

Table 1.2
Microwave Bands

Band	Frequency range
L	1–2 GHz
S	2–4 GHz
C	4–8 GHz
X	8–12 GHz
K_u	12–18 GHz
K	18–27 GHz
K_a	27–40 GHz

1.3 Radio, Television, Mobile Telephony, and Wireless Networks

Different wireless communication systems often imply different propagation scenarios and, hence, different design tools.

Amplitude modulation (AM) long-wave radio broadcasting was popular in the first half of the last century due to the ability of long waves to propagate far over the horizon, but it required very high transmitted power and the use of large and inefficient antennas. As a result, by now it has been almost completely abandoned.

AM medium-wave radio broadcasting has a shorter range but uses more efficient (although still very large) monopole transmitting antennas, so that it is still used, basically for the simplicity of the receiver devices. However, it is being dismissed in many countries. It relies on ground-wave propagation, which is dealt with in Chapter 4.

Shortwave radio broadcasting is also capable of over-the-horizon transmissions thanks to ionospheric reflection; that is *sky-wave propagation* (see Chapter 4). It is still in use, although it has mostly been abandoned for higher-frequency technologies that support higher bandwidths and transmission rates.

FM radio broadcasting, which operates in the VHF band at about 100 MHz, is currently widely used, as is television broadcasting in the VHF and UHF bands. The latter has recently moved to digital transmissions, according to the digital video broadcasting terrestrial standards (DVB-T and DVB-T2). For both radio and television broadcasting in the VHF and UHF bands, omnidirectional or wide-beam (*sector*) transmitting antennas are placed at elevated sites, most often outside urban areas, in order to obtain links in line of sight or near line of sight for large areas, whose radii are on the order of up to tens of kilometers. Concepts and basic tools presented in Chapters 2–4 can be used to predict the coverage area of these broadcasting stations. Very often, nationwide coverage is obtained by using a network of broadcast relay stations, connected via point-to-point radio links. For such links, the simple Friis formula (see Chapter 2) can be used, provided that no obstacle intersects a volume around the line of sight known as the *first Fresnel ellipsoid* (see Chapter 3). The Friis formula can also be used for the design of satellite television broadcasting systems and satellite links.

Mobile phone cellular systems were developed in the last decades of the twentieth century. First-generation (1G) analog systems, developed in the 1980s, were only able to transmit voice signals. Mobile phones became ubiquitous in the 1990s with the introduction of digital systems able to transmit not only voice, but also short text messages [second generation (2G)]. Around 2000, the transition from circuit switch to packet switch led to the third-generation (3G) systems, which could connect to the internet and hence exchange data and images. The first video calls were also enabled by 3G. In the 2010s, with the fourth-generation (4G) systems, transmission rates grew at a rate that allowed for online gaming and high-definition video streaming. In spite of this very fast evolution of system performances, the propagation scenario has not significantly changed in the moving from 1G to 4G: Employed frequencies are in the UHF band, and omnidirectional or sector antennas, often placed within an urban area, are used, each one covering its cell (i.e., its assigned area). Chapters 5 to 7 consider this propagation scenario, discussing the main tools for the design of the electromagnetic system.

The situation is now changing with the transition to fifth-generation (5G) systems. In order to allow for an even higher data rate and to connect a larger number of nodes (not only human users, but also various kinds of devices, to form the so-called *Internet of Things* (IoT)), 5G systems use the SHF and millimeter-wave bands in addition to the usual UHF band. Furthermore, transmitting and receiving base-station antennas have beamforming capabilities (i.e., they are narrow-beam antennas able to electronically steer their beam). As illustrated in Chapter 8, using higher frequencies leads to a higher transmission rate and allows for easy implementation of beamforming antennas; on the other hand, however, it reduces the coverage area of each antenna, and hence smaller cells must be designed.

Finally, parallel to the development of mobile telephony, wireless LANs (WLANs) based on the IEEE 802.11 family of standards (the so-called Wi-Fi technology) have evolved, originally to connect portable personal computers or laptops to the internet. However, currently available mobile phone devices (smartphones) embed a Wi-Fi chipset, too, so that they can be connected both to the usual cellular telephonic networks and to Wi-Fi *access points*. The latter incorporate small transmitting and receiving antennas working at microwave frequencies usually placed inside buildings, so that an indoor propagation scenario is obtained; see Chapter 6.

1.4 Challenges in the (Electromagnetic) Design of Modern Wireless Networks

Engineers involved in planning wireless communication networks have to deal with two contrasting needs: On one hand, they have to guarantee that the field levels in the area to be served are high enough to obtain a good signal-to-noise (and signal-to-interference) ratio; and on the other hand, they need to ensure that field levels comply with regulations governing human exposure to electromagnetic fields, which are described in Chapter 9. Therefore, models and practical tools to predict field levels produced by a transmitter in a complex environment are needed. When the urban scenario is considered, the statistical approach outlined in Chapter 5, complemented by empirical propagation models described in the first part of Chapter 6, can often be employed. Ray-tracing deterministic algorithms (discussed in the last part of Chapter 6 and in Chapter 7) can be more conveniently used when detailed radio coverage of a specific site is required.

However, the situation is rapidly changing with the development and deployment of 5G networks and with the currently envisioned ways to evolve toward next-generation systems (*beyond 5G* and *6G*; see Chapter 10). The main novelties of 5G (see Chapter 8), from the electromagnetic viewpoint, are the use of higher frequencies, up to the millimeter-wave band; the use of *smart antennas* able to perform digital beamforming; and the higher number of interconnected devices implied by the IoT (i.e., the higher number of transmitters). These novelties pose significant challenges in the electromagnetic design of wireless networks. For instance, the propagation of electromagnetic waves at such higher frequencies is more subject to attenuation by rain and, in the millimeter-wave band, even to attenuation by atmospheric gases (see Chapter 4). Accordingly, electromagnetic solver tools must consider such effects. Still more important, a very strong attenuation is experienced at these frequencies for through-the-wall propagation (see Chapter 2 and Chapter 5); even a rather thin glass window may produce a nonnegligible attenuation at very short wavelengths. Again, this should be properly accounted for by propagation models. Note that empirical models developed for propagation in the UHF bands are not applicable and that new empirical models are being developed for higher frequencies (see Chapter 8). In any case, the use of beamforming, with antennas able to radiate a narrow beam whose direction can be dynamically changed, makes it difficult, in practice, to use the results of such empirical models. Thus, in these cases, it becomes mandatory to use ray-tracing algorithms, provided that they are able to properly incorporate the (variable) antenna patterns (see Chapter 2) and that a sufficiently accurate description of the environment is available.

Finally, the expected higher number of transmitters leads to a higher computational load for software tools implementing ray-tracing algorithms, and this calls for the use of high-performance parallel computation architectures that can be remotely accessed by interested users (*cloud computing*).

Overall, these challenges only strengthen our opinion that it is important to know how to use design tools. To properly employ them, however, it is also necessary to have at least a basic knowledge of the mathematical and physical principles on which they rely.

References

[1]   Maxwell, J. C., "A Dynamical Theory of the Electromagnetic Field," *Philosophical Transactions of the Royal Society of London*, Vol. 155, 1865, pp. 459–512.

[2]   Hertz, H., *Electric Waves: Research on the Propagation of Electric Action With Finite Velocity Through Space*, Macmillan and Company, 1893. (English translation of the original work in German *Untersuchungen über die Ausbreitung der elektrische Kraft*.)

[3]   Khorov, E., and I. Levitsky, "Current Status and Challenges of Li-Fi: IEEE 802.11bb," *IEEE Communications Standards Magazine*, Vol. 6, No. 2, June 2022, pp. 35–41.

[4]   *Nomenclature of the frequency and wavelength bands used in telecommunications,* document ITU-R V.431-8, International Telecommunication Union (ITU), Vocabulary and related subjects V Series, Aug. 2015.

[5]   *IEEE Standard Letter Designations for Radar-Frequency Bands,* IEEE Std 521-2019, Institute of Electrical and Electronics Engineers (IEEE), Nov. 2019.

[6]   *Radio regulations*, International Telecommunication Union (ITU), 2020.

2

Fundamentals of Electromagnetic Propagation and Radiation

This chapter summarizes the theory of electromagnetic propagation and radiation. In particular, the chapter reviews the foundations of the theory, covering the basic concepts that are used in this book. For a more complete and in-depth discussion of electromagnetic theory, the reader is referred to available textbooks on electromagnetic fields and antennas [1–10].

2.1 Maxwell's Equations

All macroscopic electromagnetic phenomena are described by Maxwell's equations,

$$
\begin{cases}
\nabla \times \mathbf{e} = -\dfrac{\partial \mathbf{b}}{\partial t} \\
\nabla \times \mathbf{h} = \dfrac{\partial \mathbf{d}}{\partial t} + \mathbf{j} \\
\nabla \cdot \mathbf{d} = \rho \\
\nabla \cdot \mathbf{b} = 0
\end{cases}
\tag{2.1}
$$

Maxwell's equations govern the space-time evolution of the following fields (five vector and one scalar), each one being a function of the position vector **r** and of the time instant t:

- The *electric field* **e**($\mathbf{r}$, t);
- The *magnetic field* **h**($\mathbf{r}$, t);
- The *electric induction* **d**($\mathbf{r}$, t);
- The *magnetic induction* **b**($\mathbf{r}$, t);
- The *electric current density* **j**($\mathbf{r}$, t);
- The *electric charge density* ρ($\mathbf{r}$, t).

Mechanical actions of the fields over the charges are described by the Lorentz force density

$$\mathbf{f} = \rho\mathbf{e} + \mathbf{j} \times \mathbf{b} \tag{2.2}$$

Current and charge densities are related via the charge velocity **v** as $\mathbf{j} = \rho\mathbf{v}$. In addition, by applying the divergence to both sides of the second equation of (2.1) and using the third equation of (2.1), the current density *continuity equation* is obtained as

$$\nabla \cdot \mathbf{j} + \frac{\partial \rho}{\partial t} = 0 \tag{2.3}$$

which states the conservation of charge.

Finally, we recall that charge and current densities are defined in such a way that the charge q enclosed in a finite volume V [see Figure 2.1(a)] and

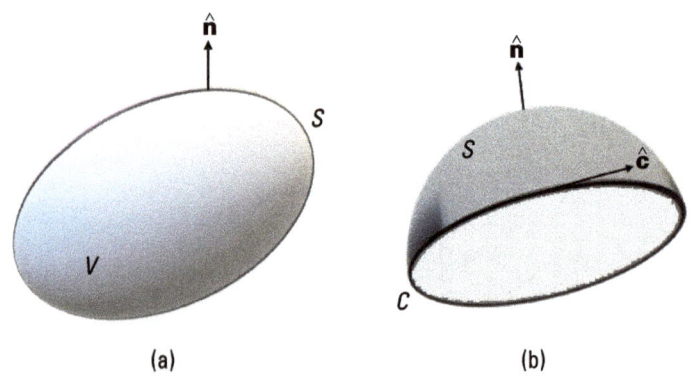

(a)   (b)

Figure 2.1 (a) Volume V bounded by a closed surface S with normal unit vector $\hat{\mathbf{n}}$, and (b) surface S with normal unit vector $\hat{\mathbf{n}}$ bounded by a closed curve C with tangent unit vector $\hat{\mathbf{c}}$.

the current i flowing through a surface S with normal unit vector $\hat{\mathbf{n}}$, [see Figure 2.1(b)] are

$$q = \iiint_V \rho \, dV \quad \text{and} \quad i = \iint_S \mathbf{j} \cdot \hat{\mathbf{n}} \, dS \tag{2.4}$$

respectively. The primary electromagnetic quantities introduced in this chapter are listed in Table 2.1 along with their measurement units.

Table 2.1
Main Electromagnetic Quantities and Their Measurement Units in the International System (SI)

Observable	Units
Current i	ampere [A]
Charge q	coulomb [C] = [A×s]
Voltage v	volt [V] = [joule/C] = [watt/A]
Magnetic flux $\phi_b = \iint_S \mathbf{b} \cdot \hat{\mathbf{n}} dS$	weber [Wb] = [V×s]
Electric field $\mathbf{e}$	volt/m [V/m]
Magnetic field $\mathbf{h}$	ampere/m [A/m]
Electric induction $\mathbf{d}$	coulomb/m² [C/m²]
Magnetic induction $\mathbf{b}$	weber/m² [Wb/m²] = tesla [T]
Current density $\mathbf{j}$	ampere/m² [A/m²]
Charge density ρ	coulomb/m³ [C/m³]
Lorentz force density $\mathbf{f}$	newton/m³ [N/m³]
Surface current density $\mathbf{j}_s$	ampere/m [A/m]
Surface charge density ρ_s	coulomb/m² [C/m²]
Capacity C	farad [F] = [C/V]
Dielectric constant (or permittivity) ε	farad/m [F/m]
Inductance L	henry [H] = [Wb/A]
Magnetic permeability μ	henry/m [H/m]
Impedance Z (resistance R, reactance X)	ohm [Ω] = [V/A]
Admittance Y (conductance G, susceptance B)	siemens [S] = [1/Ω]
Conductivity σ	siemens/m [S/m]
Poynting vector $\mathbf{s}$	watt/m² [W/m²]
Intrinsic impedance ζ	ohm [Ω] = [V/A]
Propagation constant k	[m⁻¹]

In the static case, when fields, charge, and currents are constant, time derivatives in (2.1) are zero, and Maxwell's equations become

$$\begin{cases} \nabla \times \mathbf{e} = 0 \\ \nabla \cdot \mathbf{d} = \rho \end{cases} ; \quad \begin{cases} \nabla \times \mathbf{h} = \mathbf{j} \\ \nabla \cdot \mathbf{b} = 0 \end{cases} \qquad (2.5)$$

The equations in (2.5) show that electrostatic and magnetostatic phenomena are decoupled; in addition, in this case the electric field is conservative, and the voltage of a point B with respect to a point A can be defined as

$$v_{AB} = -\int_{A}^{B} \mathbf{e} \cdot \hat{\mathbf{l}} \, dl \qquad (2.6)$$

where the line integral from A to B is independent of the integration path. In (2.6), $\hat{\mathbf{l}}$ is the unit vector tangent to the integration path.

Conversely, in the dynamic case, electric and magnetic phenomena are coupled via the terms with time derivatives in (2.1). In particular, the term $\partial \mathbf{d}/\partial t$ (*displacement current density*) was introduced by Maxwell on the basis of purely theoretical considerations, and it makes the equations symmetric, and hence more *beautiful*; in addition, with the inclusion of this term, (2.1) predicts that the electromagnetic field can propagate at a long distance as a wave; see Section 2.3.

It is sometimes useful to distinguish between current and charge densities induced by the electromagnetic fields in physical media (*induced densities*, still indicated with $\mathbf{j}$ and ρ), which are unknown functions, and current and charge densities produced by generators via, for example, mechanical or chemical processes (*impressed densities*, or *sources*, indicated with $\mathbf{j}_0$ and ρ_0), which are known functions.

Maxwell's equations must be complemented by the constitutive relationships that link inductions and induced current density to electric and magnetic fields and macroscopically describe the electromagnetic properties of materials:

$$\mathbf{d} = \mathcal{D}\{\mathbf{e},\mathbf{h}\}, \quad \mathbf{b} = \mathcal{B}\{\mathbf{e},\mathbf{h}\}, \quad \mathbf{j} = \mathcal{J}\{\mathbf{e},\mathbf{h}\} \qquad (2.7)$$

In vacuum, we have:

$$\mathbf{d} = \varepsilon_0 \mathbf{e}, \quad \mathbf{b} = \mu_0 \mathbf{h}, \quad \mathbf{j} = 0 \qquad (2.8)$$

where $\varepsilon_0 \cong 8.85 \times 10^{-12}$ farad/m and $\mu_0 \cong 4\pi \cdot 10^{-7}$ henry/m are the vacuum *dielectric constant* (or *permittivity*) and *magnetic permeability*, respectively. Constitutive relationships for some materials are presented in Section 2.2.

2.1.1 Maxwell's Equations in the Frequency Domain

We recall that an arbitrarily time-varying function $f(t)$ can be expressed as a superposition of complex exponential functions with different *angular frequencies* ω:

$$f(t) = \frac{1}{2\pi} \int_{-\infty}^{+\infty} F(\omega) \exp(j\omega t) \, d\omega \tag{2.9}$$

where

$$F(\omega) = \int_{-\infty}^{+\infty} f(t) \exp(-j\omega t) \, dt \tag{2.10}$$

is the Fourier transform (FT) of $f(t)$ and is a complex-valued function.

An important property of the FT is that if $g(t) = df(t)/dt$, then its FT is

$$G(\omega) = \int_{-\infty}^{+\infty} \frac{d\,f(t)}{dt} \exp(-j\omega t) \, dt$$

$$= \left[f(t) \exp(-j\omega t) \right]_{t=-\infty}^{t=+\infty} - \int_{-\infty}^{+\infty} -j\omega f(t) \exp(-j\omega t) \, dt = j\omega F(\omega) \tag{2.11}$$

where we have integrated by parts and assumed that $f(t)$ vanishes at $t = \pm\infty$.

Another important property of the FT is the convolution theorem:

if $u(t) = \int_{-\infty}^{+\infty} g(t-t') f(t') \, dt'$, then

$$U(\omega) = \int_{-\infty}^{+\infty} \int_{-\infty}^{+\infty} g(t-t') f(t') \, dt' \exp(-j\omega t) \, dt = \int_{-\infty}^{+\infty} \int_{-\infty}^{+\infty} g(\tau) f(t') \exp\left[-j\omega(\tau + t') \right] d\tau dt'$$

$$= \int_{-\infty}^{+\infty} g(\tau) \exp(-j\omega \tau) \, d\tau \int_{-\infty}^{+\infty} f(t') \exp(-j\omega t') \, dt' = G(\omega) F(\omega) \tag{2.12}$$

where we have set $\tau = t - t'$.

Finally, by using (2.10), it is easy to verify that if $f(t)$ is a real-valued function, and its FT is Hermitian, [i.e., $F(-\omega) = F^*(\omega)$]. Therefore, we get

$$\int_{-\infty}^{0} F(\omega) \exp(j\omega t) \, d\omega = -\int_{\infty}^{0} F(-\omega) \exp(-j\omega t) \, d\omega = \int_{0}^{\infty} F(-\omega) \exp(-j\omega t) \, d\omega$$

$$= \int_{0}^{\infty} F^*(\omega) \exp(-j\omega t) \, d\omega = \left(\int_{0}^{\infty} F(\omega) \exp(j\omega t) \, d\omega \right)^*$$

so that (2.9) can be rewritten as

$$f(t) = \frac{1}{2\pi}\left(\int_{0}^{+\infty} F(\omega)\exp(j\omega t)\,d\omega + \int_{-\infty}^{0} F(\omega)\exp(j\omega t)\,d\omega\right)$$

$$= \frac{1}{\pi}\mathrm{Re}\left\{\int_{0}^{+\infty} F(\omega)\exp(j\omega t)\,d\omega\right\} \tag{2.13}$$

If $F(\omega)$ is appreciably different from zero only for angular frequencies from $\pm\omega_0 - \Delta\omega/2$ to $\pm\omega_0 + \Delta\omega/2$, then $f(t)$ is said to have a *bandwidth* $\Delta\omega$ (see Figure 2.2). If this bandwidth reduces to zero, then

$$F(\omega) = \pi\left[F\delta(\omega - \omega_0) + F^*\delta(\omega + \omega_0)\right] \tag{2.14}$$

where $\delta(\omega)$ is the Dirac pulse (see Appendix B), and the function $f(t)$ is purely sinusoidal at the fixed angular frequency ω_0 (and period $T = 2\pi/\omega_0$). In fact, the use of (2.14) and of (B.6) in (2.13) gives

$$f(t) = \mathrm{Re}\left\{F\exp(j\omega_0 t)\right\} = |F|\cos(\omega_0 t + \arg\{F\}) \tag{2.15}$$

The complex number F is the *phasor* of the sinusoidal function $f(t)$, with $|F|$ and $\arg\{F\}$ being its modulus and its argument (or phase), respectively.

Definitions and properties [(2.9) to (2.15)] can be easily generalized to the case of vector functions. Therefore, all the fields appearing in Maxwell's equations (2.1) can be expanded according to (2.9): For the electric field

$$e(\mathbf{r},t) = \frac{1}{2\pi}\int_{-\infty}^{+\infty} \mathbf{E}(\mathbf{r},\omega)\exp(j\omega t)\,d\omega \tag{2.16}$$

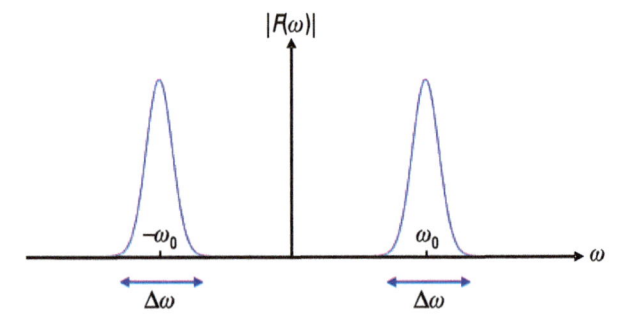

Figure 2.2 Modulus of the Fourier transform of a real function with bandwidth $\Delta\omega$.

and analogous expansions hold for all other fields. By substituting these expansions in (2.1) and (2.3) and equating the integrands, we get Maxwell's equations and the current density continuity equation in the frequency domain

$$\begin{cases} \nabla \times \mathbf{E} = -j\omega\mathbf{B} \\ \nabla \times \mathbf{H} = j\omega\mathbf{D} + \mathbf{J} \\ \nabla \cdot \mathbf{D} = \rho \\ \nabla \cdot \mathbf{B} = 0 \end{cases} \qquad (2.17a)$$

$$\nabla \cdot \mathbf{J} + j\omega\rho = 0 \qquad (2.17b)$$

2.1.2 Sinusoidal Vector Fields, Phasor Vectors, and Polarization

In radio propagation problems, it is often useful to assume that field sources, and then, in view of Maxwell's equations linearity,[1] all fields, are sinusoidal time-varying functions. In fact, in most communication applications, transmitted signals have a narrow *relative bandwidth* (i.e., the ratio between their bandwidth $\Delta\omega$ and their central frequency ω_0 is much smaller than unity), so that in the design of radio links, and to characterize propagation, the sinusoidal approximation is sufficiently accurate for most of the purposes.

For a sinusoidal time-varying electric field $\mathbf{e}(\mathbf{r}, t)$ with a fixed angular frequency ω, by using (2.15) and employing a Cartesian reference system, we can write

$$e(\mathbf{r},t) = \mathrm{Re}\left\{\mathbf{E}(\mathbf{r})\exp(j\omega t)\right\} \qquad (2.18)$$

where

$$\mathbf{E}(\mathbf{r}) = E_x(\mathbf{r})\hat{\mathbf{x}} + E_y(\mathbf{r})\hat{\mathbf{y}} + E_z(\mathbf{r})\hat{\mathbf{z}} = \mathbf{E}_R(\mathbf{r}) + j\mathbf{E}_I(\mathbf{r}) \qquad (2.19)$$

is the complex *phasor vector* of $\mathbf{e}(\mathbf{r}, t)$. In (2.19), $\mathbf{E}_R(\mathbf{r})$ and $\mathbf{E}_I(\mathbf{r})$ are real vectors that represent the real and imaginary parts of the complex vector $\mathbf{E}(\mathbf{r})$.

For all other fields appearing in (2.1), expressions analogous to (2.18) can be written. By using these expressions in (2.1), equations formally identical to (2.17) are obtained (although in this case ω is a constant rather than a variable, so that (2.17) are also the Maxwell's equations in the phasor domain.

---

1. We are here also assuming linearity of the constitutive relationships.

Equation (2.18) can be more explicitly written in terms of field components as

$$\mathbf{e}(\mathbf{r},t) = e_x(\mathbf{r},t)\hat{\mathbf{x}} + e_y(\mathbf{r},t)\hat{\mathbf{y}} + e_z(\mathbf{r},t)\hat{\mathbf{z}} \tag{2.20}$$

with

$$\begin{cases} e_x(\mathbf{r},t) = \mathrm{Re}\left\{E_x(\mathbf{r})\exp(j\omega t)\right\} = \left|E_x(\mathbf{r})\right|\cos\left(\omega t + \arg\left\{E_x(\mathbf{r})\right\}\right) \\ e_y(\mathbf{r},t) = \mathrm{Re}\left\{E_y(\mathbf{r})\exp(j\omega t)\right\} = \left|E_y(\mathbf{r})\right|\cos\left(\omega t + \arg\left\{E_y(\mathbf{r})\right\}\right) \\ e_z(\mathbf{r},t) = \mathrm{Re}\left\{E_z(\mathbf{r})\exp(j\omega t)\right\} = \left|E_z(\mathbf{r})\right|\cos\left(\omega t + \arg\left\{E_z(\mathbf{r})\right\}\right) \end{cases} \tag{2.21}$$

By using (2.19) in (2.18), we get

$$\mathbf{e}(\mathbf{r},t) = \mathbf{E}_R(\mathbf{r})\cos(\omega t) - \mathbf{E}_I(\mathbf{r})\sin(\omega t) \tag{2.22}$$

Equation (2.22) shows that for any t the vector $\mathbf{e}(\mathbf{r}, t)$ lies in the plane defined by the vectors $\mathbf{E}_R$ and $\mathbf{E}_I$, which is called *polarization plane* of the vector. If we choose the xy plane of the reference system coincident with the polarization plane, then the z component of the field is zero, and we can write

$$\begin{cases} e_x(\mathbf{r},t) = \left|E_x(\mathbf{r})\right|\cos\left(\omega t + \arg\left\{E_x(\mathbf{r})\right\}\right) \\ e_y(\mathbf{r},t) = \left|E_y(\mathbf{r})\right|\cos\left(\omega t + \arg\left\{E_y(\mathbf{r})\right\}\right) \end{cases} \tag{2.23}$$

If we interpret e_x and e_y as the coordinates of the tip of the vector, the results of (2.23) can be read as the parametric equations of an ellipse. Therefore, as time varies, the tip of the vector describes an ellipse (see Figure 2.3), and the vector is said to be *elliptically polarized*. The ellipse degenerates into a straight-line segment, and the vector is said to be *linearly polarized*, when either E_x or E_y vanishes, or when the phase difference $\arg\{E_y(\mathbf{r})\} - \arg\{E_x(\mathbf{r})\}$ is $n\pi$, with n integer. In fact, in this last case by taking the ratio of (2.23) we get $e_y/e_x = (-1)^n |E_y|/|E_x|$, which is the equation of a straight line. Conversely, the ellipse degenerates into a circumference, and the vector is said to be *circularly polarized*, when $|E_x| = |E_y| = a$ and $\arg\{E_y(\mathbf{r})\} - \arg\{E_x(\mathbf{r})\} = \pi/2 + n\pi$. In fact, in this case by squaring and summing the equations in (2.23) we get $e_x^2 + e_y^2 = a^2$.

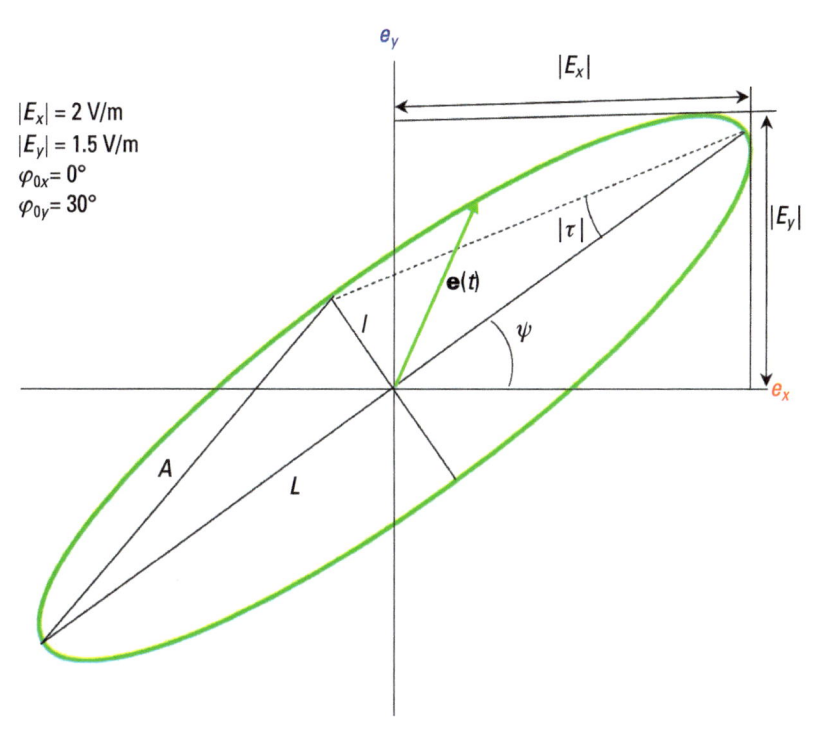

$|E_x| = 2$ V/m
$|E_y| = 1.5$ V/m
$\varphi_{0x} = 0°$
$\varphi_{0y} = 30°$

Figure 2.3 Polarization ellipse. L and I are major and minor semi-axes, respectively; ψ is the orientation angle; τ is the eccentricity angle; and $A = \sqrt{L^2 + I^2} = \sqrt{|E_x|^2 + |E_y|^2}$.

2.2 Electromagnetic Properties of Materials

The electromagnetic properties of media are macroscopically described by the constitutive relationships (2.7). For most material media of interest in propagation problems, the constitutive relationships can be accurately approximated as follows:

$$\mathbf{d}(\mathbf{r},t) = \int_{-\infty}^{+\infty} g_\varepsilon(\mathbf{r}, t - t')\mathbf{e}(\mathbf{r},t')\,dt' \tag{2.24}$$

$$\mathbf{b}(\mathbf{r},t) = \mu(\mathbf{r})\mathbf{h}(\mathbf{r},t) \cong \mu_0 \mathbf{h}(\mathbf{r},t) \tag{2.25}$$

$$\mathbf{j}(\mathbf{r},t) = \sigma(\mathbf{r})\mathbf{e}(\mathbf{r},t) \tag{2.26}$$

with $g_\varepsilon(\mathbf{r}, t) = 0$ for $t < 0$. In (2.24) to (2.26), $g_\varepsilon(\mathbf{r}, t)$ is the medium *dielectric Green function*, $\mu(\mathbf{r})$ is the medium magnetic permeability, and $\sigma(\mathbf{r})$ is the medium *conductivity*. Materials for which σ is null (or negligibly small) are said to be *insulators* or *dielectrics*, whereas materials for which σ is nonnull are said to be *conductors*.

Equation (2.24) implies that, with regard to the relation between **d** and **e**, the medium is *linear*; *local* (i.e., the value of the induction at a given point **r** depends on the value of the field at the same point **r** only); *stationary* (i.e., any time delay of the field implies the same time delay of the corresponding induction); *isotropic* (i.e., any rotation of the field implies the same rotation of the corresponding induction); and *dispersive* (i.e., *with memory*, so that the value of the induction at a given time t depends not only on the value of the field at time t, but also on the values of the field at times t' preceding t). Equations (2.25) and (2.26) imply that, with regard to the relations between **b** and **h**, and between **j** and **e**, the same medium properties hold, except for the fact that the medium is *nondispersive* (i.e., *with no memory*, so that the value of the induction, or current density, at a given time t depends on the value of the field at the same time t only). In addition, (2.25) shows that the magnetic permeability of most materials relative to our focus here is very close to the one of vacuum, so that they are *nonmagnetic* materials.

Examples of materials showing all the properties mentioned above for a wide range of field values are dielectrics listed in Table 2.2 (see Section 2.2.2) and conductors and semiconductors listed in Table 2.3 (see Section 2.2.3). Noticeable exceptions are *magnetic materials* (e.g., iron, ferrite, and magnetite), which have a magnetic permeability much higher than that of a vacuum and are often strongly *nonlinear*.

The integral relation (2.24) simplifies into an algebraic relation if we move to the phasor (or to the frequency) domain and make use of the convolution theorem (2.12). We then obtain the following expressions of the constitutive relationships in the phasor domain:

$$\mathbf{D}(\mathbf{r}) = \varepsilon(\mathbf{r},\omega)\mathbf{E}(\mathbf{r}) \tag{2.27}$$

$$\mathbf{B}(\mathbf{r}) = \mu(\mathbf{r})\mathbf{H}(\mathbf{r}) \cong \mu_0\mathbf{H}(\mathbf{r}) \tag{2.28}$$

$$\mathbf{J}(\mathbf{r}) = \sigma(\mathbf{r})\mathbf{E}(\mathbf{r}) \tag{2.29}$$

In (2.27), $\varepsilon(\mathbf{r}, \omega) = \varepsilon'(\mathbf{r},\omega) - j\varepsilon''(\mathbf{r},\omega)$ is the Fourier transform of $g_\varepsilon(\mathbf{r},t)$ evaluated at the angular frequency of the sinusoidal fields, and it is called the medium *complex dielectric constant* or *complex permittivity*.

Equations (2.28) and (2.29) imply that **b** and **h** are aligned, as well as **j** and **e**. Conversely, (2.27) ensures that **d** and **e** share the same polarization state, but they may be not aligned.

If the medium is *homogeneous*, then $\varepsilon(\omega)$, μ, and σ are independent of **r**. In addition, some media, at least in a limited range of frequencies, can be approximated as nondispersive even with regard to the relation between **d** and **e**. For nondispersive media, $\mathbf{d} = \varepsilon\mathbf{e}$, and in the phasor domain the dielectric constant is real and independent of frequency, as opposed to dispersive media, whose dielectric constant is complex and frequency dependent, as shown above.

Finally, in some particular cases, *anisotropic* media should be considered. Their permittivity is a matrix:

$$\mathbf{D}(\mathbf{r}) = \boldsymbol{\varepsilon}(\mathbf{r},\omega)\mathbf{E}(\mathbf{r}) \ \text{ or } \ \mathbf{D}(\mathbf{r}) = \boldsymbol{\varepsilon}(\mathbf{r})\mathbf{E}(\mathbf{r}) \tag{2.30}$$

depending on whether the medium is dispersive or nondispersive. Therefore, while for isotropic nondispersive media, electric field and induction are aligned, for anisotropic media they are not. Noticeable examples of anisotropic media are magnetized ferrite, which is used in some microwave circuit components, and magnetized plasma (see Section 2.2.5).

It is usually set (both for isotropic and anisotropic media) $\varepsilon(\mathbf{r}, \omega) = \varepsilon_r(\mathbf{r}, \omega)\varepsilon_0$, where $\varepsilon_r(\mathbf{r}, \omega)$ is called *relative dielectric constant* or *relative permittivity*.

2.2.1 Power Losses in Materials, Power Flux, and Energy Conservation

The Poynting vector **s** is defined as

$$\mathbf{s} = \mathbf{e} \times \mathbf{h} \tag{2.31}$$

By computing the divergence of **s** via (A.3), using Maxwell's equations (2.1) with the constitutive relationships of a nondispersive medium, and integrating over a fixed volume V bounded by the surface S with normal unit vector $\hat{\mathbf{n}}$, we get

$$-\iiint_V \mathbf{e} \cdot \mathbf{j}_0 \, dV = \iiint_V \sigma |\mathbf{e}|^2 \, dV + \frac{d}{dt} \iiint_V \left(\frac{1}{2}\varepsilon|\mathbf{e}|^2 + \frac{1}{2}\mu|\mathbf{h}|^2 \right) dV + \oiint_S \mathbf{s} \cdot \hat{\mathbf{n}} \, dS$$

$$\tag{2.32}$$

This result, known as *Poynting's theorem*, is a statement of energy conservation. In fact, it can be demonstrated [1] that the left-hand side of (2.32) is the power delivered by the sources to the electromagnetic field in the volume V, whereas the three terms at the right-hand side of (2.32) represent the power dissipated in the conducting medium due to the current flowing through it (*Joule effect*), the time derivative of the electromagnetic energy stored in

V, and the power flowing outside of the volume V through its boundary S, respectively. Therefore, the Poynting vector can be interpreted as the electromagnetic power flux density, measured in watt/m^2. In addition, $\sigma|\mathbf{e}|^2$ is the power loss per volume unit (watt/m^3) in the conducting material, $\frac{1}{2}\varepsilon|\mathbf{e}|^2$ is the electric energy density (joule/m^3), and $\frac{1}{2}\mu|\mathbf{h}|^2$ is the magnetic energy density (joule/m^3). Note that these expressions of the energy densities only hold for nondispersive media.

An equation similar to (2.32) can also be obtained for dispersive media if sinusoidal fields are considered and we move to the phasor domain. In fact, the Poynting vector $\mathbf{S}$ in the phasor domain is defined as

$$\mathbf{S} = \frac{1}{2}\mathbf{E} \times \mathbf{H}^* \tag{2.33}$$

By computing the divergence of $\mathbf{S}$ via (A.3), using Maxwell's equations in the phasor domain (2.17) with the constitutive relationships of a dispersive medium [(2.27) to (2.29)], and integrating over a fixed volume V bounded by the surface S with normal unit vector $\hat{\mathbf{n}}$, we get a complex equality whose real and imaginary parts are (*Poynting's theorem in the phasor domain*)

$$\mathrm{Re}\left\{-\iiint_V \frac{1}{2}\mathbf{E} \cdot \mathbf{J}_0^* \, dV\right\} = \iiint_V \frac{1}{2}\sigma|\mathbf{E}|^2 \, dV + \iiint_V \frac{1}{2}\omega\varepsilon''|\mathbf{E}|^2 \, dV + \mathrm{Re}\left\{\oiint_S \mathbf{S} \cdot \hat{\mathbf{n}} \, dS\right\} \tag{2.34}$$

$$\mathrm{Im}\left\{-\iiint_V \frac{1}{2}\mathbf{E} \cdot \mathbf{J}_0^* \, dV\right\} = 2\omega\iiint_V \left(\frac{1}{4}\mu|\mathbf{H}|^2 - \frac{1}{4}\varepsilon'|\mathbf{E}|^2\right) dV + \mathrm{Im}\left\{\oiint_S \mathbf{S} \cdot \hat{\mathbf{n}} \, dS\right\} \tag{2.35}$$

To interpret these expressions, recall that the time average over one period $T = 2\pi/\omega$ of the product of two sinusoidal functions $v(t)$ and $i(t)$ is

$$<v(t)i(t)>_T = \frac{1}{T}\int_0^T v(t)i(t)\,dt = \frac{1}{2}|V||I|\cos(\arg\{V\} - \arg\{I\}) = \mathrm{Re}\left\{\frac{1}{2}VI^*\right\} \tag{2.36}$$

and that if v and i are the input voltage and current of a passive device, then $VI^*/2$ is defined as the *complex power* delivered to the device. The real part of the complex power is named *active power*, and, in view of (2.36), it is the time average of power, whereas its imaginary part is named *reactive power*. For a fixed modulus $|V||I|/2$ of the complex power, a higher reactive power implies a lower active power, and hence a lower efficiency of the power transfer to the device, since a lower average power is obtained with the same peak values, $|V|$ and $|I|$, of voltage and current.

In view of these results, we can state that the left-hand side of (2.34) is the average power delivered by the sources to the electromagnetic field in V, whereas the three terms at the right-hand side of (2.34) represent the average power dissipated in the conducting medium due to the Joule effect, the average power dissipated in the medium due to *dielectric hysteresis*, and the average power flowing outside of the volume V through its boundary S, respectively. Analogously, the left-hand side of (2.35) is the reactive power delivered by the sources to the electromagnetic field in V, and the second term of the right-hand side of (2.35) is the reactive power flowing outside of the volume V through its boundary S. With regard to the first term, if the medium is nondispersive it can be interpreted as 2ω-times the difference between the average magnetic and electric energies stored in the volume V; otherwise, it has no direct physical meaning.

In conclusion, the Poynting vector in the phasor domain can be interpreted as the electromagnetic complex power flux density. In addition, $\frac{1}{2}\sigma|\mathbf{E}|^2$ is the average power loss per volume unit due to the Joule effect, and $\frac{1}{2}\omega\varepsilon''|\mathbf{E}|^2$ is the average power loss per volume unit due to dielectric hysteresis. The latter is only present if the medium is dispersive ($\varepsilon'' \neq 0$) and in the dynamic case ($\omega \neq 0$), whereas it vanishes if the medium is nondispersive ($\varepsilon'' = 0$) or in the static case ($\omega = 0$). It is important to highlight that the imaginary part (with changed signum) ε'' of the dielectric constant is related to power loss in the material. Therefore, dispersive media are also lossy.

2.2.2 Dielectric Materials

Strictly speaking, all dielectric media, except vacuums, are dispersive and, hence, lossy. The physical reason is that propagation of the electromagnetic fields in any material implements mechanical actions, and some energy is dissipated. Different models to describe the dependence of the dielectric constant on the frequency (*relaxation models*) are available. For materials made of polar molecules (e.g., water) the Debye model can be used

$$\varepsilon_r(\omega) = \varepsilon_{rb} + \frac{\varepsilon_{rl} - \varepsilon_{rb}}{1 + j\omega/\omega_0} = \varepsilon_{rb} + \frac{\varepsilon_{rl} - \varepsilon_{rb}}{1 + (\omega/\omega_0)^2} - j\frac{(\varepsilon_{rl} - \varepsilon_{rb})(\omega/\omega_0)}{1 + (\omega/\omega_0)^2} \quad (2.37)$$

where ε_{rl} and ε_{rh} are the relative dielectric constants at low ($\omega \ll \omega_0$) and high ($\omega \gg \omega_0$) frequencies, respectively, and ω_0 is a characteristic angular frequency, at which the real part of dielectric constant shows a rapid variation and its imaginary part has a peak [see Figure 2.4(a)]. Accordingly, this frequency corresponds to a peak of losses in the material.

For water, $\omega_0 \cong 2\pi \cdot 22$ GHz. For materials made of simple nonpolar molecules (e.g., oxygen, nitrogen, some building materials, and dry soil) the driven harmonic oscillator model can be used.

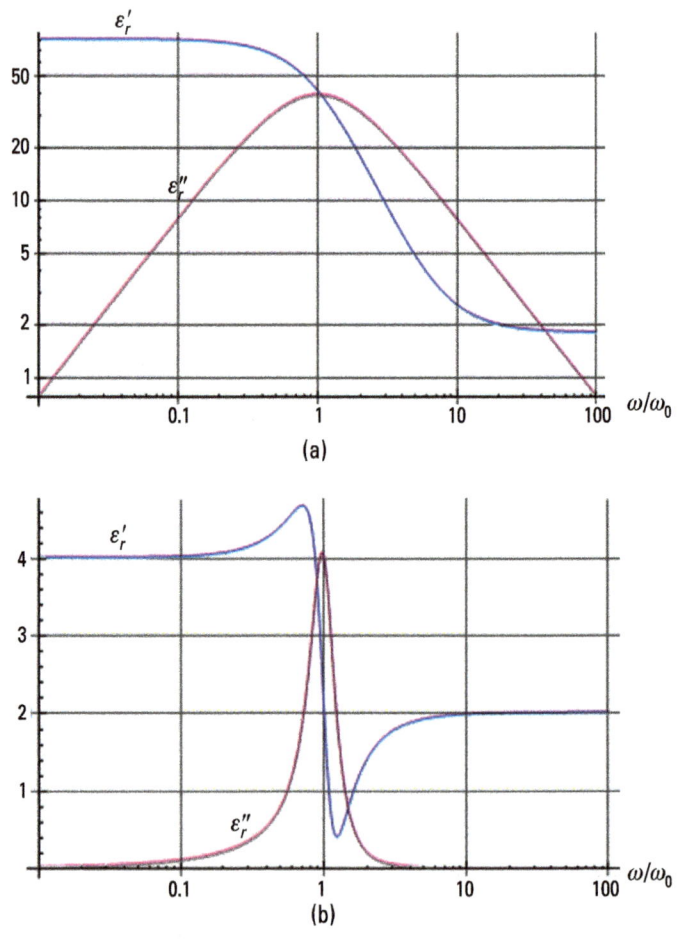

Figure 2.4   Complex permittivity versus normalized frequency for (a) a material made of polar molecules (Debye relaxation model) and (b) a material made of nonpolar molecules (driven harmonic oscillator model).

$$\varepsilon_r(\omega) = \varepsilon_{rb} + \frac{\varepsilon_{rl} - \varepsilon_{rb}}{1 - \omega^2/\omega_0^2 + j\gamma\omega/\omega_0}$$

$$= \varepsilon_{rb} + \frac{(\varepsilon_{rl} - \varepsilon_{rb})(1 - \omega^2/\omega_0^2)}{(1 - \omega^2/\omega_0^2)^2 + (\gamma\omega/\omega_0)^2} - j\frac{(\varepsilon_{rl} - \varepsilon_{rb})(\gamma\omega/\omega_0)}{(1 - \omega^2/\omega_0^2)^2 + (\gamma\omega/\omega_0)^2}$$

$$(2.38)$$

where γ is the *damping parameter*, a dimensionless constant related to losses, and the other constants have the same meaning as in (2.37) [see Figure 2.4(b)]. For oxygen, $\omega_0 \cong 2\pi \cdot 60$ GHz, whereas for the other aforementioned materials, ω_0 is much higher, beyond the microwave and millimeter-wave ranges of frequencies that are of interest in this book, so that $\varepsilon_r(\omega) \cong \varepsilon_{rl}$.

For polymers and biological tissues, the Cole-Cole model [11] can be employed:

$$\varepsilon_r(\omega) = \varepsilon_{rb} + \frac{\varepsilon_{rl} - \varepsilon_{rb}}{1 + (j\omega/\omega_0)^{1-\alpha_c}} \qquad (2.39)$$

When the parameter α_c is zero, the Cole-Cole model reduces to the Debye model.

Note that parameters appearing in (2.37) to (2.39) are, in general, dependent on temperature. In addition, for gases, ε_{rl} and ε_{rb} are also dependent on pressure. The low-frequency relative dielectric constants of some materials at 20°C are reported in Table 2.2.

2.2.3 Conductors

Conductivity of different materials has a wide variability. Metals have very high conductivity, with values of σ on the order of 10^7 S/m, but also ionic solutions (e.g., salt water) are rather good conductors. Conductivity is, in general, dependent on temperature, and for metals it decreases with temperature. However, for some substances, called *semiconductors* (e.g., silicon, germanium), conductivity increases with temperature over a wide range of values. This is explained by considering that, in both metals and semiconductors, the current conduction is determined by the presence of a fraction of atoms' electrons (*conduction electrons*) free to move in a dense crystal lattice of positive ions. Conductivity is directly proportional to the number of conduction electrons per volume unit and to the average time interval between electron-lattice collisions. The latter decreases with temperature, while the former is independent of temperature in metals and increases with temperature in semiconductors.

Table 2.2
Low-Frequency Relative Dielectric Constants of Some
Materials at a Temperature of 20°C

Material	Low-Frequency Relative Dielectric Constant at 20°C
Air (at sea level)	1.00054
Polyethylene	2.25
Polystyrene	2.4–2.7
Paper	3.5
Silicon dioxide	3.7
Concrete	4.5
Pyrex (Glass)	4.7 (3.7–10)
Rubber	7
Diamond	5.5–10
Salt	3–15
Graphite	10–15
Silicon	11.68
Methanol	30
Water	80.1

Table 2.3
Conductivities of Some Materials at a Temperature
of 20°C

Material	Conductivity at 20°C (siemens/m, $S \cdot m^{-1}$)
Silver	63.01×10^{6}
Copper	59.6×10^{6}
Annealed copper	58.0×10^{6}
Gold	45.2×10^{6}
Aluminum	37.8×10^{6}
Germanium	2.17
Silicon	4.35×10^{-4}
Sea water	5 (for a salinity of 35g/kg)
Fresh water	0.0005–0.05
Distilled water	5.5×10^{-6}

Therefore, the metals conductivity decreases with temperature, whereas in semiconductors the effect of an increased number of conduction electrons is usually prevailing, so that semiconductors' conductivity usually increases with temperature.

Conductivities of some materials at 20°C are reported in Table 2.3.

In the phasor domain, the conductivity can be embedded in a complex equivalent dielectric constant:

$$\varepsilon_{eq} = \varepsilon + \frac{\sigma}{j\omega} = \varepsilon_0 \left(\varepsilon_r - j\frac{\sigma}{\omega\varepsilon_0} \right) = \varepsilon_0 \varepsilon_{req} \tag{2.40}$$

In fact, accounting for the constitutive relationships (2.27–29) and for the continuity equation (2.17b), the second and third Maxwell's equations in the phasor domain can be written as

$$\nabla \times \mathbf{H} = j\omega\varepsilon\mathbf{E} + \sigma\mathbf{E} + \mathbf{J}_0 = j\omega\left(\varepsilon + \frac{\sigma}{j\omega} \right)\mathbf{E} + \mathbf{J}_0 = j\omega\varepsilon_{eq}\mathbf{E} + \mathbf{J}_0$$

$$\nabla \cdot \varepsilon\mathbf{E} - \rho = \nabla \cdot \varepsilon\mathbf{E} + \frac{\nabla \cdot \mathbf{J}}{j\omega} = \nabla \cdot \left(\varepsilon + \frac{\sigma}{j\omega} \right)\mathbf{E} = \nabla \cdot \varepsilon_{eq}\mathbf{E} = \rho_0 \tag{2.41}$$

Therefore, induced current and charge densities do not appear explicitly in this convenient formulation of Maxwell's equations, and the imaginary part of ε_{eq} accounts for losses due to both dielectric hysteresis and the Joule effect. If $\sigma/\omega\varepsilon_0 \gg 1$, then the material is said to be a *good conductor*. Note that this definition is frequency-dependent. Metals are good conductors up to frequencies of the order of 10^{16} Hz (i.e., up to ultraviolet frequencies), and sea water is a good conductor up to microwave frequencies. Other conductors, with conductivity on the order of fractions of siemens/m, such as fresh water and soil, are good conductors at low frequencies, but they are not for frequencies on the order of megahertz or higher. For a good conductor, $\varepsilon_{eq} \cong \sigma/j\omega$.

If, on the contrary, $\sigma/\omega\varepsilon_0 \ll 1$, then, rather than a conductor, the material should be considered as a dielectric with small losses. For example, this is the case of soil and many building materials at frequencies on the order of gigahertz.

2.2.4 Perfect Electric Conductors (PECs)

A PEC is defined as a material with $\sigma \to \infty$ (i.e., with infinite conductivity). This is an ideal material, although there are materials, named *superconductors*, that, operating at extremely low temperatures, are very near to being PEC. In addition, good conductors are well-approximated by PEC in most propagation and radiation problems (see also Section 2.3.5).

In view of (2.26), or (2.29), a finite nonzero electric field in a PEC would imply an infinite value of the current density; accordingly, in a PEC, the electric field must be equal to zero. The first Maxwell equation implies that, at nonzero frequency, also the magnetic induction, and hence the magnetic field, must be zero in a PEC.

2.2.5 Plasma

A plasma is a neutral ionized gas in which electrons and positive ions are free to move. Since the mass of positive ions is much greater than the mass of the electrons, ions acceleration is negligible with respect to the electrons acceleration, and the electromagnetic properties of a plasma are substantially determined by the movement of electrons.

The constitutive relationships of a plasma can be obtained by using a simple model of the electrons' motion under the action of an electromagnetic field (see, e.g., [1]). The resulting relative dielectric constant is

$$\varepsilon_r(\omega) = 1 - \frac{\omega_p^2}{\omega^2\left(1 - j\frac{\nu}{\omega}\right)} = 1 - \frac{\omega_p^2}{\omega^2 + \nu^2}\left(1 + j\frac{\nu}{\omega}\right) \cong 1 - \frac{\omega_p^2}{\omega^2} - j\frac{\omega_p^2 \nu}{\omega^3} \qquad (2.42)$$

with

$$\omega_p^2 = \frac{N_e q_e^2}{m_e \varepsilon_0} \qquad (2.43)$$

where ν is the average number of collisions per second among electrons and ions (or neutral, nonionized atoms); ω_p is called *plasma angular frequency*; q_e and m_e are the electron charge and mass, respectively; and N_e is the number of electrons per volume unit. The collision rate ν is usually very small due to the low density of plasma, so that, except for extremely low frequencies, the imaginary part of the dielectric constant can be neglected.

An examination of (2.42) is in order: If the frequency is smaller than the plasma frequency, the dielectric constant of the plasma is negative, whereas it is very close to the one of vacuum when the frequency is much larger than the plasma frequency. In addition, q_e, m_e, and ε_0 are universal constants, so that the plasma frequency only depends on the number of electrons per volume unit: $\omega_p = p\sqrt{N_e}$, with $p = q_e / \sqrt{m_e \varepsilon_0} \cong 2\pi \cdot 9 \text{ m}^{1.5}/\text{s}$.

If a uniform constant (i.e., static) magnetic field $\mathbf{H}_0$ is present, the ionized gas is said to be *magnetized plasma*, and it becomes an anisotropic medium. Accordingly, its relative dielectric constant is a matrix. By using a Cartesian reference system whose z-axis coincides with the direction of $\mathbf{H}_0$ (i.e., $\mathbf{H}_0 = H_0\hat{z}$), and neglecting collisions, it turns out that

$$\boldsymbol{\varepsilon}_r = \begin{pmatrix} \varepsilon_1 & -j\varepsilon_2 & 0 \\ j\varepsilon_2 & \varepsilon_1 & 0 \\ 0 & 0 & \varepsilon_3 \end{pmatrix} \tag{2.44}$$

where

$$\varepsilon_1 = 1 - \frac{\omega_p^2}{\omega^2 - \omega_c^2}$$

$$\varepsilon_2 = \frac{\omega_p^2 \omega_c}{(\omega^2 - \omega_c^2)\omega}, \text{ with } \omega_c = \frac{\mu_0 H_0 q_e}{m_e} \tag{2.45}$$

$$\varepsilon_3 = 1 - \frac{\omega_p^2}{\omega^2}$$

The parameter ω_c is called the *cyclotron angular frequency.*

The ionosphere, a layer of the Earth's upper atmosphere extending from about 60 to about 1,000 km altitude (see Section 4.5), consists of ionized gases, and, considering the Earth's magnetic field, it can be modeled as a magnetized plasma.

For the ionosphere, $\omega_c \cong 2\pi \cdot 1.4$ MHz, and the maximum value of N_e is on the order of 10^{12} m^{-3}.

2.2.6 Boundary Between Two Media and Boundary Condition on PEC's Surface

Let us consider a smooth surface S, with normal unit vector $\hat{n}$, separating two media of different characteristics, medium 1 and medium 2. On such a surface, the electromagnetic fields are discontinuous and nondifferentiable with respect to space variables $\mathbf{r}$, so that Maxwell's equations (2.1) or (2.17) cannot be used on S. Therefore, they must be solved separately in the two media, and then proper matching conditions must be imposed on S. The latter are obtained by resorting to the integral form of Maxwell's equations (see, e.g., [1]) that do not require field continuity and differentiability. If $\mathbf{e}_1$, $\mathbf{h}_1$, $\mathbf{d}_1$, and $\mathbf{b}_1$ are the fields and inductions in medium 1, and $\mathbf{e}_2$, $\mathbf{h}_2$, $\mathbf{d}_2$, and $\mathbf{b}_2$ are the fields and inductions in medium 2, it turns out that

$$\hat{\mathbf{n}} \times \left(\mathbf{e}_2 - \mathbf{e}_1 \right) \Big|_S = 0$$
$$\hat{\mathbf{n}} \times \left(\mathbf{h}_2 - \mathbf{h}_1 \right) \Big|_S = \mathbf{j}_s$$
$$\hat{\mathbf{n}} \cdot \left(\mathbf{d}_2 - \mathbf{d}_1 \right) \Big|_S = \rho_s \qquad (2.46)$$
$$\hat{\mathbf{n}} \cdot \left(\mathbf{b}_2 - \mathbf{b}_1 \right) \Big|_S = 0$$

where $\mathbf{j}_s$ is the *surface electric current (linear) density* possibly flowing on the surface S, measured in amperes per meter; and ρ_s is the *surface electric charge (surface) density* possibly laying on the surface S, measured in coulomb/m^2. Equation (2.46) indicates that, on the boundary between two media, the electric field tangential component and the magnetic induction normal component are continuous; the magnetic field tangential component and the electric induction normal component are continuous, too, unless surface currents and charges are present on S, in which case the discontinuities of the magnetic field tangential component and the electric induction normal component are equal to the surface current density and the surface charge density, respectively.

On the boundary between two nonideal media, surface current and charges cannot be induced by the electromagnetic field, so that tangential components of both electric and magnetic fields are continuous. However, if medium 1 is a PEC, then $\mathbf{e}_1 = 0$, and from the first of (2.46) we understand that the tangential component of the electric field on S is always zero. Therefore, we can state that the boundary condition on the surface of a PEC is

$$\hat{\mathbf{n}} \times \mathbf{e} \Big|_S = 0, \quad \text{or} \quad \hat{\mathbf{n}} \times \mathbf{E} \Big|_S = 0 \qquad (2.47)$$

in the time and phasor domains, respectively. In addition, from the second of (2.46), by recalling that in a PEC at nonzero frequency also the magnetic field is null, the tangential component of the dynamic magnetic field on S is

$$\hat{\mathbf{n}} \times \mathbf{H} \Big|_S = \mathbf{J}_s \qquad (2.48)$$

so that the electromagnetic field induces surface currents on the surface of a PEC to support the discontinuity of the magnetic field tangential component.

2.3 Plane-Wave Propagation, Reflection, and Transmission

Let us consider the phasor-domain Maxwell's equations with no sources for a linear, local, stationary, isotropic, and homogeneous medium:

$$\begin{cases} \nabla \times \mathbf{E} = -j\omega\mu\mathbf{H} \\ \nabla \times \mathbf{H} = j\omega\varepsilon\mathbf{E} \\ \nabla \cdot \mathbf{E} = 0 \\ \nabla \cdot \mathbf{H} = 0 \end{cases} \tag{2.49}$$

By applying the curl to both sides of the first of (2.49) and then using the second of (2.49) we get

$$\nabla \times \nabla \times \mathbf{E} = -j\omega\mu\nabla \times \mathbf{H} = \omega^2\varepsilon\mu\mathbf{E} \tag{2.50}$$

By using (A.9) and the third of (2.49), we obtain

$$\nabla^2\mathbf{E} + k^2\mathbf{E} = 0 \tag{2.51}$$

where we have set

$$k = \omega\sqrt{\varepsilon\mu} \tag{2.52}$$

with k being the *propagation constant* of the considered homogeneous medium. Equation (2.52) implies that the propagation constant is real for nondispersive media and complex for dispersive media (because in this case ε is complex), so that, in general, $k = \beta - j\alpha$.

Equation (2.51) is the *Helmholtz homogeneous equation*. A solution of (2.51) is

$$\mathbf{E}(\mathbf{r}) = \mathbf{E}_0\exp(-j\mathbf{k}\cdot\mathbf{r}) = \mathbf{E}_0\exp\left[-j\left(k_x x + k_y y + k_z z\right)\right] \tag{2.53}$$

with

$$\mathbf{k}\cdot\mathbf{k} = k^2 \quad \text{i.e.,} \quad k_x^2 + k_y^2 + k_z^2 = k^2 \tag{2.54}$$

where $\mathbf{k}$ is the *propagation vector* (or *wavenumber vector*), that in general is complex: $\mathbf{k} = \mathbf{k}_R - j\mathbf{k}_I$.

To verify that (2.53) to (2.54) is a solution of (2.51), let us use (2.53) in (2.51). Noting that application of the operator ∇ to the expression (2.53) corresponds to a multiplication by $-j\mathbf{k}$, we have

$$\left(-\mathbf{k}\cdot\mathbf{k}+k^2\right)\mathbf{E}_0\exp\left(-j\mathbf{k}\cdot\mathbf{r}\right)=0 \tag{2.55}$$

which, in view of (2.54), is satisfied for any value of $\mathbf{E}_0$.

By using (2.53) in the first and third Maxwell's equations (2.49) we get

$$\mathbf{H}=\frac{\mathbf{k}\times\mathbf{E}_0}{\omega\mu}\exp\left(-j\mathbf{k}\cdot\mathbf{r}\right)$$
$$\mathbf{k}\cdot\mathbf{E}_0=0 \tag{2.56a}$$

Equivalently, we can apply the curl to both sides of the second of Maxwell's equations (2.49), obtain an expression for $\mathbf{H}$ analogous to (2.53) and (2.54) and then use it in the second and fourth of (2.49), getting

$$\mathbf{E}=\frac{\mathbf{H}_0\times\mathbf{k}}{\omega\varepsilon}\exp\left(-j\mathbf{k}\cdot\mathbf{r}\right)$$
$$\mathbf{k}\cdot\mathbf{H}_0=0 \tag{2.56b}$$

By summarizing the above results, we can state that a solution of source-free Maxwell's equations in a linear, local, stationary, isotropic, and homogeneous medium is given by

$$\mathbf{E}(\mathbf{r})=\mathbf{E}_0\exp\left(-j\mathbf{k}\cdot\mathbf{r}\right)$$
$$\mathbf{H}(\mathbf{r})=\mathbf{H}_0\exp\left(-j\mathbf{k}\cdot\mathbf{r}\right) \tag{2.57}$$

provided that the constant complex vectors $\mathbf{E}_0$, $\mathbf{H}_0$, and $\mathbf{k}$ satisfy the following relations:

$$\begin{cases}\mathbf{k}\cdot\mathbf{k}=k^2\\ \mathbf{H}_0=\dfrac{\mathbf{k}\times\mathbf{E}_0}{\omega\mu}\\ \mathbf{k}\cdot\mathbf{E}_0=0\end{cases}\quad\text{or equivalently}\quad\begin{cases}\mathbf{k}\cdot\mathbf{k}=k^2\\ \mathbf{E}_0=\dfrac{\mathbf{H}_0\times\mathbf{k}}{\omega\varepsilon}\\ \mathbf{k}\cdot\mathbf{H}_0=0\end{cases} \tag{2.58}$$

Such a solution is called a *plane wave*.

We explicitly note that plane waves cannot exist in practice, since they require infinite power to be generated. However, as discussed in Section 2.4.3, they are a good approximation of actual fields in many situations of practical interest; in addition, it can be shown that in a source-free homogeneous half-space, or also in a source-free homogeneous plane layer, any electromagnetic field can be expressed as a superposition of plane waves [1].

2.3.1 Homogeneous Plane Waves

The condition (2.54) is certainly satisfied by letting

$$\mathbf{k} = k\hat{\mathbf{k}} \tag{2.59}$$

although this is not the only possibility (see Section 2.3.2). Equation (2.59) implies that $\mathbf{k}$ is real in a lossless medium, and complex, with parallel real and imaginary parts, in a lossy one.

A plane wave whose wavenumber vector satisfies (2.59) is called *homogeneous plane wave*. The direction $\hat{\mathbf{k}}$ of the wavenumber vector is said *direction of propagation*.

By using (2.59) we can rewrite (2.58) as

$$\begin{aligned} \mathbf{H}_0 &= \frac{1}{\zeta}\hat{\mathbf{k}} \times \mathbf{E}_0 \\ \hat{\mathbf{k}} \cdot \mathbf{E}_0 &= 0 \end{aligned} \quad \text{or equivalently} \quad \begin{aligned} \mathbf{E}_0 &= \zeta \mathbf{H}_0 \times \hat{\mathbf{k}} \\ \hat{\mathbf{k}} \cdot \mathbf{H}_0 &= 0 \end{aligned} \tag{2.60}$$

where $\zeta = \sqrt{\mu/\varepsilon}$ is the intrinsic impedance of the medium. In vacuum, $\zeta = \zeta_0 = \sqrt{\mu_0/\varepsilon_0} \cong 377\Omega \cong 120\pi\Omega$. Equation (2.60) shows that in a homogeneous plane wave, the electric and magnetic fields are mutually perpendicular, and both are perpendicular to the direction of propagation. Therefore, a homogeneous plane wave is a transverse-electromagnetic (TEM) wave. It is also easy to verify that electric and magnetic fields of a homogeneous plane wave only vary along the direction of propagation, so that they are uniform over planes orthogonal to this direction. To show that they actually behave as a wave traveling along the direction of propagation, we choose the reference system in such a way that the z-axis coincides with the direction of propagation, so that $\mathbf{k} = k\hat{z}$ and $\mathbf{k} \cdot \mathbf{r} = kz$, and compute the x component of the time domain electric field from (2.15) and (2.53):

$$\begin{aligned} e_x(z,t) &= \mathrm{Re}\left\{E_{0x}\exp(-jkz)\exp(j\omega t)\right\} = \mathrm{Re}\left\{E_{0x}\exp\left[-j(\beta - j\alpha)z\right]\exp(j\omega t)\right\} \\ &= \left|E_{0x}\right|\exp[-\alpha z]\cos\left(\omega t - \beta z + \arg\left\{E_{0x}\right\}\right) \\ &= \left|E_{0x}\right|\exp[-\alpha z]\cos\left[-\beta\left(z - \frac{\omega}{\beta}t\right) + \arg\left\{E_{0x}\right\}\right] \end{aligned}$$

$$\tag{2.61}$$

An analogous expression is obtained for the y component.

Equation (2.61) shows that, apart from an exponential attenuation that is absent for nondispersive media, the time-varying sinusoidal fields are also space-varying sinusoidals, with a spatial period called *wavelength* $\lambda = 2\pi/\beta$. In addition, as time varies, the wave shifts along the z-axis (i.e., along the direction of propagation) with a *propagation* (or *phase*) *velocity* $v_p = \omega/\beta$ (see Figure 2.5). For a nondispersive (and hence lossless) medium, $v_p = 1/\sqrt{\varepsilon\mu}$, and it is independent of frequency. In vacuum, $v_p = 1/\sqrt{\varepsilon_0\mu_0} = c \cong 2.998 \cdot 10^8$ m/s.

By using (2.60) and (A.2), the Poynting vector of a homogeneous plane wave can be computed as

$$\mathbf{S} = \frac{1}{2}\mathbf{E} \times \mathbf{H}^* = \frac{1}{2\zeta^*}\mathbf{E} \times \left(\hat{\mathbf{k}} \times \mathbf{E}^*\right) = \frac{1}{2\zeta^*}|\mathbf{E}|^2\,\hat{\mathbf{k}} = \frac{1}{2\zeta^*}\left|\mathbf{E}_0\right|^2 \exp\left(-2\alpha\hat{\mathbf{k}} \cdot \mathbf{r}\right)\hat{\mathbf{k}}$$

$$(2.62)$$

Equation (2.62) shows that the power flux is along the direction of propagation; in addition, for a dispersive medium ($\alpha \neq 0$, complex ζ), this power flux attenuates exponentially, due to the power dissipation in the medium. Conversely, for a nondispersive medium ($\alpha = 0$, real ζ), (2.62) reduces to

$$\mathbf{S} = \frac{1}{2\zeta}\left|\mathbf{E}_0\right|^2\hat{\mathbf{k}}$$

$$(2.63)$$

and the power flux density is real (i.e., there is only a flux of active power) and uniform, with no attenuation.

2.3.2 Nonhomogeneous Plane Waves

Equation (2.54) [i.e., the first equation of (2.58)] can be satisfied also by considering a complex wavenumber vector $\mathbf{k} = \mathbf{k}_R - j\mathbf{k}_I$ with nonparallel real, $\mathbf{k}_R$

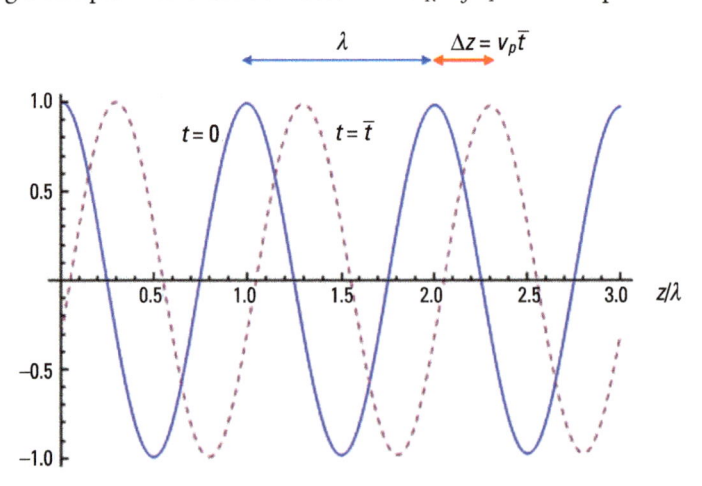

Figure 2.5 Propagation of a sinusoidal wave.

and imaginary, $-\mathbf{k}_I$, parts. In this case, the plane wave is called *nonhomogeneous plane wave*, and

$$\mathbf{E}(\mathbf{r}) = \mathbf{E}_0 \exp\left(-j\mathbf{k}_R \cdot \mathbf{r}\right)\exp\left(-\mathbf{k}_I \cdot \mathbf{r}\right) \tag{2.64}$$

which shows that the wave propagates along the direction of $\mathbf{k}_R$ and attenuates along the direction of $\mathbf{k}_I$.

By separating real and imaginary parts of (2.54) we get

$$\begin{aligned} \mathbf{k}_R \cdot \mathbf{k}_R - \mathbf{k}_I \cdot \mathbf{k}_I &= \beta^2 - \alpha^2 \\ \mathbf{k}_R \cdot \mathbf{k}_I &= \beta\alpha \end{aligned} \tag{2.65}$$

which shows that for a lossless medium ($\alpha = 0$) $\mathbf{k}_R$ and $\mathbf{k}_I$ are orthogonal. Note that in this last case attenuation along the direction of $\mathbf{k}_I$ is not related to power dissipation.

2.3.3 Plane Waves for Arbitrarily Time-Varying Fields

Up to now sinusoidal plane waves have been considered. However, obtained results can be generalized to arbitrarily time-varying fields by recalling that the phasor-domain Maxwell's equations (2.17) and constitutive relationships [(2.27) to (2.29)] are formally identical to their frequency-domain counterparts (see Section 2.1.2). Therefore, the solution of source-free Maxwell's equations in the frequency domain for a linear, local, stationary, isotropic, and homogeneous medium are still provided by (2.57) to (2.58), although with $\mathbf{E}_0(\omega)$, $\mathbf{H}_0(\omega)$, and $\mathbf{k}(\omega)$ functions of frequency. Time-domain solutions for arbitrarily time-varying fields can then be obtained by inverse-Fourier-transforming these expressions.

If we consider a homogeneous plane wave propagating along the z-axis [i.e., $\mathbf{k}(\omega) = k(\omega)\,\hat{\mathbf{z}}$,], the x component of the time domain electric field can be computed as

$$e_x(z,t) = \frac{1}{2\pi}\int_{-\infty}^{+\infty} E_x(z,\omega)\exp\left(j\omega t\right)d\omega = \frac{1}{2\pi}\int_{-\infty}^{+\infty} E_{0x}(\omega)\exp\left[-jk(\omega)z\right]\exp\left(j\omega t\right)d\omega$$

$$= \frac{1}{2\pi}\int_{-\infty}^{+\infty} E_{0x}(\omega)\exp\left\{j\left[\omega t - k(\omega)z\right]\right\}d\omega$$

$$\tag{2.66}$$

with $k(\omega) = \omega\sqrt{\varepsilon(\omega)\mu} = \beta(\omega) - j\alpha(\omega)$. For nondispersive media, $k(\omega) = \omega\sqrt{\varepsilon\mu} = \omega/v_p$ and (2.66) becomes

$$e_x(z,t) = \frac{1}{2\pi} \int_{-\infty}^{+\infty} E_{0x}(\omega)\exp\left\{ j\omega\left[t - \frac{z}{v_p} \right] \right\} d\omega = e_x\left(0, t - \frac{z}{v_p} \right) \quad (2.67)$$

so that the signal at the abscissa z is a perfect copy of the signal at the abscissa 0, delayed by the time needed by the wave to propagate from 0 to z at velocity v_p. Therefore, signals propagate undistorted through nondispersive media. For dispersive media we have

$$e_x(z,t) = \frac{1}{2\pi} \int_{-\infty}^{+\infty} E_{0x}(\omega)\exp\left[-\alpha(\omega)z - j\beta(\omega)z \right]\exp\left\{ j\omega t \right\} d\omega \quad (2.68)$$

and the signal at the abscissa z is not a delayed copy of the signal at the abscissa 0. Therefore, signals are distorted when propagating through dispersive media.

2.3.4   Narrowband Signals and Group Velocity

Signal distortion in dispersive media can be made negligible by using narrowband signals (i.e., signals for which the ratio between their bandwidth $\Delta\omega$ and their central frequency ω_0 is much smaller than unity). In this case, we can expand $\beta(\omega)$ and $\alpha(\omega)$ in Taylor series around ω_0:

$$\beta(\omega) = \beta(\omega_0) + \left.\frac{d\beta}{d\omega}\right|_{\omega=\omega_0} (\omega - \omega_0) + \frac{1}{2}\left.\frac{d^2\beta}{d\omega^2}\right|_{\omega=\omega_0} (\omega - \omega_0)^2 + \dots$$

$$\alpha(\omega) = \alpha(\omega_0) + \left.\frac{d\alpha}{d\omega}\right|_{\omega=\omega_0} (\omega - \omega_0) + \dots$$

$$(2.69)$$

If

$$\left| \left.\frac{d^2\beta}{d\omega^2}\right|_{\omega=\omega_0} \Delta\omega^2 z \right| \ll 1 \quad \text{and} \quad \left| \left.\frac{d\alpha}{d\omega}\right|_{\omega=\omega_0} \Delta\omega z \right| \ll 1 \quad (2.70)$$

we can make the following substitutions in the argument of the exponential of (2.68):

$$\beta(\omega) \cong \beta_0 + \frac{\omega - \omega_0}{v_g} \quad \text{and} \quad \alpha(\omega) \cong \alpha_0 \quad (2.71)$$

where

$$\beta_0 = \beta(\omega_0), \quad \alpha_0 = \alpha(\omega_0) \quad \text{and} \quad v_g = 1 \bigg/ \frac{d\beta}{d\omega}\bigg|_{\omega=\omega_0} \tag{2.72}$$

with v_g being the *group velocity*.

By replacing (2.71) in (2.68) and recalling that the FT of a real function is Hermitian [i.e., using (2.13)], we get

$$
\begin{aligned}
e_x(z,t) &= \frac{1}{\pi}\text{Re}\left\{ \int_{\omega_0-\Delta\omega/2}^{\omega_0+\Delta\omega/2} E_{0x}(\omega)\exp\left[-\alpha_0 z + j\left(\omega t - \beta_0 z - \frac{\omega-\omega_0}{v_g}z \right) \right]d\omega \right\} \\
&= \exp(-\alpha_0 z)\text{Re}\left\{ \exp\left[j(\omega_0 t - \beta_0 z) \right]\frac{1}{\pi}\int_{-\Delta\omega/2}^{\Delta\omega/2} E_{0x}(\omega'+\omega_0)\exp\left[j\omega'\left(t - \frac{z}{v_g} \right) \right]d\omega' \right\} \\
&= \left| \tilde{e}_x\left(t - \frac{z}{v_g} \right) \right|\exp(-\alpha_0 z)\cos\left[\omega_0 t - \beta_0 z + \arg\left\{ \tilde{e}_x\left(t - \frac{z}{v_g} \right) \right\} \right]
\end{aligned}
$$

$$\tag{2.73}$$

where $\omega' = \omega - \omega_0$ and

$$\tilde{e}_x(t) = \frac{1}{\pi}\int_{-\Delta\omega/2}^{\Delta\omega/2} E_{0x}(\omega'+\omega_0)\exp\left[j\omega't \right]d\omega' \tag{2.74}$$

is the *complex envelope* of the transmitted signal. The propagating (and atten-uating) wave described by (2.73) is then a sinusoidal signal (*carrier*) whose amplitude and phase are time-varying, since they are modulated by the com-plex envelope (*modulating signal*). The carrier travels at the phase velocity $v_p = \omega_0/\beta_0$, while the complex envelope travels undistorted at the group velocity v_g defined in (2.72) (see Figure 2.6). Due to the narrowband condition $\Delta\omega \ll \omega_0$, the time variations of the envelope are over a time scale much longer than the carrier period. Usually, either the modulus of $\tilde{e}_x(t)$ is variable and its phase is constant, or its phase is variable and its modulus is constant. In the former case, *amplitude modulation* is obtained, whereas in the latter, *phase* or *frequency modulation* is achieved. However, in some digital modulation schemes, both amplitude and phase are modulated.

Since the signal information is conveyed by the envelope, it is not dis-torted as far as the approximation (2.71) is valid. However, the validity of this approximation, given by conditions (2.70), depends on the traveled distance

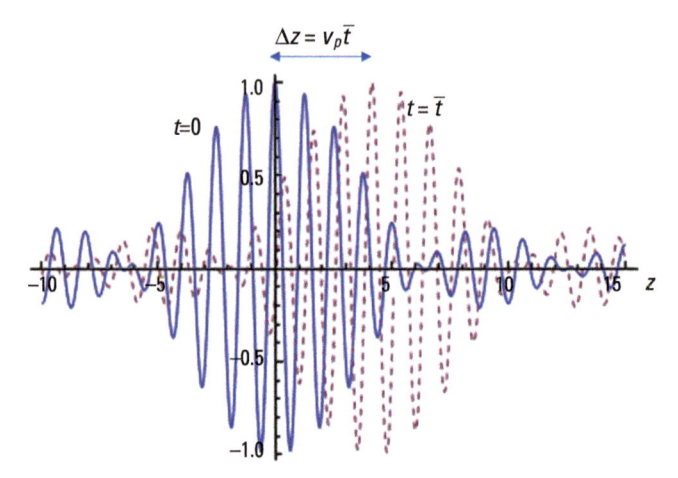

Figure 2.6   Propagation of a wave packet.

$|z|$. For sufficiently large distances, conditions (2.70) are no longer valid, and the signal is distorted. Fortunately, in usual radio links most of the distance is traveled in air, which can be considered nondispersive, and only short distances are traveled in dispersive media, such as building walls.

A plane wave traveling according to (2.73) is also called a *wave packet*. Note that group velocity cannot exceed the velocity c of light in vacuum, since it represents the speed at which electromagnetic energy, and information conveyed by the wave, propagate. Conversely, phase velocity does not represent the speed of any physical entity, so that it may exceed the velocity c of light in vacuum. Finally, it is important to note that in a nondispersive medium, phase and group velocities coincide.

2.3.5   Plane Wave Reflection and Transmission at a Plane Boundary

Let us consider a homogeneous plane wave with propagation vector $\mathbf{k}_i$ (*incident wave*) impinging on a plane boundary between two half-spaces filled with two different homogeneous media, medium 1 and medium 2, whose parameters are ε_1, μ_1, and ε_2, μ_2, respectively; see Figure 2.7(a). Accordingly, propagation constants in the two half-spaces are $k_1 = \omega\sqrt{\varepsilon_1\mu_1}$ and $k_2 = \omega\sqrt{\varepsilon_2\mu_2}$, and intrinsic impedances are $\zeta_1 = \sqrt{\mu_1/\varepsilon_1}$ and $\zeta_2 = \sqrt{\mu_2/\varepsilon_2}$. We also define the *refractive index* (or *index of refraction*) of medium 1 as $n_1 = \sqrt{\varepsilon_{r1}\mu_{r1}}$ and of medium 2 as $n_2 = \sqrt{\varepsilon_{r2}\mu_{r2}}$. Finally, we define the refractive index of medium 2 with respect to medium 1 as $n_{21} = n_2/n_1 = k_2/k_1$.

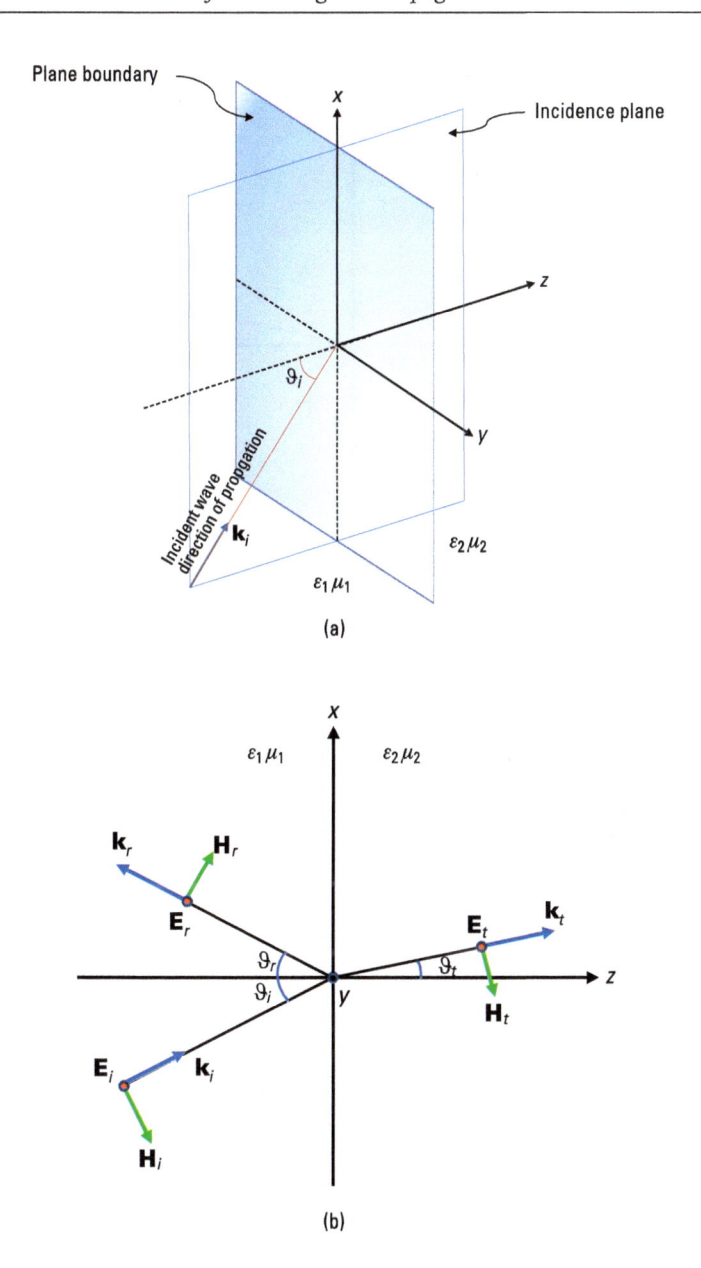

Figure 2.7 Plane waves at a plane interface between two homogeneous media: (a) problem geometry and definition of incidence plane, (b) perpendicular polarization, and (c) parallel polarization.

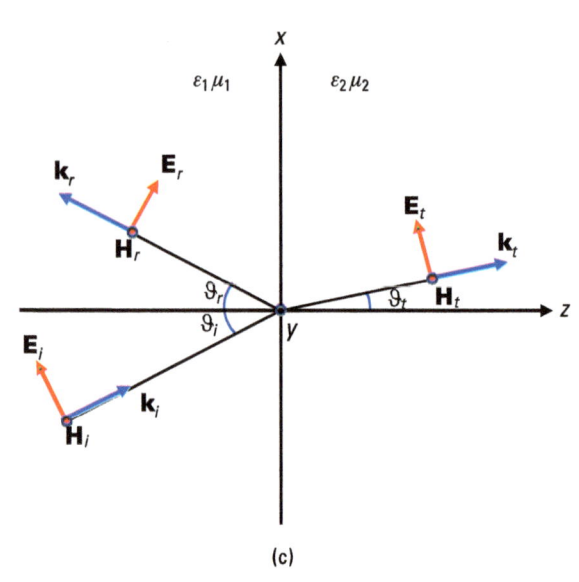

(c)

Figure 2.7 *(Continued)*

The incident wave comes from the half-space filled with medium 1; see Figure 2.7(a). Its direction of propagation and the direction orthogonal to the plane boundary form an angle ϑ_i (*incidence angle*) and define the *incidence plane*. Medium 1 is assumed to be lossless, whereas medium 2 may be lossy.

Continuity of tangential electric and magnetic fields at the boundary (see Section 2.2.5) requires that a *reflected wave*, with propagation vector $\mathbf{k}_r$, is also present in half-space 1, as well as a *transmitted* (or *refracted*) *wave*, with propagation vector $\mathbf{k}_r$, in half-space 2. We choose a reference system with the *x-y* plane coincident with the plane boundary between the two media, and the *x-z* plane coincident with the incidence plane; see Figure 2.7. Then, the propagation vector of the incident wave can be written as

$$\mathbf{k}_i = k_{ix}\hat{\mathbf{x}} + k_{iz}\hat{\mathbf{z}} = k_1 \sin\vartheta_i \hat{\mathbf{x}} + k_1 \cos\vartheta_i \hat{\mathbf{z}} \tag{2.75}$$

In the half-space 1 (i.e., for $z \leq 0$), we have

$$\mathbf{E}(\mathbf{r}) = \mathbf{E}_i(\mathbf{r}) + \mathbf{E}_r(\mathbf{r}) = \mathbf{E}_{i0}\exp(-j\mathbf{k}_i \cdot \mathbf{r}) + \mathbf{E}_{r0}\exp(-j\mathbf{k}_r \cdot \mathbf{r})$$
$$\mathbf{H}(\mathbf{r}) = \mathbf{H}_i(\mathbf{r}) + \mathbf{H}_r(\mathbf{r}) = \mathbf{H}_{i0}\exp(-j\mathbf{k}_i \cdot \mathbf{r}) + \mathbf{H}_{r0}\exp(-j\mathbf{k}_r \cdot \mathbf{r}) \tag{2.76}$$

and in the half-space 2 (i.e., for $z \geq 0$), we have

$$\mathbf{E}(\mathbf{r}) = \mathbf{E}_t(\mathbf{r}) = \mathbf{E}_{t0}\exp\left(-j\mathbf{k}_t \cdot \mathbf{r}\right)$$
$$\mathbf{H}(\mathbf{r}) = \mathbf{H}_t(\mathbf{r}) = \mathbf{H}_{t0}\exp\left(-j\mathbf{k}_t \cdot \mathbf{r}\right)$$

$$(2.77)$$

By enforcing continuity of, for instance, the x component of the electric field at the plane interface $z = 0$, we get $E_{i0x}\exp(-jk_{ix}x) + E_{r0x}\exp(-jk_{rx}x - jk_{ry}y)$ $= E_{t0x}\exp(-jk_{tx}x - jkt_yy)$. This equality must be satisfied for all values of x and y, which implies that $k_{ry} = k_{ty} = k_{iy} = 0$, and $k_{rx} = k_{tx} = k_{ix} = k_1\sin\vartheta_i$. Therefore, $\mathbf{k}_i$, $\mathbf{k}_r$, and $\mathbf{k}_t$ all lie in the incidence plane x-z. In addition, by using $k_{ry} = 0$ and $k_{rx} = k_1\sin\vartheta_i$ we can write

$$\mathbf{k}_r = k_{rx}\hat{\mathbf{x}} + k_{rz}\hat{\mathbf{z}} = k_1\sin\vartheta_r\hat{\mathbf{x}} - k_1\cos\vartheta_r\hat{\mathbf{z}} = k_1\sin\vartheta_i\hat{\mathbf{x}} - k_1\cos\vartheta_i\hat{\mathbf{z}} \quad (2.78)$$

which shows that the reflection angle ϑ_r is equal to the incidence angle (*reflection law*).

Similarly, by using $k_{ty} = 0$ and $k_{tx} = k_1\sin\vartheta_i$, we have

$$\mathbf{k}_t = k_{tx}\hat{\mathbf{x}} + k_{tz}\hat{\mathbf{z}} = k_1\sin\vartheta_i\hat{\mathbf{x}} + \sqrt{k_2^2 - k_1^2\sin^2\vartheta_i}\,\hat{\mathbf{z}} = k_1\sin\vartheta_i\hat{\mathbf{x}} + k_1\sqrt{n_{21}^2 - \sin^2\vartheta_i}\,\hat{\mathbf{z}}$$

$$(2.79)$$

where (2.54) has been used to compute k_{tz}. If the transmitted wave is homogeneous, $\mathbf{k}_t$ can be also expressed in terms of the transmission (or refraction) angle ϑ_t (see Figure 2.7):

$$\mathbf{k}_t = k_2\sin\vartheta_t\hat{\mathbf{x}} + k_2\cos\vartheta_t\hat{\mathbf{z}} \quad (2.80)$$

By equating the x components of (2.79) and (2.80), we get the *Snell refraction law*:

$$\sin\vartheta_t = \frac{n_1}{n_2}\sin\vartheta_i \quad (2.81)$$

In order to relate $\mathbf{E}_{r0}$ and $\mathbf{E}_{t0}$ to $\mathbf{E}_{i0}$, it is useful to express each arbitrarily polarized plane wave as the superposition of two orthogonally linearly polarized plane waves. We therefore consider *perpendicular polarization*, in which the electric field is perpendicular to the incidence plane and to the z direction [transverse-electric, or TEz, field (see Figure 2.7(b))] and *parallel polarization*, in which the electric field is parallel to the incidence plane and the magnetic field is perpendicular to the z direction [transverse-magnetic, or TMz, field]; see Figure 2.7(c). For perpendicular polarization we have

$$\mathbf{E}_{i0} = E_{i0y}\hat{\mathbf{y}}, \ \mathbf{E}_{r0} = E_{r0y}\hat{\mathbf{y}}, \ \mathbf{E}_{t0} = E_{t0y}\hat{\mathbf{y}} \tag{2.82}$$

and, by using the first of (2.56a),

$$\mathbf{H}_{i0} = -\frac{k_{iz}}{\omega\mu_1}E_{i0y}\hat{\mathbf{x}} + \frac{k_{ix}}{\omega\mu_1}E_{i0y}\hat{\mathbf{z}}$$

$$\mathbf{H}_{r0} = -\frac{k_{rz}}{\omega\mu_1}E_{r0y}\hat{\mathbf{x}} + \frac{k_{rx}}{\omega\mu_1}E_{r0y}\hat{\mathbf{z}} = \frac{k_{iz}}{\omega\mu_1}E_{r0y}\hat{\mathbf{x}} + \frac{k_{ix}}{\omega\mu_1}E_{r0y}\hat{\mathbf{z}} \tag{2.83}$$

$$\mathbf{H}_{t0} = -\frac{k_{tz}}{\omega\mu_2}E_{t0y}\hat{\mathbf{x}} + \frac{k_{tx}}{\omega\mu_2}E_{r0y}\hat{\mathbf{z}} = -\frac{k_{tz}}{\omega\mu_2}E_{t0y}\hat{\mathbf{x}} + \frac{k_{ix}}{\omega\mu_2}E_{r0y}\hat{\mathbf{z}}$$

By imposing continuity of tangential components of electric and magnetic fields (i.e., of E_y and H_x) on the interface plane $z = 0$,

$$\begin{cases} E_{i0y} + E_{r0y} = E_{t0y} \\ \dfrac{k_{iz}}{\omega\mu_1}\left(E_{i0y} - E_{r0y}\right) = \dfrac{k_{tz}}{\omega\mu_2}E_{t0y} \end{cases} \tag{2.84}$$

We now define the perpendicular polarization reflection coefficient $\Gamma_\perp = E_{r0y}/E_{i0y}$ and the perpendicular polarization transmission coefficient $T_\perp = E_{t0y}/E_{i0y}$. By using these definitions in (2.84) and solving with respect to $\Gamma_\perp$ and $T_\perp$ we get

$$\Gamma_\perp = \frac{k_{iz}\mu_2 - k_{tz}\mu_1}{k_{iz}\mu_2 + k_{tz}\mu_1} = \frac{\mu_2\cos\vartheta_i - \mu_1\sqrt{n_{21}^2 - \sin^2\vartheta_i}}{\mu_2\cos\vartheta_i + \mu_1\sqrt{n_{21}^2 - \sin^2\vartheta_i}} \tag{2.85}$$

$$T_\perp = 1 + \Gamma_\perp$$

The active power density flux through the interface is

$$\mathrm{Re}\{\mathbf{S}\cdot\hat{\mathbf{z}}\} = \mathrm{Re}\left\{-\frac{1}{2}\left(E_{i0y} + E_{r0y}\right)\left(H_{i0x} + H_{r0x}\right)^*\right\} = \frac{1}{2}\frac{k_{iz}}{\omega\mu_1}\left|E_{i0y}\right|^2\left(1 - \left|\Gamma_\perp\right|^2\right) \tag{2.86}$$

and since it must be nonnegative, it follows that $\left|\Gamma_\perp\right| \leq 1$.

Analogously, for parallel polarization we have

$$\mathbf{H}_{i0} = H_{i0y}\hat{\mathbf{y}}, \ \mathbf{H}_{r0} = H_{r0y}\hat{\mathbf{y}}, \ \mathbf{H}_{t0} = H_{t0y}\hat{\mathbf{y}} \tag{2.87}$$

and, by using the first of (2.56b),

$$\mathbf{E}_{i0} = \frac{k_{iz}}{\omega\varepsilon_1} H_{i0y}\hat{\mathbf{x}} - \frac{k_{ix}}{\omega\varepsilon_1} H_{i0y}\hat{\mathbf{z}}$$

$$\mathbf{E}_{r0} = \frac{k_{rz}}{\omega\varepsilon_1} H_{r0y}\hat{\mathbf{x}} - \frac{k_{rx}}{\omega\varepsilon_1} H_{r0y}\hat{\mathbf{z}} = -\frac{k_{iz}}{\omega\varepsilon_1} H_{r0y}\hat{\mathbf{x}} - \frac{k_{ix}}{\omega\varepsilon_1} H_{r0y}\hat{\mathbf{z}} \qquad (2.88)$$

$$\mathbf{E}_{t0} = \frac{k_{tz}}{\omega\varepsilon_2} H_{t0y}\hat{\mathbf{x}} - \frac{k_{tx}}{\omega\varepsilon_2} H_{r0y}\hat{\mathbf{z}} = \frac{k_{tz}}{\omega\varepsilon_2} H_{t0y}\hat{\mathbf{x}} - \frac{k_{ix}}{\omega\varepsilon_2} H_{r0y}\hat{\mathbf{z}}$$

Imposing continuity of tangential components of electric and magnetic fields (i.e., of E_x and H_y) on the interface plane $z = 0$, we get

$$\begin{cases} E_{i0x} + E_{r0x} = E_{t0x} \\ \dfrac{\omega\varepsilon_1}{k_{iz}}\left(E_{i0x} - E_{r0x}\right) = \dfrac{\omega\varepsilon_2}{k_{tz}} E_{t0x} \end{cases} \qquad (2.89)$$

We now define the *parallel polarization reflection coefficient* $\Gamma_{\|} = E_{r0x}/E_{i0x}$ and the *parallel polarization transmission coefficient* $T_{\|} = E_{t0x}/E_{i0x}$. Using these definitions in (2.89) and solving with respect to $\Gamma_{\|}$ and $T_{\|}$ yields

$$\Gamma_{\|} = \frac{k_{tz}\varepsilon_1 - k_{iz}\varepsilon_2}{k_{tz}\varepsilon_1 + k_{iz}\varepsilon_2} = \frac{\varepsilon_1\sqrt{n_{21}^2 - \sin^2\vartheta_i} - \varepsilon_2\cos\vartheta_i}{\varepsilon_1\sqrt{n_{21}^2 - \sin^2\vartheta_i} + \varepsilon_2\cos\vartheta_i} \qquad (2.90)$$

$$T_{\|} = 1 + \Gamma_{\|}$$

The active power density flux through the interface is

$$\mathrm{Re}\{\mathbf{S}\cdot\hat{\mathbf{z}}\} = \mathrm{Re}\left\{\frac{1}{2}\left(E_{i0x} + E_{r0x}\right)\left(H_{i0y} + H_{r0y}\right)^*\right\} = \frac{1}{2}\frac{\omega\varepsilon_1}{k_{iz}}\left|E_{i0x}\right|^2\left(1 - \left|\Gamma_{\|}\right|^2\right)$$

$$(2.91)$$

so that $|\Gamma_{\|}| \leq 1$.

If we assume $\mu_1 = \mu_2 = \mu_0$, see (2.28), then (2.85) and (2.90) become the *Fresnel reflection coefficients* in perpendicular and parallel polarization:

$$\Gamma_{\perp} = \frac{\cos\vartheta_i - \sqrt{n_{21}^2 - \sin^2\vartheta_i}}{\cos\vartheta_i + \sqrt{n_{21}^2 - \sin^2\vartheta_i}}$$

$$(2.92)$$

$$\Gamma_{\|} = \frac{\sqrt{n_{21}^2 - \sin^2\vartheta_i} - n_{21}^2\cos\vartheta_i}{\sqrt{n_{21}^2 - \sin^2\vartheta_i} + n_{21}^2\cos\vartheta_i}$$

If medium 2 is lossless, then n_{21} is real, and $\Gamma_\parallel$ is zero if the incidence angle is equal to the *Brewster angle* $\vartheta_B = \arctan n_{21}$. In addition, if $n_{21} > 1$ then Fresnel coefficients are both real for any incidence angle, and the transmitted wave is homogeneous. Conversely, if $n_{21} < 1$ then Fresnel coefficients are both real only for incidence angles smaller than the *critical angle* $\vartheta_C = \arcsin n_{21}$; in this case, the transmitted wave is homogeneous. For higher incidence angles, the argument of the square root in (2.92) is negative, so that Fresnel coefficients are complex, with unitary modulus. Therefore [see (2.86) and (2.91)], there is no active power flux through the interface (*total reflection* of active power); in addition, the transmitted wave is nonhomogeneous, since $k_{tz} = k_1\sqrt{n_{21}^2 - \sin^2\vartheta_i} = -j\alpha_{tz}$ is imaginary, whereas $k_{tx} = k_1\sin\vartheta_i$ is real, so that $\mathbf{k}_R$ is along x and $\mathbf{k}_I$ along z. Accordingly, the transmitted wave propagates along the x direction and attenuates exponentially along the z direction, so that it is confined within the region close to the interface. For this reason, it is called a *surface wave*.

If medium 2 is lossy, then n_{21}, $\Gamma_\perp$, and $\Gamma_\parallel$ are complex. In addition, the transmitted wave is in general nonhomogeneous, with attenuation along the z direction and propagation along an obliquus direction in the x-z plane. However, if the medium 2 is a good conductor (e.g., a metal), its refractive index is:

$$n_m \cong \sqrt{-j\frac{\sigma}{\omega\varepsilon_0}} = \frac{1-j}{\sqrt{2}}\sqrt{\frac{\sigma}{\omega\varepsilon_0}} \qquad (2.93)$$

so that $|n_m| \gg 1$. We then have, independently of the incidence angle,

$$\Gamma_\perp \cong \Gamma_\parallel \cong -1 \quad \text{and} \quad k_{tz} \cong k_0 n_{21} = (1-j)\sqrt{\frac{\sigma\omega\mu_0}{2}} = \frac{1-j}{\delta} \gg k_{tx} \qquad (2.94)$$

where $\delta = \sqrt{2/(\sigma\omega\mu)_0}$ is the *skin depth*. Therefore, for any incidence angle, the transmitted wave propagates and exponentially attenuates along the z direction (i.e., the direction perpendicular to the interface), so that it becomes negligible as it penetrates a few skin depths within the metal. Note that for metals at radio frequencies the skin depth is on the order of a fraction of millimeter, and of a few micrometers at microwaves. In addition, for the transmitted wave we have

$$\mathbf{E} = \zeta_m \mathbf{H} \times \hat{\mathbf{z}} \quad \text{where} \quad \zeta_m \cong \zeta_0\sqrt{j\frac{\omega\varepsilon_0}{\sigma}} = \frac{1+j}{\sqrt{2}}\sqrt{\frac{\omega\mu_0}{\sigma}} = \frac{1+j}{\sigma\delta} \qquad (2.95)$$

Since electric and magnetic fields are tangent to the interface, and tangential components of fields are continuous at the interface, then on the surface of the metal the electric field tangential component is

$$\mathbf{E}_{\text{tan}} = \zeta_m \mathbf{H} \times \hat{\mathbf{z}} \tag{2.96}$$

Since $|\zeta_m| << \zeta_0$, this tangential component is very small. In summary, the field only penetrates for a very small skin depth in a good conductor, such as a metal, so that the current flows only in the immediate proximity of the metal surface (*skin effect*); and the tangential component of the electric field on the metal surface is very small. Therefore, when computing the electromagnetic field outside the metal, the latter can be approximated with a PEC, in which the electric field is null, the current is confined on its surface, the tangential component of the electric field on its surface is null, and reflection coefficient is −1. However, to compute the active power dissipated in the metal, the flux of the Poynting vector through the metal surface must be computed, and the tangential electric field cannot be ignored; it can be obtained from the (PEC-approximated) magnetic field by using (2.96), called the *Leontovic boundary condition*.

2.3.6 Plane Wave Propagation in Layered Media

A procedure similar to the one described in Section 2.3.5 can be followed to analyze plane wave propagation in a layered medium (i.e., a series of parallel plane interfaces between homogeneous layers); see Figure 2.8. Using the same hypotheses and definitions regarding the incident wave as in Section 2.3.5, we choose a reference system with the z-axis perpendicular to the interfaces, the x-y plane coincident with the first interface, and the x-z plane coincident with the incidence plane. In each layer, a *progressive* ($k_z > 0$) and a *regressive*

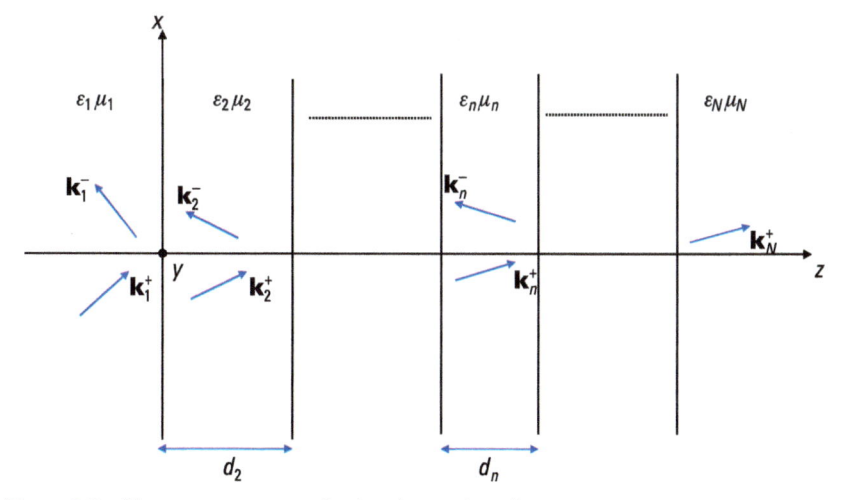

Figure 2.8 Plane wave propagation in a layered medium.

$(k_z < 0)$ wave are present, except for the last half-space, at the end of the layered medium, in which only the progressive wave is present; and continuity of tangential fields must be imposed at each interface. This leads to expressing the fields of all the waves in terms of those of the incidence one.

The general case of Figure 2.8 is not analyzed here, but we consider the simplest interesting case of two interfaces (i.e., of a single homogeneous layer of width d and (complex) dielectric constant ε_2 between two half-spaces of dielectric constants ε_1 (real) and ε_3 (real or complex), respectively); see Figure 2.9. We also assume that $\mu_1 = \mu_2 = \mu_3 = \mu_0$. This structure may schematize, for instance, the case of propagation through a homogeneous wall of width d. In this case, ε_1 and ε_3 are usually coincident (i.e., $\varepsilon_1 = \varepsilon_3 = \varepsilon_0$) where ε_0 is the free-space value.

By imposing the continuity of tangential fields at the two interfaces, A and B (see Figure 2.9) have $k_y = 0$ and $k_x = k_1 \sin\vartheta_i$ for all the plane waves in the layered structure, and $k_{z1} = k_1 \cos\vartheta_i$, $k_{z2} = k_1 \sqrt{n_{21}^2 - \sin^2 \vartheta_i} = \beta_{z2} - j\alpha_{z2}$, $k_{z3} = k_1 \sqrt{n_{31}^2 - \sin^2 \vartheta_i} = \beta_{z3} - j\alpha_{z3}$, where $n_{31} = k_3/k_1$. For parallel and perpendicular polarizations, the transverse (with respect to the z-axis) components of the electric and magnetic fields can be expressed in a unified way as:

$$\begin{cases} E = E_1^+ \exp\left(-jk_x x - jk_{z1} z\right) + E_1^- \exp\left(-jk_x x + jk_{z1} z\right) \\ H = \dfrac{1}{Z_1}\left[E_1^+ \exp\left(-jk_x x - jk_{z1} z\right) - E_1^- \exp\left(-jk_x x + jk_{z1} z\right)\right] \end{cases} \quad \text{for } z \leq 0$$

$$(2.97)$$

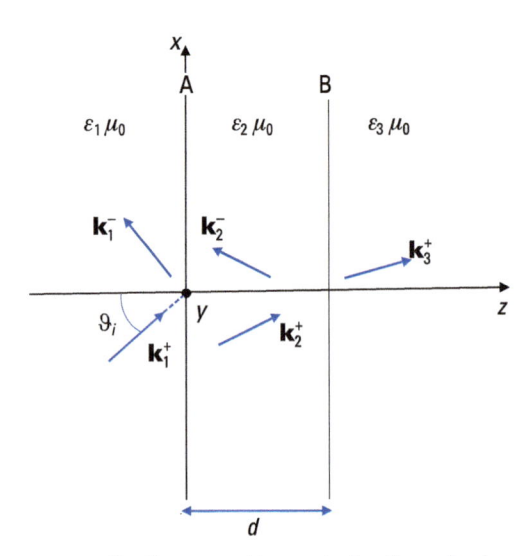

Figure 2.9 Plane-wave reflection on and transmission through a homogeneous layer.

$$\begin{cases} E = E_2^+ \exp\left(-jk_x x - jk_{z2} z\right) + E_2^- \exp\left(-jk_x x + jk_{z2} z\right) \\ H = \dfrac{1}{Z_2}\left[E_2^+ \exp\left(-jk_x x - jk_{z2} z\right) - E_2^- \exp\left(-jk_x x + jk_{z2} z\right)\right] \end{cases} \quad \text{for } 0 \leq z \leq d$$

$$(2.98)$$

$$\begin{cases} E = E_3^+ \exp\left(-jk_x x - jk_{z3} z\right) \\ H = \dfrac{1}{Z_3} E_3^+ \exp\left(-jk_x x - jk_{z3} z\right) \end{cases} \quad \text{for } z \geq d \qquad (2.99)$$

where for parallel polarization $E = E_x$, $H = H_y$, and $Z_n = k_{zn}/(\omega \varepsilon_n)$, whereas for perpendicular polarization $E = E_y$, $H = -H_x$, and $Z_n = \omega \mu_0 / k_{zn}$, with $n = 1,2,3$.

By imposing the continuity of tangential fields at $z = 0$ (interface A) we get

$$\begin{cases} E_1^+ + E_1^- = E_2^+ + E_2^- \\ \dfrac{1}{Z_1}\left(E_1^+ - E_1^- \right) = \dfrac{1}{Z_2}\left(E_2^+ - E_2^- \right) \end{cases} \qquad (2.100)$$

and by imposing continuity of tangential fields at $z = d$ (interface B) we get

$$\begin{cases} E_2^+ \exp\left(-jk_{z2} d\right) + E_2^- \exp\left(jk_{z2} d\right) = E_3^+ \exp\left(-jk_{z3} d\right) \\ \dfrac{1}{Z_2}\left[E_2^+ \exp\left(-jk_{z2} d\right) - E_2^- \exp\left(jk_{z2} d\right)\right] = \dfrac{1}{Z_3} E_3^+ \exp\left(-jk_{z3} d\right) \end{cases} \qquad (2.101)$$

We now define the reflection and transmission coefficients at the interface A, Γ_A and T_A, as

$$\Gamma_A = \frac{E_1^-}{E_1^+}, \quad T_A = \frac{E_2^+}{E_1^+} \qquad (2.102)$$

and the reflection and transmission coefficients at the interface B, Γ_B and T_B, as

$$\Gamma_B = \frac{E_2^- \exp\left(jk_{z2} d\right)}{E_2^+ \exp\left(-jk_{z2} d\right)}, \quad T_B = \frac{E_3^+ \exp\left(-jk_{z3} d\right)}{E_2^+ \exp\left(-jk_{z2} d\right)} \qquad (2.103)$$

We also define the layer transmission coefficient, T_{AB}, as

$$T_{AB} = \frac{E_3^+ \exp\left(-jk_{z3} d\right)}{E_1^+} = T_A T_B \exp\left(-jk_{z2} d\right) \qquad (2.104)$$

By using (2.103) in (2.101) and solving with respect to Γ_B and T_B, we get

$$\Gamma_B = \frac{Z_3 - Z_2}{Z_3 + Z_2}, \quad T_B = 1 + \Gamma_B = \frac{2Z_3}{Z_3 + Z_2} \tag{2.105}$$

Similarly, by using (2.103) and (2.104) in (2.100) and solving with respect to Γ_A and T_A, we get

$$\Gamma_A = \frac{\Gamma_{12} + \Gamma_B'}{1 + \Gamma_{12}\Gamma_B'}, \quad T_A = \frac{1 + \Gamma_A}{1 + \Gamma_B'} = \frac{1 + \Gamma_{12}}{1 + \Gamma_{12}\Gamma_B'} \tag{2.106}$$

where

$$\Gamma_{12} = \frac{Z_2 - Z_1}{Z_2 + Z_1}, \quad \Gamma_B' = \Gamma_B \exp\left(-j2k_{z2}d\right) \tag{2.107}$$

By using the expressions of Z_1 and Z_2 reported above for parallel and perpendicular polarizations, it can be verified that Γ_{12} is coincident with (2.85) and (2.90) for perpendicular and parallel polarizations, respectively; in other words, it is the Fresnel reflection coefficient for a single interface between two half-spaces with dielectric constants ε_1 and ε_2.

If the layer is lossy and $2\alpha_{z2}d \gg 1$, then $\exp(-j2k_{z2}d) \cong 0$ and $\Gamma_A \cong \Gamma_{12}$; see Figure 2.10(b). Conversely, if the layer is lossless, or its losses are negligible, the layer reflection coefficient Γ_A may significantly differ from the one of a single interface Γ_{12}; see Figure 2.10(a).

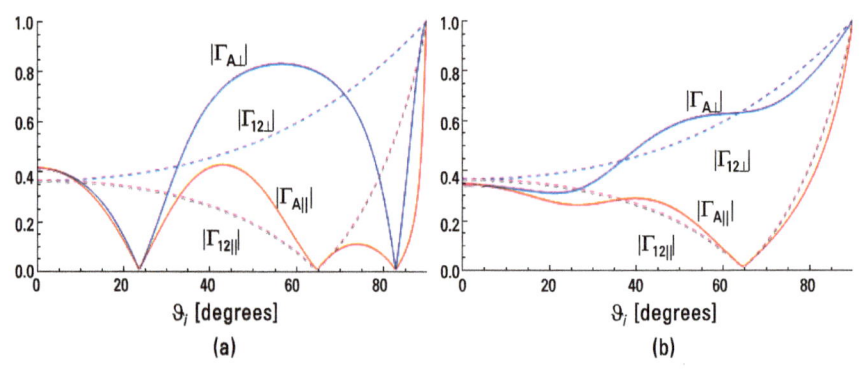

Figure 2.10 Perpendicular (blue) and parallel (red) polarization reflection coefficients Γ_A of a 30-cm-thick homogeneous wall with (a) a relative dielectric constant $\varepsilon_r = 4.5$ (lossless wall) and (b) $\varepsilon_r = 4.5 - j0.2$ (lossy wall), between two air half-spaces. The frequency is 2,400 MHz, and reflection coefficients for a homogeneous half-space (infinitely thick wall) Γ_{12} are also plotted for reference as dashed lines.

In the case that $\varepsilon_3 = \varepsilon_1$, we have $Z_3 = Z_1$, so that (2.105) and (2.106) become:

$$\Gamma_B = \frac{Z_1 - Z_2}{Z_1 + Z_2} = -\Gamma_{12}, \quad T_B = 1 - \Gamma_{12} \tag{2.108}$$

$$\Gamma_A = \frac{\Gamma_{12}\left[1 - \exp\left(-j2k_{z2}d\right)\right]}{1 - \Gamma_{12}^2 \exp\left(-j2k_{z2}d\right)}, \quad T_A = \frac{1 + \Gamma_{12}}{1 - \Gamma_{12}^2 \exp\left(-j2k_{z2}d\right)} \tag{2.109}$$

An example of use of the results of this subsection is illustrated in Section 5.1.1.

2.3.7   Plane Wave Propagation in Anisotropic Media

Plane waves can propagate also in homogeneous anisotropic media. In fact, let us consider the source-free phasor-domain Maxwell's equations for a linear, local, stationary, homogeneous, and anisotropic medium, whose relative dielectric constant $\boldsymbol{\varepsilon}_r$ is a matrix:

$$\begin{cases} \nabla \times \mathbf{E} = -j\omega\mu_0\mathbf{H} \\ \nabla \times \mathbf{H} = j\omega\varepsilon_0\boldsymbol{\varepsilon}_r\mathbf{E} \\ \nabla \cdot \boldsymbol{\varepsilon}_r\mathbf{E} = 0 \\ \nabla \cdot \mathbf{H} = 0 \end{cases} \tag{2.110}$$

Applying the curl to both sides of the first of (2.110) and then using the second of (2.110) we get

$$\nabla \times \nabla \times \mathbf{E} = -j\omega\mu_0\nabla \times \mathbf{H} = \omega^2\varepsilon_0\mu_0\boldsymbol{\varepsilon}_r\mathbf{E} \tag{2.111}$$

By using (A.9), we obtain

$$\nabla^2\mathbf{E} - \nabla\nabla \cdot \mathbf{E} + k_0^2\boldsymbol{\varepsilon}_r\mathbf{E} = 0 \tag{2.112}$$

where we have set

$$k_0 = \omega\sqrt{\varepsilon_0\mu_0} \tag{2.113}$$

with k_0 being the propagation constant in vacuum.

It can be verified that the plane wave electric field of (2.53), with a proper choice of $\mathbf{k}$, is a solution of (2.112). In fact, using (2.53) in (2.112) and recalling that application of the operator ∇ to (2.53) corresponds to a multiplication by $-j\mathbf{k}$, we have

$$\left(-\mathbf{k}\cdot\mathbf{k}I + \mathbf{k}\mathbf{k}\cdot + k_0^2\varepsilon_r\right)\mathbf{E}_0 \exp\left(-j\mathbf{k}\cdot\mathbf{r}\right) = 0 \tag{2.114}$$

where I is the identity matrix; see (A.31). The matrix equation (2.114) has nonnull solutions for the vector $\mathbf{E}_0$ if the wavenumber vector $\mathbf{k}$ satisfies

$$\mathrm{Det}\left(-\mathbf{k}\cdot\mathbf{k}I + \mathbf{k}\mathbf{k}\cdot + k_0^2\varepsilon_r\right) = 0 \tag{2.115}$$

where $\mathrm{Det}(\cdot)$ stands for the determinant of the matrix in parentheses; see (A.34). If we consider homogeneous plane waves (i.e., $\mathbf{k}= k\hat{\mathbf{k}}$,), then (2.115) becomes

$$\mathrm{Det}\left[k^2\left(I - \hat{\mathbf{k}}\hat{\mathbf{k}}\cdot\right) - k_0^2\varepsilon_r\right] = 0 \tag{2.116}$$

which, for each fixed propagation direction $\hat{\mathbf{k}}$, can be solved with respect to k. By replacing (2.53) in (2.110) and using $\mathbf{k}= k\hat{\mathbf{k}}$, we get

$$\begin{aligned} \mathbf{H}_0 &= \frac{k}{\omega\mu_0}\hat{\mathbf{k}}\times\mathbf{E}_0 \\ \hat{\mathbf{k}}\cdot\varepsilon_r\mathbf{E}_0 &= 0 \end{aligned} \quad \text{or equivalently} \quad \begin{aligned} \varepsilon_r\mathbf{E}_0 &= \frac{k}{\omega\varepsilon_0}\mathbf{H}_0\times\hat{\mathbf{k}} \\ \hat{\mathbf{k}}\cdot\mathbf{H}_0 &= 0 \end{aligned} \tag{2.117}$$

which shows that in an anisotropic medium, it is not guaranteed that the plane wave is TEM, but only that it is transverse magnetic.

An example of propagation in an anisotropic medium is propagation in the ionosphere, modeled as a magnetized plasma. This situation is analyzed in Section 4.5.

2.4 Radiation

Radiation from prescribed sources in free space can be evaluated by solving Maxwell's equations in the whole space, assumed to be filled with a linear, local, stationary, isotropic, and homogeneous medium:

$$\begin{cases} \nabla\times\mathbf{E} = -j\omega\mu\mathbf{H} \\ \nabla\times\mathbf{H} = j\omega\varepsilon\mathbf{E} + \mathbf{J} \\ \nabla\cdot\varepsilon\mathbf{E} = \rho \\ \nabla\cdot\mu\mathbf{H} = 0 \end{cases} \tag{2.118}$$

We note that ρ is related to $\mathbf{J}$ by the current density continuity equation (2.17b), so that, at nonzero frequency, it is sufficient to assign $\mathbf{J}$ to completely define the sources.

By applying the curl to both sides of the first of (2.118) and then using the second of (2.118) we get

$$\nabla \times \nabla \times \mathbf{E} = -j\omega\mu\nabla \times \mathbf{H} = \omega^2\varepsilon\mu\mathbf{E} - j\omega\mu\mathbf{J} \qquad (2.119)$$

By using (A.9) and recalling (2.52), we obtain $\nabla^2\mathbf{E} + k^2\mathbf{E} = j\omega\mu\mathbf{J} + \nabla\nabla \cdot \mathbf{E}$. Finally, by using the third of (2.118) and the current density continuity equation (2.17b), we get

$$\nabla^2\mathbf{E} + k^2\mathbf{E} = j\omega\left(I + \frac{\nabla\nabla \cdot}{k^2} \right)\mu\mathbf{J} \qquad (2.120)$$

To evaluate the electric field radiated by $\mathbf{J}$, this equation must be solved. Its solution can be made simpler by letting

$$\mathbf{E} = -j\omega\left(I + \frac{\nabla\nabla \cdot}{k^2} \right)\mathbf{A} \qquad (2.121)$$

where $\mathbf{A}$ is an auxiliary function named the *Lorentz vector potential*. By substituting (2.121) in (2.120) and noting that

$$\begin{aligned} \nabla^2(\nabla\nabla \cdot \mathbf{A}) &= \nabla\nabla \cdot (\nabla\nabla \cdot \mathbf{A}) - \nabla \times \nabla \times (\nabla\nabla \cdot \mathbf{A}) \\ &= \nabla\nabla \cdot (\nabla^2\mathbf{A} - \nabla \times \nabla \times \mathbf{A}) = \nabla\nabla \cdot (\nabla^2\mathbf{A}) \end{aligned} \qquad (2.122)$$

we obtain

$$-j\omega\left(I + \frac{\nabla\nabla \cdot}{k^2} \right)(\nabla^2\mathbf{A} + k^2\mathbf{A}) = j\omega\left(I + \frac{\nabla\nabla \cdot}{k^2} \right)\mu\mathbf{J} \qquad (2.123)$$

In deriving (2.122), we have used (A.9), (A.6), and (A.7).

Equation (2.123) is satisfied if

$$\nabla^2\mathbf{A} + k^2\mathbf{A} = -\mu\mathbf{J} \qquad (2.124)$$

which is a *Helmholtz vector equation*.

In conclusion, solving Maxwell's equations (2.118) for $\mathbf{E}$ and $\mathbf{H}$ is equivalent to solving the Helmholtz equation (2.124) for the vector potential $\mathbf{A}$, then using (2.121) to obtain **E**, and finally inserting it in the first of (2.118) to obtain **H**:

$$\mathbf{H} = \frac{1}{-j\omega\mu}\nabla \times \mathbf{E} = \frac{1}{\mu}\nabla \times \mathbf{A} \tag{2.125}$$

where (A.6) has been also used.

2.4.1 Elementary Source

Let us consider a very short, thin current filament placed at the origin of the reference system and oriented along the z-axis; see Figure 2.11. In this case, the current density can be expressed as

$$\mathbf{J}(\mathbf{r}) = \mathbf{J}_0\delta(\mathbf{r}) = J_0\delta(\mathbf{r})\hat{\mathbf{z}} \tag{2.126}$$

By inserting (2.126) in (2.124), the latter is solved by $A_x = A_y = 0$ and A_z satisfying the Helmholtz scalar equation

$$\nabla^2 A_z + k^2 A_z = -\mu J_0\delta(\mathbf{r}) \tag{2.127}$$

The problem defined by (2.127) is invariant for rotations around the origin (i.e., the problem has spherical symmetry), so that its solution A_z only depends on the distance r from the origin, and it is independent of the polar

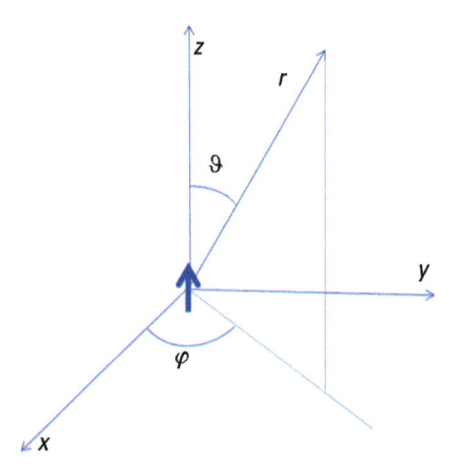

Figure 2.11   Current element placed at the origin of the reference system and oriented along the z-axis.

angle ϑ and of the azimuth angle φ: $A_z(r, \vartheta, \varphi) = A_z(r)$. For values of r different from zero, the right-hand side of (2.127) is zero. By expressing the Laplace operator ∇^2 in spherical coordinates [see (A.27)], and noting that the derivatives with respect to ϑ and φ are zero, (2.127) for r different from zero becomes

$$\frac{1}{r^2}\frac{\partial}{\partial r}r^2\frac{\partial A_z}{\partial r} + k^2 A_z = 0 \tag{2.128}$$

By letting $A_z(r) = f(r)/r$, (2.128) becomes

$$\frac{\partial^2 f}{\partial r^2} + k^2 f = 0 \tag{2.129}$$

whose general solution is $f(r) = C_1\exp(-jkr) + C_2\exp(jkr)$, so that

$$A_z(r) = C_1\frac{\exp(-jkr)}{r} + C_2\frac{\exp(jkr)}{r} \tag{2.130}$$

The first and second terms of the solution (2.130) are an outward- and an inward-propagating spherical wave, respectively. The latter is not physical, since it violates causality, so that we have to set $C_2 = 0$. The constant C_1 can be determined by imposing that (2.127) is satisfied in a small spherical volume V, centered in the origin, of radius r_0 and bounded by a spherical surface S:

$$\iiint_V \nabla^2 A_z\, dV + k^2 \iiint_V A_z\, dV = -\mu J_0 \iiint_V \delta(\mathbf{r})\, dV = -\mu J_0 \tag{2.131}$$

If the radius r_0 tends to zero, the right-hand side of (2.31) does not change, and the second term at the left-hand side vanishes, since $V \sim r_0^3$ and $A_z \sim 1/r_0$. The first term at the left-hand side of (2.131) can be more explicitly expressed by using (A.8), the divergence theorem (A.10), and (2.130) with $C_2 = 0$:

$$\iiint_V \nabla^2 A_z\, dV = \iiint_V \nabla \cdot \nabla A_z\, dV = \oiint_S \nabla A_z \cdot \hat{r}\, dS = \oiint_S \frac{\partial A_z}{\partial r}\bigg|_{r=r_0} dS$$

$$= 4\pi r_0^2 \frac{dA_z}{dr}\bigg|_{r=r_0} = -4\pi C_1\left(1 + jkr_0\right)\exp\left(-jkr_0\right) \tag{2.132}$$

Therefore, for r_0 approaching zero, this term tends to $-4\pi C_1$. We then have $-4\pi C_1 = -\mu J_0$, from which C_1 is easily obtained. In conclusion, the vector potential produced by the elementary source defined by (2.126) is given by

$$\mathbf{A} = \frac{\mu J_0}{4\pi r}\exp(-jkr)\hat{\mathbf{z}} \tag{2.133}$$

Therefore, the vector potential is parallel to the elementary source, and it only depends on the distance from the source itself. By exploiting homogeneity and isotropy, we can state that these properties are maintained for any position and orientation of the source, so that the vector potential of a short-current filament placed in an arbitrary point $\mathbf{r}'$ and arbitrarily oriented, $\mathbf{J}(\mathbf{r}) = \mathbf{J}_0\delta(\mathbf{r}-\mathbf{r}')$, can be expressed as

$$\mathbf{A}(\mathbf{r}) = \frac{\mu J_0}{4\pi|\mathbf{r}-\mathbf{r}'|}\exp(-jk|\mathbf{r}-\mathbf{r}'|) \tag{2.134}$$

2.4.2 Radiation from an Arbitrary Current Distribution

Let us consider an arbitrary current density distribution, different from zero in a volume V enclosed in a sphere of finite radius d, centered in the origin. By using the properties of the Dirac pulse [see (B.11)], the current density can be expressed as a superposition of elementary sources as follows:

$$\mathbf{J}(\mathbf{r}) = \iiint\limits_{\mathbf{r}'\in V} \mathbf{J}(\mathbf{r}')\delta(\mathbf{r}-\mathbf{r}')d\mathbf{r}' \tag{2.135}$$

The linearity of the Helmholtz equation (2.124) leads to the expression of the vector potential of this arbitrary current density as the linear superposition of solutions (2.134):

$$\mathbf{A}(\mathbf{r}) = \iiint\limits_{\mathbf{r}'\in V} \frac{\mu \mathbf{J}(\mathbf{r}')}{4\pi|\mathbf{r}-\mathbf{r}'|}\exp(-jk|\mathbf{r}-\mathbf{r}'|)d\mathbf{r}' \tag{2.136}$$

The electromagnetic field radiated by an arbitrary current density can be then computed by using (2.136) in (2.121) and (2.125):

$$\mathbf{E} = -\frac{j\omega\mu}{4\pi}\left(I + \frac{\nabla\nabla\cdot}{k^2}\right)\iiint\limits_{\mathbf{r}'\in V} \frac{\exp(-jk|\mathbf{r}-\mathbf{r}'|)}{|\mathbf{r}-\mathbf{r}'|}\mathbf{J}(\mathbf{r}')d\mathbf{r}'$$

$$\mathbf{H} = \frac{1}{4\pi}\nabla\times\iiint\limits_{\mathbf{r}'\in V} \frac{\exp(-jk|\mathbf{r}-\mathbf{r}'|)}{|\mathbf{r}-\mathbf{r}'|}\mathbf{J}(\mathbf{r}')d\mathbf{r}' \tag{2.137}$$

2.4.3 Far Field

The expressions in (2.137) can be analytically evaluated only for very few particular geometries, and therefore they are usually computed numerically. However, if the point in which the fields are evaluated is sufficiently far from the sources, these expressions simplify and show that all radiated fields share interesting common features that are independent of the particular source. In fact, if $r \gg d$ then $r'/r \ll 1$, and we can expand $|\mathbf{r} - \mathbf{r}'|$ as follows,

$$|\mathbf{r} - \mathbf{r}'| = \sqrt{(\mathbf{r} - \mathbf{r}') \cdot (\mathbf{r} - \mathbf{r}')} = \sqrt{r^2 - 2\mathbf{r} \cdot \mathbf{r}' + r'^2} = r\sqrt{1 - 2\frac{\hat{\mathbf{r}} \cdot \mathbf{r}'}{r} + \frac{r'^2}{r^2}}$$

$$\cong r\left[1 - \frac{\hat{\mathbf{r}} \cdot \mathbf{r}'}{r} + \frac{r'^2}{2r^2} - \frac{r'^2(\hat{\mathbf{r}} \cdot \hat{\mathbf{r}}')^2}{2r^2}\right] = r - \hat{\mathbf{r}} \cdot \mathbf{r}' + \frac{r'^2 \sin^2 \alpha}{2r} \qquad (2.138)$$

where α is the angle between $\mathbf{r}$ and $\mathbf{r}'$, and the expansion $\sqrt{1 + x} \cong 1 + x/2 - x^2/8$ has been used. Therefore, at the denominators of (2.137) we can set $|\mathbf{r} - \mathbf{r}'| \cong r$; in addition, if $kd^2/r \ll 1$ we can assume that $\exp(-jk|\mathbf{r} - \mathbf{r}'|) \cong \exp(-jkr)\exp(jk\hat{\mathbf{r}} \cdot \mathbf{r}')$, and (2.137) can be rewritten as

$$\mathbf{E} \cong -\frac{j\omega\mu}{4\pi}\left(I + \frac{\nabla\nabla \cdot}{k^2}\right)\exp(-jkr)\iiint_{\mathbf{r}' \in V} \frac{\exp(jk\hat{\mathbf{r}} \cdot \mathbf{r}')}{r} \mathbf{J}(\mathbf{r}')d\mathbf{r}'$$

$$\mathbf{H} \cong \frac{1}{4\pi}\nabla \times \exp(-jkr)\iiint_{\mathbf{r}' \in V} \frac{\exp(jk\hat{\mathbf{r}} \cdot \mathbf{r}')}{r} \mathbf{J}(\mathbf{r}')d\mathbf{r}' \qquad (2.139)$$

Finally, if $kr \gg 1$, then the integral in (2.139) is very slowly spatially varying compared to $\exp(-jkr)$, so that application of the operator ∇ in (2.139) can be approximated as multiplication by $-jk\hat{\mathbf{r}}$, and the neglected terms are infinitesimal of order greater than one for $r \to \infty$.

In conclusion, if the point $\mathbf{r}$ is in the *far zone* (or *Fraunhofer region*), or, in other words, if the following conditions

$$r \gg d, \quad r \gg \frac{d^2}{\lambda}, \quad r \gg \lambda \qquad (2.140)$$

are satisfied,[2] then the radiated electromagnetic field is expressed as (*far field*, or *radiation field*)

---

2. In writing these conditions, we are assuming that the medium is lossless, so that $k = 2\pi/\lambda$. Although the field expressions obtained in this section also hold for lossy media, the far region is of interest in practice only if losses are negligible, or anyway very small ($k = \beta - j\alpha$ with $\alpha \ll \beta$).

$$\mathbf{E} \cong -\frac{j\omega\mu}{4\pi r}\exp(-jkr)(I - \hat{\mathbf{r}}\hat{\mathbf{r}}\cdot)\iiint_{\mathbf{r}'\in V}\exp(jk\hat{\mathbf{r}}\cdot\mathbf{r}')\mathbf{J}(\mathbf{r}')\,d\mathbf{r}'$$

$$\mathbf{H} \cong -\frac{jk}{4\pi r}\exp(-jkr)\hat{\mathbf{r}} \times \iiint_{\mathbf{r}'\in V}\exp(jk\hat{\mathbf{r}}\cdot\mathbf{r}')\mathbf{J}(\mathbf{r}')\,d\mathbf{r}'$$

$$(2.141)$$

It is easy to realize that the operator $I - \hat{\mathbf{r}}\hat{\mathbf{r}}\cdot$ removes the component along the direction $\hat{\mathbf{r}}$, so that the electric field is perpendicular to the radial direction, $\hat{\mathbf{r}}\cdot\mathbf{E} \cong 0$, and it has only components along ϑ and φ. In addition, the second of (2.141) can be written as

$$\mathbf{H} \cong \frac{1}{\zeta}\hat{\mathbf{r}} \times \mathbf{E} \qquad (2.142)$$

Finally, the integral in (2.141) is independent of r, and it only depends on ϑ and φ.

Therefore, independently of the source, the electromagnetic field in the far zone behaves locally as a TEM plane wave propagating outward along the radial direction with velocity $v_p = 1/\sqrt{\varepsilon\mu}$, and attenuating as $1/r$. The source distribution dictates the field behavior as a function of the direction of radiation (i.e., the field dependence on ϑ and φ) and the field intensity and polarization.

By using (2.142), the Poynting vector of the far field is expressed as

$$\mathbf{S} \cong \frac{1}{2\zeta}|\mathbf{E}|^2\,\hat{\mathbf{r}} \qquad (2.143)$$

Its modulus attenuates as $1/r^2$, so that the power flux through a spherical surface centered in the origin is independent of the radius of the sphere, coherently with energy conservation (in a lossless medium).

2.4.4 Equivalent Problems and Magnetic Sources

In the real world, as far as we know today, only electric charges and currents exist, whereas no magnetic charges or currents have ever been observed. However, in some cases, a real radiation problem can be solved by considering a simpler "equivalent" radiation problem in which magnetic current and charge densities $\mathbf{j}_m$ and ρ_m may be present. If this is the case, Maxwell's equations (2.1) must be replaced by the following ones:

$$\begin{cases} \nabla \times \mathbf{e} = -\dfrac{\partial \mathbf{b}}{\partial t} - \mathbf{j}_m \\[4pt] \nabla \times \mathbf{h} = \dfrac{\partial \mathbf{d}}{\partial t} \\[4pt] \nabla \cdot \mathbf{d} = 0 \\[4pt] \nabla \cdot \mathbf{b} = -\rho_m \end{cases} \tag{2.144}$$

Note that (2.144) can be obtained from (2.1) with the substitutions $\mathbf{e} \leftrightarrow \mathbf{h}$, $\mathbf{d} \leftrightarrow -\mathbf{b}$, $\mathbf{j} \to -\mathbf{j}_m$, $\rho \to \rho_m$. Solutions of (2.144) can be then obtained from solutions of (2.1) by making the same substitutions. With regard to far fields produced by magnetic sources, based on (2.141) we obtain:

$$\mathbf{H} \cong -\frac{j\omega\varepsilon}{4\pi r}\exp(-jkr)\left(I - \hat{\mathbf{r}}\hat{\mathbf{r}}\cdot\right)\iiint_{\mathbf{r}'\in V}\exp\left(jk\hat{\mathbf{r}}\cdot\mathbf{r}'\right)\mathbf{J}_m(\mathbf{r}')d\mathbf{r}'$$

$$\mathbf{E} \cong \frac{jk}{4\pi r}\exp(-jkr)\hat{\mathbf{r}}\times\iiint_{\mathbf{r}'\in V}\exp\left(jk\hat{\mathbf{r}}\cdot\mathbf{r}'\right)\mathbf{J}_m(\mathbf{r}')d\mathbf{r}' \tag{2.145}$$

A first example of the approach described above is the *Ampere's equivalence principle*, which states that a current loop I with a radius that is small with respect to the wavelength, enclosing an area S with a normal unit vector $\hat{\mathbf{n}}$, radiates the same electromagnetic field as a magnetic current element

$$\mathbf{J}_m(\mathbf{r}) = \mathbf{J}_{m0}\delta(\mathbf{r}) = j\omega\mu SI\hat{\mathbf{n}}\delta(\mathbf{r}) \tag{2.146}$$

Another interesting example is the *equivalence theorem* (or *Love's field-equivalence principle*). Let us consider a source distribution $\mathbf{J}_0$ generating the electromagnetic field $\mathbf{E}$, $\mathbf{H}$. A smooth closed surface S with normal unit vector $\hat{\mathbf{n}}$ surrounds the sources. The equivalence theorem states that the original sources $\mathbf{J}_0$ can be removed and substituted by *equivalent sources* $\mathbf{J}_s = \hat{\mathbf{n}} \times \mathbf{H}_s$ and $\mathbf{J}_{ms} = -\hat{\mathbf{n}} \times \mathbf{E}_s$; that is, electric $\mathbf{J}_s$ and magnetic $\mathbf{J}_{ms}$ surface current densities distributed over the surface S, where $\mathbf{E}_s$ and $\mathbf{H}_s$ are the fields produced by $\mathbf{J}_0$ over S; see Figure 2.12. These equivalent sources generate a field $\mathbf{E}'$, $\mathbf{H}'$ coincident with $\mathbf{E}$, $\mathbf{H}$ in the volume outside S, and identically equal to zero in the volume inside S.

The equivalence theorem can be considered as a rigorous statement of the *Huygens principle*: Each point over S, hit by the radiation, in turn becomes a source of radiation for the space outside S. This result also clearly shows that, in principle, the electromagnetic field propagation from the source to any arbitrary point involves the entire space. However, in practice, only a

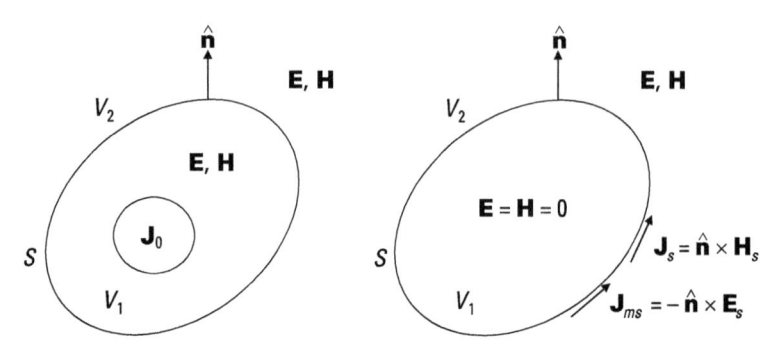

Figure 2.12 Equivalence theorem: actual problem (left) and equivalent problem (right).

finite volume is appreciably involved; it can be determined via a method that is illustrated in Section 3.2.

As a last example of use of an equivalent problem, we discuss the *image theorem*. Let us consider an electric current element radiating in a homogeneous half-space limited by a perfectly conducting plane. The image theorem states that the field produced by the current element in the half-space is equal to the field radiated in free space by the original current element and by its *image*. The latter is a current element placed symmetrically with respect to the plane, and such that its component orthogonal to the plane is equal to the one of the original current element, whereas its component parallel to the plane is equal to the opposite of the one of the original current element; see Figure 2.13.

2.5 Transmitting and Receiving Antennas

Antennas are (usually metallic) devices designed to efficiently radiate or collect electromagnetic signals. A transmitting antenna is powered by a generator, to which it is connected via a transmission line (usually a coaxial cable) or a waveguide. Electric currents are excited in the antenna, which radiates the

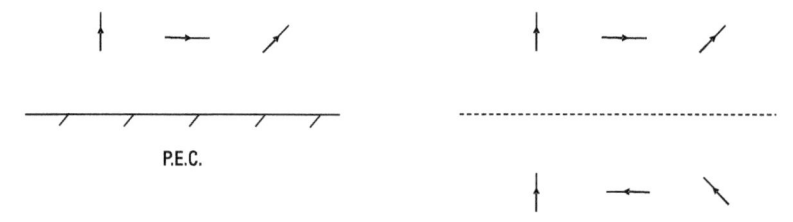

Figure 2.13 Image theorem: actual problem (left) and equivalent problem (right).

electromagnetic wave. Once the current distribution on the antenna is known, the radiated field can be computed as explained in Section 2.4. However, such a current distribution is actually unknown; it must be evaluated once the power supply circuit, possibly schematized by an ideal voltage generator with its series internal impedance, is assigned. In general, this is not an easy task, and the computation can be carried out analytically in some cases, with some form of approximation. More often, it is performed numerically. Similar considerations are valid for receiving antennas; see Section 2.5.2.

2.5.1 Parameters of a Transmitting Antenna

From an engineering viewpoint, the behavior of a transmitting antenna can be fully characterized by assigning a few parameters. First of all, the antenna *input impedance* is defined as the ratio between input voltage and input current:

$$Z_{in} = R_{in} + jX_{in} = \frac{V_0}{I_0} \tag{2.147}$$

Its real part R_{in} is the *input resistance*, and its imaginary part X_{in} is the *input reactance*. If the input antenna terminals are physically present, as in Figure 2.14, input voltage and current are those measured at the antenna terminals. Otherwise, a reference transversal section can be defined on the feeding transmission line, and V_0 and I_0 are the voltage and current at that reference section. The input impedance of the antenna must be matched to the impedance of the feeding line so that all the power provided by the feeding line is actually delivered to the antenna. The antenna *input power* P_{in} is expressed as

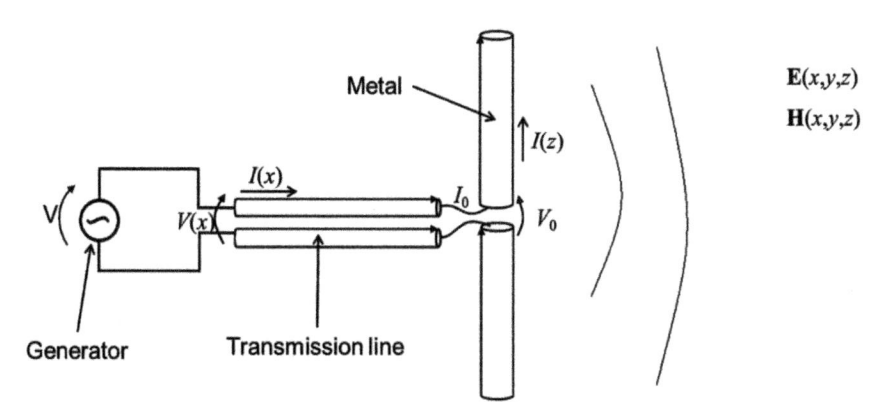

Figure 2.14 Transmitting antenna.

$$P_{\text{in}} = \frac{1}{2} R_{\text{in}} |I_0|^2 \tag{2.148}$$

In an ideal, perfectly conducting antenna, the input power is equal to the *radiated power* P_{rad}; however, in a real antenna, part of the input power is dissipated on the antenna itself due to the Joule effect, so that $P_{\text{rad}} < P_{\text{in}}$. We can then define the antenna *radiation resistance* R_{rad} as

$$R_{\text{rad}} = \frac{2 P_{\text{rad}}}{|I_0|^2} \tag{2.149}$$

so that the radiated power can be expressed in terms of the radiation resistance according to a formula analogous to (2.148). The ratio $\eta = P_{\text{rad}}/P_{\text{in}} = R_{\text{rad}}/R_{\text{in}}$ is the antenna efficiency. For an ideal antenna, $P_{\text{in}} = P_{\text{rad}}$, $R_{\text{in}} = R_{\text{rad}}$, and antenna efficiency is unitary; otherwise, $\eta < 1$.

By exploiting (2.141), and noting that $\omega\mu = k\zeta$, the radiated field in the far zone can be expressed as

$$\mathbf{E}(r,\vartheta,\varphi) = \frac{jk\zeta I_0}{4\pi r} \exp(-jkr)\mathbf{l}(\vartheta,\varphi)$$
$$\mathbf{H}(r,\vartheta,\varphi) = \frac{1}{\zeta}\hat{\mathbf{r}} \times \mathbf{E}(r,\vartheta,\varphi) \tag{2.150}$$

where

$$\mathbf{l}(\vartheta,\varphi) = (I - \hat{\mathbf{r}}\hat{\mathbf{r}}\cdot)\iiint\limits_{\mathbf{r}'\in V} \frac{\exp(jk\hat{\mathbf{r}}\cdot\mathbf{r}')\mathbf{J}(\mathbf{r}')}{-I_0} d\mathbf{r}' \tag{2.151}$$

The vector function $\mathbf{l}(\vartheta,\varphi)$ is termed antenna *effective length*, and $\mathbf{l}(\vartheta,\varphi)$ normalized to its maximum modulus (i.e., $\mathbf{l}(\vartheta,\varphi)/|\mathbf{l}(\vartheta,\varphi)|_{\text{max}}$), is called the antenna *radiation pattern*.

Input impedance and effective length completely characterize the antenna behavior with regard to the radiated field in the far zone. In fact, given the generator, the input current can be computed from the input imped-ance, and then, given the input current, the far field can be computed from the effective length. Indeed, these two parameters are partially related, since the radiation resistance can be expressed in terms of the effective length via the radiated power. In fact, the latter can be computed integrating (2.143), in which (2.150) is used, over a spherical surface centered in the origin; then the radiation resistance can be computed via (2.149):

$$R_{rad} = \frac{2P_{rad}}{|I_0|^2} = \frac{2\oiint_S \frac{1}{2\zeta}|\mathbf{E}|^2 \, dS}{|I_0|^2} = \frac{\zeta}{4\lambda^2} \int_0^{2\pi}\int_0^{\pi} |l(\vartheta,\varphi)|^2 \sin\vartheta \, d\vartheta \, d\varphi \quad (2.152)$$

Other useful functions can be defined that facilitate relating the radiated power density (i.e., radiated power flux per unit area) to the radiated or input power. The *directivity function* is defined as

$$D(\vartheta,\varphi) = \frac{\text{radiated power density}}{\text{average radiated power density}} = \frac{\frac{1}{2\zeta}|\mathbf{E}(r,\vartheta,\varphi)|^2}{\frac{P_{rad}}{4\pi r^2}} \quad (2.153)$$

The *maximum directivity D*, often referred to simply as *directivity*, is a measure of the ability of the antenna to concentrate the radiated power in a given direction. Therefore, antennas with high directivity (*directive antennas*) are typically used for point-to-point radio links, while antennas with low directivity are usually employed for signal broadcasting or in mobile devices.

The definition of *antenna gain function* is similar to the one of directivity:

$$G(\vartheta,\varphi) = \frac{\text{radiated power density}}{\text{average radiated power density for lossless antenna}}$$

$$= \frac{\frac{1}{2\zeta}|\mathbf{E}(r,\vartheta,\varphi)|^2}{\frac{P_{in}}{4\pi r^2}} = \eta D(\vartheta,\varphi) \quad (2.154)$$

The antenna *maximum gain G*, or simply *gain*, has the same meaning as the maximum directivity, but it is often more useful in practice, since it makes it possible to evaluate the maximum radiated power density from knowledge of the input power, which can be more easily measured than radiated power:

$$|\mathbf{S}|_{max} = \frac{P_{in}G}{4\pi r^2} = \frac{\text{EIRP}}{4\pi r^2} \quad (2.155)$$

where EIRP $= P_{in}G$ is the *effective isotropic radiated power*.

Gain (or directivity) can be also expressed in terms of effective length and input (or radiation) resistance. In fact, by using (2.148) [or (2.149)], and (2.150) in (2.154) [or (2.153)], we get

$$G(\vartheta,\varphi) = \frac{\pi\zeta\left|\mathbf{l}(\vartheta,\varphi)\right|^2}{\lambda^2 R_{in}} \quad \text{and} \quad D(\vartheta,\varphi) = \frac{\pi\zeta\left|\mathbf{l}(\vartheta,\varphi)\right|^2}{\lambda^2 R_{rad}} \quad (2.156)$$

If we use (2.152) in (2.156), approximate the antenna radiation pattern as equal to its maximum value within the 3-dB beam and equal to zero outside, and the direction of maximum radiation is in the horizontal (x-y) plane, we obtain a simple approximate relation between maximum directivity and beamwidth:

$$D = \frac{\pi\zeta\left|\mathbf{l}(\vartheta_{max},\varphi_{max})\right|^2}{\lambda^2 R_{rad}} = \frac{4\pi\left|\mathbf{l}(\pi/2,0)\right|^2}{\int_0^{2\pi}\int_0^{\pi}\left|\mathbf{l}(\vartheta,\varphi)\right|^2 \sin\vartheta \, d\vartheta \, d\varphi} \cong \frac{4\pi}{BW_{xy}BW_{xz}}$$

$$(2.157)$$

where BW_{xy} and BW_{xz} are the beamwidths, in radians, in the horizontal (x-y) and vertical (x-z) planes.

2.5.2 Parameters of a Receiving Antenna

A receiving antenna is connected, via a transmission line or a waveguide, to a receiver. Currents are excited on the antenna by an incident, locally plane wave, and in turn they generate voltage and current on the receiving circuit. The latter is schematized by a load impedance Z_L, which is the ratio between voltage and current at the receiving antenna (real or virtual (see Section 2.5.1)) terminals. The receiving antenna is schematized by an ideal voltage generator with its internal impedance. It can be shown that the internal impedance is equal to the antenna input impedance, whereas the equivalent generator voltage is the open-circuit voltage V_0 at the receiving antenna terminals. The latter is linearly dependent on the electric field of the incident wave:

$$V_0 = \mathbf{E}_i \cdot \mathbf{l}_r(\vartheta,\varphi) \quad (2.158)$$

where $\mathbf{l}_r$ is the *receiving effective length* of the antenna, and ϑ and φ individuate the direction of propagation of the incident wave. A fundamental result of antenna theory is that the receiving effective length is equal to the transmitting one: $\mathbf{l}_r = \mathbf{l}$. Accordingly, the same parameters $\mathbf{l}$ and Z_{in} specify both the transmitting and the receiving properties of an antenna. The final result is that if an antenna is able to efficiently radiate along a certain direction, then it is able to efficiently receive from that direction, too.

It can be noted that, for given receiving antenna and open-circuit voltage V_0 at its terminals, the power delivered to the receiver depends on the receiver impedance (i.e., the load impedance). This power is maximum if $Z_L = Z_{in}^*$ (*power-matching condition*), in which case $P_R = |V_0|^2/(8R_{in})$. Similarly, for a given receiving antenna and power density of the incident wave, the modulus of the open-circuit voltage, and hence the power delivered to the receiver, depends on the polarization of the incidence wave. The received power is maximum if $\mathbf{E}_i = a\mathbf{l}^*$, with a being a scalar constant, so that the modulus of the scalar product in (2.158) is maximum (*polarization-matching condition*). If $\mathbf{l}$ is real, this condition implies that the incident field is linearly polarized along the direction of $\mathbf{l}$.

Once power and polarization matching are achieved, the received power is directly proportional to the incident power density S_i via a constant—called the *antenna effective area* A_{eff}—that characterizes the antenna and only depends on the direction of propagation of the incident wave:

$$P_R = S_i A_{eff}(\vartheta, \varphi) \tag{2.159}$$

In general, the effective area may significantly differ from the geometric area of the antenna; however, the latter is a good approximation of the former if it is much larger than the wavelength squared.

It can be shown that gain and effective area are related by

$$G(\vartheta, \varphi) = \frac{4\pi}{\lambda^2} A_{eff}(\vartheta, \varphi) \tag{2.160}$$

This relation shows that a high directivity is obtained by using *electrically large* antennas (i.e., antennas much larger than wavelength).

2.5.3 Some Commonly Used Antennas

The simplest and perhaps most popular radiating and receiving devices are wire antennas. They consist of two coaxial thin cylinders of length l and radius a, separated by a small gap Δ, with $\Delta \ll a \ll \lambda$, and $a \ll l$; see Figure 2.15(a). Wire antennas have low directivity (in the range of 1.5 to 1.65), and the most efficient among them is the halfwave dipole, whose length $2l$ is equal to $\lambda/2$. In fact, this antenna has a purely resistive input impedance equal to about 75Ω, easily matched to available coaxial cables, and its efficiency is nearly unitary. When size constraints prevent the use of a halfwave dipole, a short dipole can be used. It is characterized by the condition $l \ll \lambda$. However, it has a small radiation resistance (see Table 2.4), a low efficiency, and a large frequency-depending capacitive input reactance that must be compensated for by using a coil to match the antenna to the feeding coaxial cable. The coil further increases ohmic losses.

Another simple, small, low-directivity and low-efficiency antenna is the small current loop, consisting of N turns wrapped into a circular coil of radius $a \ll \lambda$; see Figure 2.15(b). This antenna has a large frequency-depending inductive input reactance and high ohmic losses. To increase radiation resistance and hence antenna efficiency, the coil may be wrapped on a ferrite core with a high relative magnetic relative permittivity μ_r. In any case, due to its very low efficiency, the small loop antenna is used only as a receiving antenna.

Short-dipole, halfwave-dipole, and small-loop antennas have the maximum of radiation along the directions in the plane perpendicular to their axis; and they, like all wire antennas, radiate a null far field along the direction of their axis.

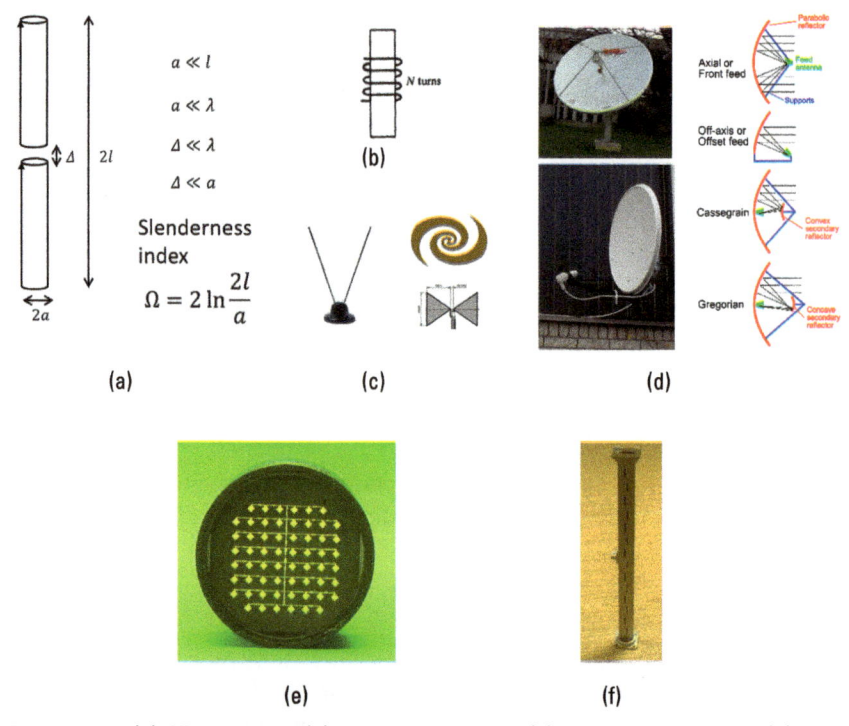

Figure 2.15 (a) Wire antenna, (b) small-loop antenna, (c) broadband antennas, (d) paraboloid reflector antennas, (e) patch antennas, and (f) slot antennas.

The aforementioned antennas can only radiate (or receive) in a narrow band around their nominal working frequency; in fact, the power provided by the feeding line is entirely delivered to the antenna only if the frequency-depending matching condition is met. When the frequency significantly deviates from the nominal one, matching is lost, and most of the power provided by the feeding line is reflected back toward the transmitter. However, broadband antennas exist, such that their input impedance, and radiation pattern, are practically constant with frequency over a wide band. Among these antennas we mention the biconical antenna, antennas derived from it (e.g., the rabbit-ear antenna and bow-tie antenna), and antennas with a scale-invariant geometry (e.g., the logarithmic-spiral antenna and fractal antennas). Some broadband antennas are depicted in Figure 2.15(c).

All the antennas mentioned so far have a low directivity. High directivity can be obtained by either using arrays of antennas, described in Section 2.5.4, or employing paraboloidal reflector antennas. A paraboloidal reflector is illuminated by a primary source (*feed*) usually consisting of a horn antenna (i.e., a waveguide whose terminal part is flared into a horn). The feed is placed in the paraboloid focus, so that radiation is reflected by the paraboloid mainly along its symmetry axis; see Figure 2.15(d). Double-reflector geometries can be also used to simplify the feeding circuit implementation. The effective area of an electrically large reflector is equal to its geometric area multiplied by its efficiency. A well-designed paraboloid reflector has an efficiency η of about 0.8. Its gain is obtained via (2.160).

Finally, patch antennas [see Figure 2.15(e)] are useful if feeding is provided by printed circuits, whereas slot antennas [see Figure 2.15(f)] are sometimes employed when feeding is provided by waveguides. Both are more commonly used as elements of arrays rather than as individual antennas.

Table 2.4 lists the effective length, radiation resistance, and gain of some of the antennas discussed to this point. The effective length is computed according to (2.151), or (2.145) and (2.146) for the loop antenna; radiation resistance is determined via (2.152); and gain is calculated via (2.154) or (2.160).

2.5.4 Arrays of Antennas, Phased Arrays, and Beamforming

As shown above, simple antennas, except large reflectors, have a low directivity. To increase directivity, simple radiating elements can be clustered to form an *antenna array*. The overall radiated field is the coherent summation of the fields radiated by the different elements, and radiation is reinforced around the direction (or directions) for which the different contributions sum up in

Table 2.4

Parameters of Some Commonly Used Antennas

Antenna	Effective length (I)	Radiation Resistance (R_{rad})	Gain (G)
Wire antenna: short dipole	$l \sin \vartheta \hat{\vartheta}$	$\frac{2}{3} \pi \zeta_0 \left(\frac{l}{\lambda} \right)^2 \cong 800 \left(\frac{l}{\lambda} \right)^2 \Omega$	$\eta 1.5$
Wire antenna: halfwave dipole	$\frac{\lambda}{\pi} \dfrac{\cos\left(\frac{\pi}{2} \cos \vartheta \right)}{\sin \vartheta} \hat{\vartheta}$	$\cong 75\Omega$	$\cong 1.65$
Small-loop antenna	$-j \mu_r N k_0 \pi a^2 \sin \vartheta \hat{\varphi}$	$\frac{2}{3} \pi \zeta_0 \left(\frac{\mu_r N k_0 \pi a^2}{\lambda} \right)^2 \cong 32000 \left(\frac{\mu_r N \pi a^2}{\lambda^2} \right)^2 \Omega$	$\eta 1.5$
Paraboloidal reflector antenna	Depending on feed geometry	Depending on feed geometry	$\frac{4\pi}{\lambda^2} \eta A$

The employed reference system has the z-axis coincident with the antenna symmetry axis.

phase, while it is weakened in the directions where contributions are out of phase. The result is an increased directivity. If radiating elements are aligned along a specific direction, a *linear array* is obtained; if they are distributed over a plane, a *planar array* is achieved. The elements of *conformal arrays* are distributed over nonplanar surfaces.

The most popular arrays are Yagi-Uda antennas [see Figure 2.16(a)] commonly used to receive terrestrial television broadcasting signals. Broadband arrays can be obtained either by using broadband antennas as radiating elements, or by arranging the elements according to scale-invariant geometries; a common example is the log-periodic array of Figure 2.16(b). Antennas employed for mobile telephony base stations are usually linear [Figure 2.16(c)] or less frequently planar [Figure 2.16(d)] arrays, often enclosed in a *radome* (i.e., a box made of a material transparent at radio frequencies that protects the antenna from dangerous atmospheric phenomena).

If, as it is usually the case, all the radiating elements are equal and with the same orientation, the field radiated in the far zone is

$$\mathbf{E}(r,\vartheta,\varphi) = \frac{jk\zeta I_0}{4\pi r} \exp(-jkr) \mathbf{l}_e(\vartheta,\varphi) F(\vartheta,\varphi)$$

$$\text{with } F(\vartheta,\varphi) = \sum_{n=0}^{N-1} a_n \exp\left(jk\mathbf{r}'_n \cdot \hat{\mathbf{r}} \right)$$

(2.161)

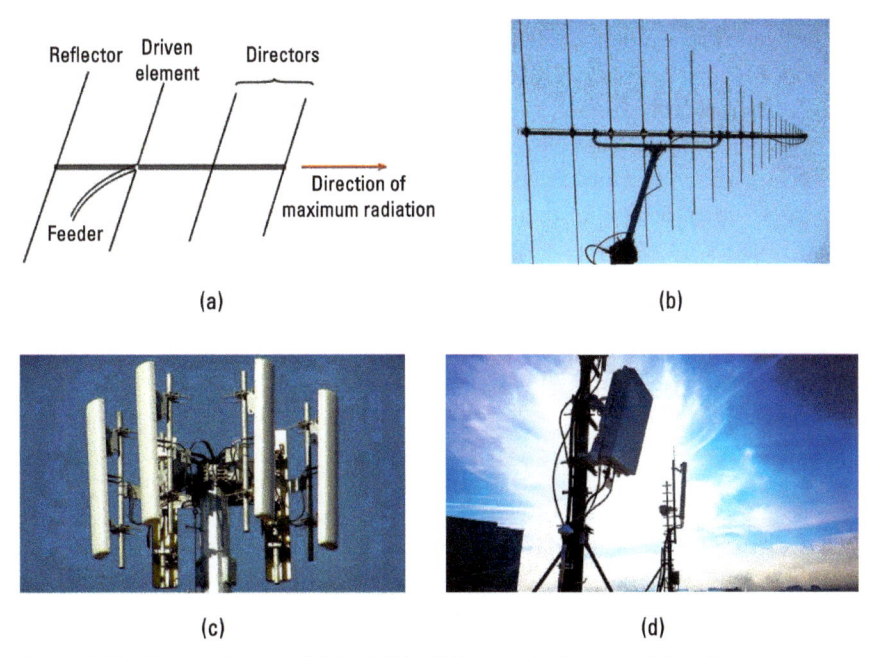

Figure 2.16 Types of arrays: (a) Yagi-Uda, (b) log-periodic array, (c) six linear arrays (with radome) for mobile telephony base stations, and (d) planar array (with radome) for 5G mobile telephony base stations.

where $\mathbf{l}_e$ is the effective length of the radiating elements (*element factor*), F is the *array factor*, $\mathbf{r}'_n$ is the position vector of the nth element, I_0 is the input current of the element $n = 0$ (*reference current*), and a_n is the normalized input current of the nth element. If the elements are aligned along the z direction and are equally spaced (*uniform array*) with spacing d, then $\mathbf{r}'_n = nd\hat{\mathbf{z}}$ and

$$F(\vartheta,\varphi) = F(\vartheta) = \sum_{n=0}^{N-1} a_n \exp\left(jnkd\cos\vartheta\right) \tag{2.162}$$

If the input currents have all the same modulus, and their phases vary progressively [i.e., $a_n = a\exp(-jnu_0)$, with u_0 being the phase difference between two adjacent elements], then

$$\left|F(\vartheta)\right| = a\left|\sum_{n=0}^{N-1} \exp\left[jn\left(kd\cos\vartheta - u_0\right)\right]\right| = a\left|\frac{1 - \exp\left[jN\left(kd\cos\vartheta - u_0\right)\right]}{1 - \exp\left[j\left(kd\cos\vartheta - u_0\right)\right]}\right|$$

$$= a\left|\frac{\sin\left[\frac{N}{2}\left(kd\cos\vartheta - u_0\right)\right]}{\sin\left[\frac{1}{2}\left(kd\cos\vartheta - u_0\right)\right]}\right| \tag{2.163}$$

Equation (2.163) shows that by varying u_0 it is possible to change the direction of maximum radiation of the array (i.e., the pointing direction of the radiated beam). In fact, the maximum of the array factor is obtained for $\vartheta = \vartheta_0$, with ϑ_0 such that $kd\cos\vartheta_0 = u_0$. If $u_0 = 0$ (i.e., all the elements are in phase), the maximum is obtained for $\vartheta = \pi/2$, and the array is called a *broadside array*. If $u_0 = kd$ then the maximum is obtained for $\vartheta = 0$, and the array is called an *end-fire array*.

Figure 2.17 shows the plot of $|F|$ for a broadside array. The effective width $\Delta\vartheta$ of the radiated beam can be evaluated by computing the position of the first null of $|F|$, which is obtained for $Nkd\cos(\pi/2 - \Delta\vartheta)/2 = \pi$, so that, recalling that $\cos(\pi/2 - \Delta\vartheta) = \sin(\Delta\vartheta)$,

$$\sin\Delta\vartheta \cong \Delta\vartheta = \frac{2\pi}{Nkd} = \frac{\lambda}{Nd} \tag{2.164}$$

Therefore, the beamwidth reduces (and hence the directivity increases) as the array length Nd increases. An analogous result is obtained for other beam-pointing directions. However, the spacing d cannot be arbitrarily increased, to avoid the appearance of undesired additional maxima of radiation (*grating lobes*) due to the periodicity of (2.163). It can be verified that grating lobes are always avoided if d does not exceed $\lambda/2$. Therefore, to increase the array length, the number of elements N must be increased.

If electronically controlled phase shifters are inserted at the array element inputs, it is possible to electronically steer the radiated beam (*phased arrays*). If a planar, instead of linear, array is used, a narrow pencil beam can be obtained, that can be electronically steered in any direction. In more general terms, by

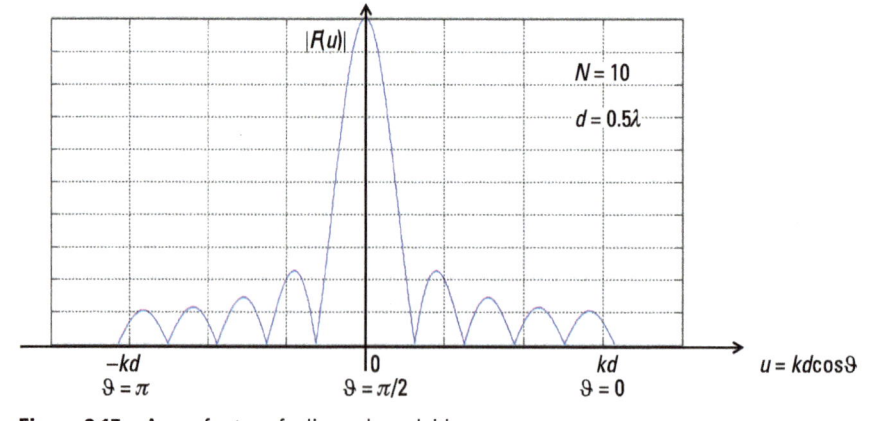

Figure 2.17   Array factor of a linear broadside array.

properly designing array element input currents (selecting both their moduli and their phases), desired beam shape and pointing direction can be obtained (*beamforming*). In any case, the beamwidth in radians cannot be smaller than a value of the order of λ/L, where L is the linear size of the array, and directivity cannot be larger than $4\pi A/\lambda^2$, where A is the area covered by the array.

Phased arrays and beamforming have been used in radars for military applications for some decades, but now that the costs of this technology have decreased, it is being employed in antennas for the base stations of the new 5G mobile phone systems.

2.6 Friis Formula for Free-Space Radio Links

Let us consider a transmitting antenna with input power P_T and gain G_T, radiating in free space; and let us assume that at distance r, in the far zone, along the direction of the maximum radiation of the transmitting antenna, a receiving antenna is placed, with effective area A_R and gain G_R, pointed toward the transmitting antenna and oriented in such a way that polarization matching is achieved. The power density of the wave radiated by the transmitting antenna and impinging on the receiving one can be computed via (2.155):

$$S_i = \frac{P_T G_T}{4\pi r^2} \tag{2.165}$$

and the power P_R delivered by the receiving antenna to a matched load (the *received power*) can be computed via (2.159) and (2.160):

$$P_R = S_i A_R = S_i \frac{\lambda^2}{4\pi} G_R \tag{2.166}$$

By combining (2.165) and (2.166) we obtain the *Friis formula*[3] (or *radio-link formula*), which directly relates the received power to the transmitted one:

$$P_R = P_T G_T G_R \left(\frac{\lambda}{4\pi r}\right)^2 \tag{2.167}$$

The Friis formula is often expressed in decibels, where the value in decibels of a power P is $[P]_{dB} = 10\,\text{Log}(P)$. Accordingly, (2.167) can be written as

---

3. Danish-American radio engineer Harald T. Friis derived the formula in 1946.

$$\left[P_R \right]_{dB} = \left[P_T \right]_{dB} + \left[G_T \right]_{dB} + \left[G_R \right]_{dB} + 20 \operatorname{Log}\left(\frac{c}{4\pi r f} \right)$$
$$= \left[P_T \right]_{dB} + \left[G_T \right]_{dB} + \left[G_R \right]_{dB} - L_{FS}$$

(2.168)

where $f = \omega/(2\pi) = ck/(2\pi) = c/\lambda$ is the frequency and L_{FS} is the *free-space loss*:

$$L_{FS} = -20 \operatorname{Log}\left(\frac{c}{4\pi r f} \right) = 20 \operatorname{Log}\left(\frac{4\pi}{c} \right) + 20 \operatorname{Log}(f) + 20 \operatorname{Log}(r)$$

$$\cong -147.6 + 20 \operatorname{Log}(f) + 20 \operatorname{Log}(r) = 32.4 + 20 \operatorname{Log}\left(f_{MHz} \right) + 20 \operatorname{Log}\left(r_{km} \right)$$

(2.169)

This formula is also provided when megahertz and kilometers are used as measurement units, because these are often applied in telecommunication design and applications. Further loss terms may be needed to account for pointing and polarization mismatches, but they can be easily made negligible by adjusting antennas' pointing and orientations.

The Friis formula is a very easy-to-use tool for the design of radio links. Keep in mind, however, that it refers to transmitting and receiving antennas operating in free space. In the real world, antennas operate in a complex environment in which obstacles are present. Therefore, this formula can be used only when at least one of the antennas is very directive, and the obstacles are sufficiently far from the line of sight (LOS) connecting the two antennas. The minimum clearance from the LOS can be determined via a method illustrated in Section 3.2. These conditions are usually met in radio or microwave links between directive antennas placed at elevated sites, and in satellite microwave links. Conversely, for links between nondirective antennas in a complex environment, such as an urban scenario, the Friis formula cannot be used as it is, and methods presented in the following chapters must be employed.

The received power, possibly computed via the Friis formula, must be compared to the noise power at the receiver, to verify that a sufficient signal-to-noise ratio (SNR) is obtained.

2.6.1 Antenna Noise Temperature and Receiver Noise Figure

Any material body that has a temperature greater than the absolute zero radiates an electromagnetic field (*thermal radiation*), due to the thermal agitation of the microscopic charged particles that compose matter. For the same reason, any electrical or electronic device generates at its terminals a random voltage (*thermal noise*) that, in telecommunication systems, disturbs the signal. In particular, a resistor at temperature T, measured in kelvin (K), generates a thermal noise whose power spectral density (i.e., power per frequency unit, measured in watts per hertz or joules) is KT, where $K = 1.38 \cdot 10^{-23}$ joules per kelvin (J/K) is the *Boltzmann's constant*. The noise power spectral density produced by a generic electrical or electronic device can be then expressed as KT_n, where T_n is the *noise temperature* of the device (i.e., the temperature that a resistor should have in order to produce the same noise power spectral density as the considered device). Therefore, in general, the noise temperature of a device may differ from its physical temperature, and it may be frequency-dependent.

Let us now consider a telecommunication system that is used to transmit signals of bandwidth $\Delta f = \Delta \omega / (2\pi)$. Spectral noise components outside this bandwidth can be filtered out, so that only spectral components within this bandwidth contribute to the noise power. Therefore, according to the definitions above, the noise power at the receiver of this telecommunication system is $N = KT_n \Delta f$, where T_n is the system noise temperature. In order to evaluate T_n, we have to consider that the noise at the receiver is the sum of two contributions: the noise picked up by the receiving antenna and the thermal noise generated by the electronic components of the receiver itself. The former is due to the thermal radiation of bodies placed around the receiving antenna (ground, atmosphere, and cosmic thermal radiation), and, at lower frequency (up to a few megahertz), to electrical discharges, such as lightning that could occur at any moment somewhere in the Earth's atmosphere. The power of the noise received by the antenna can be computed as $N_a = KT_a \Delta f$, where T_a is the *antenna noise temperature*, which is usually determined experimentally. It depends on the antenna gain function and pointing direction, and it varies with frequency: At frequencies below 1 MHz it is extremely high, ranging from thousands to billions of kelvin; conversely, in the UHF band (300 MHz–3 GHz) and at microwave frequencies it is not larger than about 100K, being as low as about 10K for antennas pointing at high elevation angles.

Similarly, the power of thermal noise generated by the electronic components of the receiver can be computed as $N_r = KT_r \Delta f$, where T_r is the *receiver noise temperature*. The latter is sometimes expressed in terms of the *receiver noise figure F* via the relation $T_r = (F - 1)T_0$, where T_0 is a reference temperature usually set to 290K.

Since the two noise contributions are uncorrelated, the total noise power N is the sum of the two powers, so that

$$N = N_a + N_r = K\left(T_a + T_r\right)\Delta f = K\left[T_a + (F-1)T_0\right]\Delta f = KT_n \Delta f \quad (2.170)$$

2.6.2 Example: Downlink in a Satellite Communication System

Let us consider a geostationary telecommunication satellite transmitting toward a receiving ground station. The parameters of the link are listed in Table 2.5. Let us compute the SNR at the receiver.

By using (2.169), the free-space loss is computed as

$$L_{FS} = 32.4 + 20\,\mathrm{Log}(12000) + 20\,\mathrm{Log}(37132) = 32.4 + 81.6 + 91.4 = 205.4 \text{ dB}$$

The received power can be then evaluated by the Friis formula (2.168):

$$\left[P_R\right]_{dB} = 23 + 36 + 53 - 205.4 = -93.4 \text{ dB}$$

The received power is then −93.4 dB, or −63.4 dBm (decibels over 1 mW) (i.e., 4.6×10^{-10} W).

The system noise temperature is $T_n = T_a + (F - 1)T_0 = 50 + 14 \cdot 290 = 50 + 4060 = 4110\mathrm{K}$, and this shows that at microwave frequencies the noise produced by the receiver is usually dominant with respect to noise received by the antenna. By using (2.170), we get the following noise power: $N = KT_n\Delta f$

Table 2.5
Parameters of a Satellite Telecommunication Downlink

Parameter	Value
Frequency f	12 GHz
Transmitted power (P_T)	200W (or 23 dB, or 53 dBm)
Transmitting antenna gain (G_T)	4,000 (or 36 dB)
Receiving antenna gain (G_R)	200,000 (or 53 dB)
Range r	37,132 km
Receiving antenna noise temperature (T_a)	50K
Receiver bandwidth (Δf)	10 MHz
Receiver noise figure (F)	15 (or 11.8 dB)

$= 1.38 \cdot 10^{-23} \cdot 4110 \cdot 10^7 \cong 5.7 \cdot 10^{-13}$W, or -122.4 dB. Therefore, the SNR is $[P_R]_{\text{dB}} - [N]_{\text{dB}} = -93.4 + 122.4 = 29$ dB.

References

[1]   Franceschetti, G., *Electromagnetics: Theory, Techniques, and Engineering Paradigms*, New York: Plenum Press, 1997.

[2]   Stratton, J. A., *Electromagnetic Theory*, New York: McGraw-Hill, 1941.

[3]   Harrington, R. F., *Time-Harmonic Electromagnetic Fields*, New York: McGraw-Hill, 1961.

[4]   Jones, D. S., *The Theory of Electromagnetism*, Oxford: Pergamon Press, 1964.

[5]   Papas, C. H., *Theory of Electromagnetic Wave Propagation*, New York: McGraw-Hill, 1965.

[6]   Van Bladel, J., *Electromagnetic Fields*, Washington: Hemisphere, 1985.

[7]   Kong, J. A., *Electromagnetic Wave Theory*, New York: John Wiley and Sons, 1990.

[8]   Felsen, L. B., and N. Marcuvitz, *Radiation and Scattering of Waves*, New York: IEEE Press, 1994.

[9]   Balanis, C. A., *Antenna Theory: Analysis and Design*, New York: Harper and Row, 1982.

[10]   Collin, E. R., *Antennas and Radiowave Propagation*, New York: McGraw-Hill, 1985.

[11]   Cole, K. S., and R. H. Cole, "Dispersion and Absorption in Dielectrics I—Alternating Current Characteristics," *J. Chem. Phys.*, Vol. 9, No. 4, 1941, pp. 341–352.

3

Asymptotic Techniques

This chapter illustrates a set of mathematical tools and techniques that are needed for the design of radio links in complex environments, such as urban scenarios. In fact, the Friis formula [see (2.67)], is a very useful and simple tool for the design of radio links. However, it is obtained by assuming that transmitting and receiving antennas operate in free space (i.e., when there are no obstacles, and the medium is lossless and homogeneous). Therefore, it can be safely employed only to design radio or microwave links between directive antennas placed at elevated sites, and satellite microwave links. Conversely, for links between nondirective antennas in a complex environment, such as an urban scenario, the Friis formula cannot be used as it is, and different procedures must be employed. These procedures can rely on the solution of Maxwell's equations in a nonhomogeneous space, with obstacles showing different, possibly space-varying, complex dielectric constants. In this more general scenario, exact analytical solutions of Maxwell's equations are not available. Two different approaches can be then used in this case.

The first one is to employ *numerical methods*: discretization of space (and time, if time domain Maxwell's equations are considered) is performed, so that space (and time) derivatives appearing in Maxwell's equations are approximated by corresponding difference quotients, and finite difference equations are obtained. These can be solved by resorting to available methods for the solution of algebraic linear systems of equations. This approach leads to some widespread methods, including the finite-difference time domain (FDTD) method, the finite element method (FEM), and the method of moments (MoM) [1]. However, numerical methods can be fruitfully used in practice only if the linear size of the volume of interest is not very large compared to wavelength. In fact, in order to obtain a sufficient accuracy, the spatial sampling step must be a small fraction of wavelength. Considering that the unknowns of the problem are the values of electric and magnetic vector fields (i.e., six complex scalar values) at each point of the sampling *grid* (or *mesh*), the number of unknowns (and of equations to solve) becomes too high if the linear size of the considered volume is much larger than wavelength, as is the case for radio propagation in urban areas. As a matter of fact, in this case, wavelength is on the order of tens of centimeters, whereas the linear size of the volume of interest is on the order of at least a few hundred meters, so that grid points are on the order of at least 10^{12}. Such a huge number of grid points and, hence, of equations and unknowns leads to unacceptable processing times. Therefore, in this book numerical methods are not considered, and the interested reader is referred to available textbooks on this specific topic; see, for example, [1].

The second approach, more suitable for studying radio wave propagation in complex scenarios where the size is much larger than the wavelength, such as urban areas, is to employ *asymptotic techniques*; the solution of Maxwell's equations is obtained by performing proper asymptotic expansions either of the equations themselves, or of their solutions expressed in integral form. This leads to some useful methods, which are illustrated in this chapter. The limiting parameter of the asymptotic expansion is usually the frequency, and its limiting value is usually infinity. Therefore, these techniques are also termed *high-frequency techniques*.

Sections 3.1 to 3.6 illustrate the main concepts of such techniques, focusing on the results that are useful to describe propagation in urban areas. A more in-depth discussion of these techniques can be found in [2, 3].

3.1 Geometrical Optics

Let us consider the source-free Maxwell's equations in the phasor domain for a linear, local, isotropic, stationary, nonhomogeneous medium:

$$\begin{cases} \nabla \times \mathbf{E} = -j\omega\mu_0\mu_r \mathbf{H} \\ \nabla \times \mathbf{H} = j\omega\varepsilon_0\varepsilon_r \mathbf{E} \\ \nabla \cdot \varepsilon_r \mathbf{E} = 0 \\ \nabla \cdot \mu_r \mathbf{H} = 0 \end{cases} \qquad (3.1)$$

where $\varepsilon_r = \varepsilon_r(\mathbf{r})$ and $\mu_r = \mu_r(\mathbf{r})$ are the medium space-varying relative permittivity and permeability, respectively. We now assume that the functions governing the fields can be expressed as

$$\mathbf{E}(\mathbf{r}) = \mathbf{E}_0(\mathbf{r})\exp\left[-jk_0 L(\mathbf{r})\right]$$
$$\mathbf{H}(\mathbf{r}) = \mathbf{H}_0(\mathbf{r})\exp\left[-jk_0 L(\mathbf{r})\right]$$

(3.2)

where $k_0 = \omega\sqrt{\varepsilon_0 \mu_0}$ is the free-space propagation constant, and $L(\mathbf{r})$ is called the *eikonal function*. By substituting (3.2) in (3.1) and using (A.4) and (A.5), we obtain

$$\begin{cases} \nabla L \times \mathbf{E}_0 - \dfrac{\nabla \times \mathbf{E}_0}{jk_0} = \zeta_0 \mu_r \mathbf{H}_0 \\[2mm] \nabla L \times \mathbf{H}_0 - \dfrac{\nabla \times \mathbf{H}_0}{jk_0} = -\dfrac{1}{\zeta_0}\varepsilon_r \mathbf{E}_0 \\[2mm] \nabla L \cdot \mathbf{E}_0 - \dfrac{\nabla \cdot \varepsilon_r \mathbf{E}_0}{jk_0 \varepsilon_r} = 0 \\[2mm] \nabla L \cdot \mathbf{H}_0 - \dfrac{\nabla \cdot \mu_r \mathbf{H}_0}{jk_0 \mu_r} = 0 \end{cases}$$

(3.3)

where $\zeta_0 = \sqrt{\mu_0/\varepsilon_0}$ is the free-space intrinsic impedance. If the frequency tends to infinity (i.e., $\omega \to \infty$, so that $k_0 \to \infty$ and $\lambda \to 0$), and if we assume that the fields and their derivatives do not diverge, then the second terms at the left-hand sides of (3.3) vanish, and the equations in (3.3) become

$$\begin{cases} \nabla L \times \mathbf{E}_0 = \zeta_0 \mu_r \mathbf{H}_0 \\[2mm] \nabla L \times \mathbf{H}_0 = -\dfrac{1}{\zeta_0}\varepsilon_r \mathbf{E}_0 \\[2mm] \nabla L \cdot \mathbf{E}_0 = 0 \\[2mm] \nabla L \cdot \mathbf{H}_0 = 0 \end{cases}$$

(3.4)

Fields expressed by (3.2) with $\mathbf{E}_0$, $\mathbf{H}_0$, and L satisfying (3.4) are the *geometrical optics* (GO) fields. By vectorially multiplying both sides of the first of (3.4) by ∇L, and then using the second of (3.4), we get

$$\nabla L \times \left(\nabla L \times \mathbf{E}_0\right) = -n^2 \mathbf{E}_0$$

(3.5)

where $n = \sqrt{\varepsilon_r \mu_r}$ is the medium space-varying refractive index. By using (A.2) and then the third of (3.4), we finally obtain

$$\nabla L \cdot \nabla L = n^2 \qquad (3.6)$$

which is the eikonal equation. It can be satisfied by

$$\nabla L = n\hat{\mathbf{s}} \qquad (3.7)$$

so that, for a lossless medium, ∇L is a vector field whose amplitude is the refractive index and whose direction is provided by the unit vector $\hat{\mathbf{s}}$.

In view of (3.4), ∇L is perpendicular to the electric and magnetic fields; its *field lines* (i.e., the curves that are tangent to ∇L at each point along their length) are called *rays*. Therefore, rays are perpendicular to the surfaces on which L is constant, which are the *wavefronts*, and the tangent unit vector of a ray is $\hat{\mathbf{s}}$. If rays in a volume V have no mutual intersection (i.e., if one and only one ray passes through each point of V), then they are said to form a *ray congruence*; see Figure 3.1.

It is easy to verify that electromagnetic power flux is along the rays; in fact, by using (3.4), (A.2), and (3.7), the Poynting vector can be expressed as

$$S = \frac{1}{2}\mathbf{E} \times \mathbf{H}^* = \frac{1}{2\zeta_0 \mu_r}\mathbf{E}_0 \times \left(\nabla L \times \mathbf{E}_0^*\right) = \frac{n\hat{\mathbf{s}}}{2\zeta_0 \mu_r}\left|\mathbf{E}_0\right|^2 = \frac{1}{2\zeta}\left|\mathbf{E}_0\right|^2\hat{\mathbf{s}} \qquad (3.8)$$

where ζ is the medium space-varying intrinsic impedance. In addition, it can be shown that if the electromagnetic field is known at a point O of a given ray, then it can be computed at any point P along the same ray. In fact, the eikonal function L along a ray γ can be computed by performing the line integral of its gradient ∇L along γ, and then using (3.7):

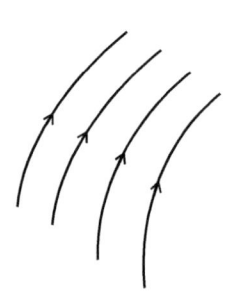

Figure 3.1 Ray congruence.

$$L(\mathrm{P}) = L(\mathrm{O}) + \gamma \int_{\mathrm{O}}^{\mathrm{P}} \nabla L \cdot \hat{\mathbf{s}}\, dl = L(\mathrm{O}) + \gamma \int_{\mathrm{O}}^{\mathrm{P}} n\, dl \qquad (3.9)$$

Note that in (3.9) and (3.12), the notation $c \int_{A}^{B}$ stands for "line integral from point A to point B along the curve c."

The integral of the refractive index from O to P along a curve c, $c \int_{\mathrm{O}}^{\mathrm{P}} n\, dl$, is called the *optical length* of c from O to P. Accordingly, (3.9), which is called the *phase transport equation*, states that the variation of the eikonal function from O to P along a ray is equal to the optical length of the ray path from O to P. The phase variation is then easily obtained by multiplying the optical length by k_0; see (3.2).

The *amplitude transport equation* can be obtained by using energy conservation as follows. Let us define a *flux tube* as the surface formed by all the rays passing through a closed contour. In view of (3.8), there is no power flux through such a surface. Let us now consider a point O on a ray γ, and let us consider, on the wavefront including O, a vanishingly small surface element $dS(\mathrm{O})$ whose contour defines a flux tube; see Figure 3.2. This flux tube intersects a second wavefront, including the point P of γ, so defining the (vanishingly small) surface element $dS(\mathrm{P})$. If the medium is lossless, the power flux through $dS(\mathrm{O})$ must be equal to the one through $dS(\mathrm{P})$, because there is no power dissipation and no power flux through the tube surface. We can then write, also recalling that rays are perpendicular to wavefronts, and hence to $dS(\mathrm{O})$ and $dS(\mathrm{P})$,

$$\frac{1}{2\zeta(\mathrm{P})}\left|\mathbf{E}_0(\mathrm{P})\right|^2 dS(\mathrm{P}) = \frac{1}{2\zeta(\mathrm{O})}\left|\mathbf{E}_0(\mathrm{O})\right|^2 dS(\mathrm{O}) \qquad (3.10)$$

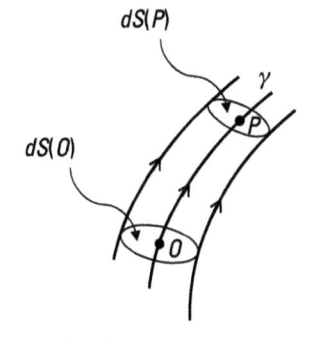

Figure 3.2 Flux tube and power density.

The amplitude transport equation is then obtained as

$$\left|\mathbf{E}_0(P)\right| = \left|\mathbf{E}_0(O)\right|\sqrt{\frac{dS(O)\zeta(P)}{dS(P)\zeta(O)}} \tag{3.11}$$

A more general transport equation for the vector $\mathbf{E}_0$ that also makes it possible to compute how the wave polarization state changes along a ray, is available [2–5], but its expression in the general nonhomogeneous medium case is beyond the scope of this book.

In conclusion, the results in (3.9) and (3.11) show that in the GO limit (i.e., if the frequency tends to infinity), the value of the electromagnetic field at a point P only depends on the field values along the ray passing through P, so that propagation is along the rays; GO propagation from a point O to a point P only involves points on the ray connecting O and P.

The GO solution can be used to approximate electromagnetic propagation at a finite frequency when the corresponding wavelength is much smaller than the obstacles, and the spatial variations of $\mathbf{E}_0$ and $\mathbf{H}_0$ are not too fast, so that the second terms at the left-hand side of (3.3) are negligible.

3.1.1 Fermat's Principle

As shown above, simple formulas can be used to compute the GO fields, once the ray congruence is known. The latter can be obtained by solving (3.6), the eikonal equation, to determine the scalar function L. A useful result that often helps to determine the ray geometry is the *Fermat's principle*, detailed as follows.

Let us consider a volume V in which a single ray congruence is present. The Fermat's principle states that among all the curves joining two points A and B in V, the ray is the one with the smallest optical length. This can be demonstrated by recalling that ∇L is a conservative field (i.e., its line integral from A to B is independent of the integration path). In fact, by indicating with γ the ray and with c a different, generic line joining A and B, whose tangent unit vector is $\hat{\mathbf{t}}$ (see Figure 3.3), we can write

$$c\int_A^B n\,dl > c\int_A^B n\hat{\mathbf{s}}\cdot\hat{\mathbf{t}}\,dl = c\int_A^B \nabla L\cdot\hat{\mathbf{t}}\,dl = \gamma\int_A^B \nabla L\cdot\hat{\mathbf{s}}\,dl = \gamma\int_A^B n\,dl \tag{3.12}$$

where we have used $\hat{\mathbf{s}}\cdot\hat{\mathbf{t}} < 1$ and (3.7).

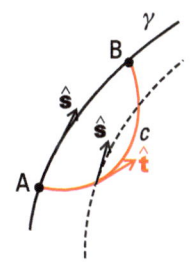

Figure 3.3 Relevant to the Fermat's principle.

3.1.2 GO in Homogeneous Media

In a homogeneous medium, the refractive index n is a constant, so that the optical length of a curve is simply n times its geometrical length. And since the shortest curve joining two points is the straight line, the Fermat's principle implies that, in a homogeneous medium, rays are straight lines. Accordingly, in this case (3.9) simplifies as $L(P) = L(O) + nl$, where l is the length of the straight-line segment OP. With regard to (3.11), in this case $\zeta(P) = \zeta(O) = \zeta$; in addition (see Figure 3.4), $dS(O) = AB \cdot CD = r_M \vartheta_1 r_m \vartheta_2$ and $dS(P) =$

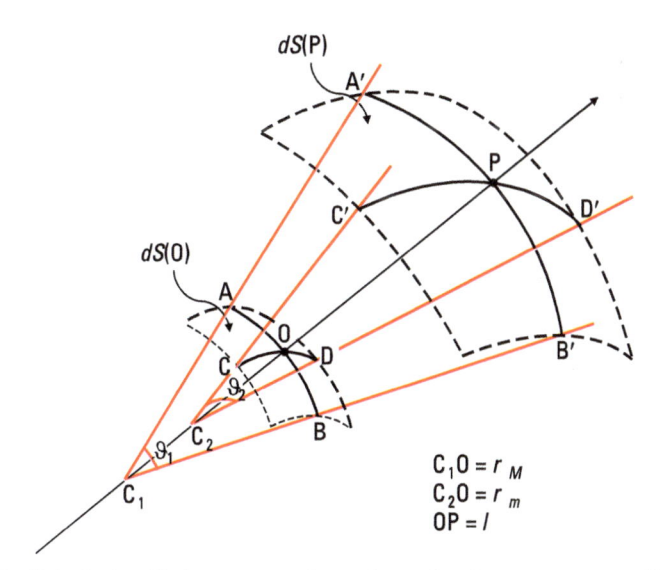

$$C_1 O = r_M$$
$$C_2 O = r_m$$
$$OP = l$$

Figure 3.4 Principal radii of curvature of wavefronts in a homogeneous medium; r_M and r_m are the principal radii of curvature of $dS(O)$ and $r_M + l$ and $r_m + l$ are the principal radii of curvature of $dS(P)$.

A'B' · C'D' = $(r_M + l)\vartheta_1(r_m + l)\vartheta_2$, where r_M and r_m are the principal radii of curvature of the (vanishingly small) surface element $dS(O)$, and all other quantities are defined in Figure 3.4. Therefore, in a homogeneous medium, (3.11) can be rewritten as

$$\left|\mathbf{E}_0(P)\right| = \left|\mathbf{E}_0(O)\right| \sqrt{\frac{r_M r_m}{(r_M + l)(r_m + l)}} \tag{3.13}$$

Finally, in a homogeneous medium the polarization state does not change along the ray. In conclusion, if the straight line segment joining O and P lays in a homogeneous medium, then according to GO the electric field in P is related to the electric field in O as follows:

$$\mathbf{E}(P) = \mathbf{E}(O)\exp(-jk_0 nl) \sqrt{\frac{r_M r_m}{(r_M + l)(r_m + l)}} = \mathbf{E}(O)\exp(-jkl)A(l) \tag{3.14}$$

where $k = k_0 n$ and

$$A(l) = \sqrt{\frac{r_M r_m}{(r_M + l)(r_m + l)}} \tag{3.15}$$

The attenuation function $A(l)$ depends on the wavefront radii of curvature (i.e., on the wavefront shape), and it simplifies in some important particular cases.

In the most general case, r_M and r_m are different, and the wavefront is said to be *astigmatic*. In this case, rays forming a flux tube do not intersect all at a single point.

If $r_M \to \infty$ and $r_m \to \infty$, then the wavefront is a plane (*plane ray congruence*) and $A(l) = 1$; together with (3.4), this implies that the electromagnetic field is a plane wave.

If $r_M \to \infty$, and $r_m = r$ is finite, then the wavefront is a cylinder with radius r (*cylindrical ray congruence*) and

$$A(l) = \sqrt{\frac{r}{r + l}} \tag{3.16a}$$

A cylindrical ray congruence is generated by a linear source of infinite length, which is the axis of the cylindrical wavefront, and the field amplitude decays as the inverse of the square root of the distance from the source.

Finally, if $r_M = r_m = r$, then the wavefront is a sphere of radius r (*spherical ray congruence*) and

$$A(l) = \frac{r}{r+l} \tag{3.16b}$$

A spherical ray congruence is generated by a point-like source, which is the center of the spherical wavefront, and the field amplitude decays as the inverse of the distance from the source. Therefore, in the GO limit, the field produced by a point-like source is coincident with the far field produced by a finite source [see (2.150)], and it is locally approximated by a plane wave. Therefore, in this limit, if no obstacle is present along the straight line joining transmitting and receiving antennas (i.e., the LOS), then the free-space formula (2.150) can be used to compute the LOS field at the receiver. However, the LOS field might be not the only contribution to the field at the receiver; in fact, rays emitted by the transmitting antenna may reach the receiving antenna after one or more reflections on obstacles. This phenomenon is illustrated in Section 3.1.3.

3.1.3 Interface Between Homogeneous Media: Reflected and Transmitted Ray Congruences

Let us consider a (curved, in general) boundary between two different homogeneous media, medium 1 and medium 2, with refractive indexes n_1 and n_2, respectively, and let us consider a ray, radiated by a source placed in T, that impinges onto this boundary at a point S, coming from medium 1 (see Figure 3.5). Since propagation is along the ray, and the electromagnetic field associated to a ray is a locally plane wave, the results of Section 2.3.5 on plane wave incidence onto a plane boundary can be applied to compute the fields in the vicinity of point S. The curved boundary is approximated by its local tangent plane, and *reflected* and *refracted* (or *transmitted*) *rays* arise, whose directions obey the Snell reflection and refraction laws [see (2.78) and (2.81)]. In addition, the reflected and transmitted electric fields at S are related to the incident electric field at S via the reflection and transmission coefficients [(2.85), (2.90), and (2.92)]. Accordingly, if a spherical ray congruence, radiated by a source placed in T, impinges onto the considered boundary, a reflected ray congruence arises in medium 1, and a refracted ray congruence is transmitted in medium 2. The total field at a point R of medium 1 will be the (coherent) sum of the direct (LOS) contribution $\mathbf{E}_i(R)$, computed via (2.150), and of the

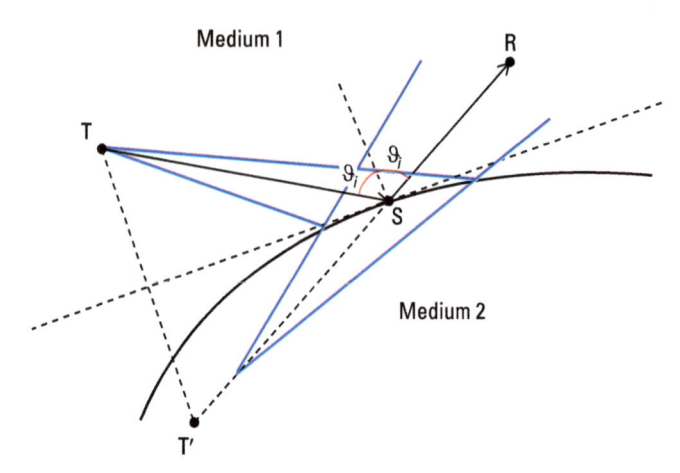

Figure 3.5 Ray reflection on a curved boundary.

reflected field $\mathbf{E}_r(R)$, whose perpendicular and parallel polarization components $E_{r\perp}(R)$ and $E_{r\parallel}(R)$ are computed from those of the incident field at S, $\mathbf{E}_i(S)$, as:

$$E_{r\perp,\parallel}(R) = E_{i\perp,\parallel}(S)\Gamma_{\perp,\parallel}(\vartheta_i)\exp(-jkr')\sqrt{\frac{r'_M r'_m}{(r'_M + r')(r'_m + r')}} \qquad (3.17)$$

where, see Figure 3.5, ϑ_i is the incidence angle, r' is the length of SR, r'_M and r'_m are the principal radii of the curvature of the reflected wavefront, and $\Gamma_\perp(\vartheta_i)$ and $\Gamma_\parallel(\vartheta_i)$ are the Fresnel reflection coefficients at perpendicular and parallel polarization, respectively. By using (2.150) to evaluate $\mathbf{E}_i(S)$, we get:

$$E_{r\perp,\parallel}(R) = \frac{jk\zeta I_0}{4\pi r}\exp\left[-jk(r+r')\right]l_{\perp,\parallel}\Gamma_{\perp,\parallel}(\vartheta_i)\sqrt{\frac{r'_M r'_m}{(r'_M + r')(r'_m + r')}} \qquad (3.18)$$

where I_0 is the input current of the transmitting antenna placed in T, r is the length of TS, and $l_\perp$ and $l_\parallel$ are the perpendicular and parallel polarization components of the transmitting antenna effective length. The principal radii of curvature of the reflected wavefront, r'_M and r'_m, can be related, with purely geometrical considerations, to r, ϑ_i and to the shape of the boundary surface; if the latter is a sphere of radius a, we get [4, Section 5.3]:

$$r'_M = r\frac{a}{a + 2r\cos\vartheta_i}$$

$$r'_m = r\frac{a\cos\vartheta_i}{a\cos\vartheta_i + 2r} \qquad (3.19)$$

If $a \to \infty$ (i.e., if the boundary is a plane surface), then $r'_M = r'_m = r$, and (3.18) becomes

$$E_{r\perp,\|}(R) = \frac{jk\zeta I_0}{4\pi(r+r')}\exp\left[-jk(r+r')\right]\ell_{\perp,\|}\Gamma_{\perp,\|}(\vartheta_i) \qquad (3.20)$$

so that the reflected field is equal to Γ times the field that would be radiated in free space by an image source placed in T′; see Figure 3.5.

Equation (3.20), which is obtained for a plane boundary, can be safely used also for a spherical boundary when $a \cdot \cos\vartheta_i \gg r$, as it is the case for an antenna radiating in the vicinity of the Earth surface. (In this case, a is the Earth radius.)

3.1.4 Example of Inhomogeneous Media: Stratified Medium

Let us consider a ray propagating in a nonhomogeneous medium, $\hat{s} = \hat{s}(x, y, z)$ being its tangent unit vector at a generic point $P \equiv (x, y, z)$. Let us define a ray local spherical reference system [see Figure 3.6(a)], in which $\vartheta = \vartheta(x, y, z)$ is the angle between $\hat{s}$ and the z-axis, and $\varphi = \varphi(x, y, z)$ is the angle between the $\hat{s}$-z and x-z local planes. Accordingly, $\hat{s}$ can be written as

$$\hat{s} = \sin\vartheta\cos\varphi\,\hat{x} + \sin\vartheta\sin\varphi\,\hat{y} + \cos\vartheta\,\hat{z} \qquad (3.21)$$

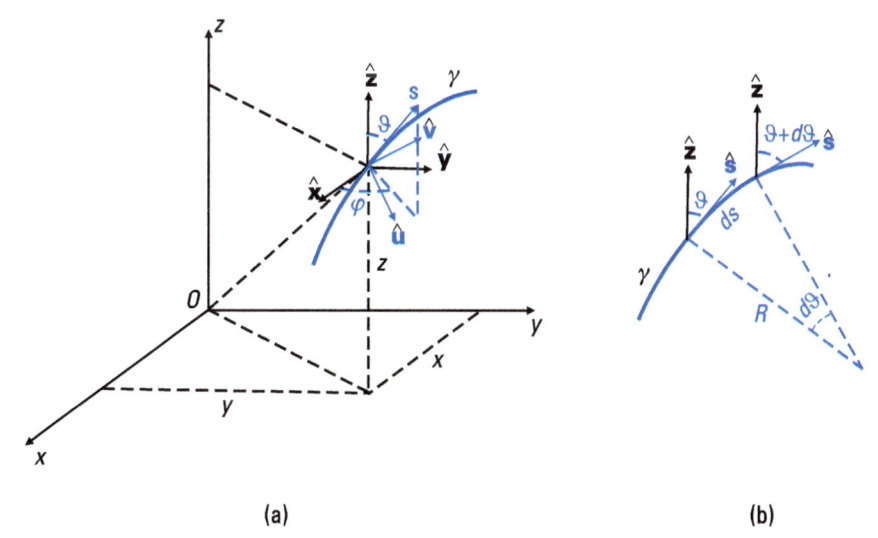

(a)   (b)

Figure 3.6 (a) Ray local reference system and (b) ray radius of curvature R.

We also define the following unit vectors:

$$\hat{\mathbf{v}} = \frac{\hat{\mathbf{z}} \times \hat{\mathbf{s}}}{|\hat{\mathbf{z}} \times \hat{\mathbf{s}}|} = -\sin\varphi\,\hat{\mathbf{x}} + \cos\varphi\,\hat{\mathbf{y}}$$

$$\hat{\mathbf{u}} = \hat{\mathbf{v}} \times \hat{\mathbf{s}} = \cos\vartheta\cos\varphi\,\hat{\mathbf{x}} + \cos\vartheta\sin\varphi\,\hat{\mathbf{y}} - \sin\vartheta\,\hat{\mathbf{z}}$$

(3.22)

so that $\hat{\mathbf{u}}, \hat{\mathbf{v}}$, and $\hat{\mathbf{s}}$ are mutually orthogonal unit vectors. Note that they may vary along the ray, since they depend on x, y, and z.

By using (A.14) and (3.21) and (3.22), and after some tedious but simple algebraic manipulations, it can be verified that

$$\nabla \times \hat{\mathbf{s}} \cdot \hat{\mathbf{u}} = -\sin\vartheta\left(\sin\vartheta\cos\varphi\frac{\partial\varphi}{\partial x} + \sin\vartheta\sin\varphi\frac{\partial\varphi}{\partial y} + \cos\vartheta\frac{\partial\varphi}{\partial z}\right)$$

$$= -\sin\vartheta\nabla\varphi \cdot \hat{\mathbf{s}} = -\sin\vartheta\frac{\partial\varphi}{\partial s}$$

(3.23)

$$\nabla \times \hat{\mathbf{s}} \cdot \hat{\mathbf{v}} = \sin\vartheta\cos\varphi\frac{\partial\vartheta}{\partial x} + \sin\vartheta\sin\varphi\frac{\partial\vartheta}{\partial y} + \cos\vartheta\frac{\partial\vartheta}{\partial z} = \nabla\vartheta \cdot \hat{\mathbf{s}} = \frac{\partial\vartheta}{\partial s}$$

$$\nabla \times \hat{\mathbf{s}} \cdot \hat{\mathbf{s}} = -\nabla\vartheta \cdot \hat{\mathbf{v}} + \sin\vartheta\nabla\varphi \cdot \hat{\mathbf{u}}$$

The ray tangent unit vector can be related to the medium index of refraction by noting that (3.7) and (A.6) imply that

$$\nabla \times n\hat{\mathbf{s}} = 0$$

(3.24)

which, by using (A.5), can be written as

$$\nabla n \times \hat{\mathbf{s}} + n\nabla \times \hat{\mathbf{s}} = 0 \Leftrightarrow \nabla \times \hat{\mathbf{s}} = -\frac{1}{n}\nabla n \times \hat{\mathbf{s}} = -\nabla(\ln n) \times \hat{\mathbf{s}}$$

(3.25)

Let us now consider, as a particular case, a *stratified medium* (i.e., a medium whose index of refraction varies only along a single direction such as the z-axis direction): $n(x, y, z) = n(z)$. In this case, (3.25) becomes

$$\nabla \times \hat{\mathbf{s}} = -\frac{d\ln n}{dz}\hat{\mathbf{z}} \times \hat{\mathbf{s}} = -\frac{d\ln n}{dz}\sin\vartheta\,\hat{\mathbf{v}}$$

(3.26)

which, projected onto the $\hat{\mathbf{u}}$ and $\hat{\mathbf{v}}$ directions, in view of (3.23) gives

$$\sin\vartheta\frac{d\varphi}{ds} = 0$$

$$\frac{d\vartheta}{ds} = -\frac{d\ln n}{dz}\sin\vartheta$$

(3.27)

The first of (3.27) states that φ is constant along the ray, so that the $\hat{s}$-z plane (i.e., the vertical plane to which the ray tangent belongs) is constant along the ray, and the ray is a plane curve. The *osculating circle* locally approximating the ray can be then defined, whose radius is the radius of curvature R of the ray, with $ds = R d\vartheta$; see Figure 3.6(b). Its inverse $1/R = d\vartheta/ds$ is the *ray curvature*, which is a measure of the ray bending. It can be obtained from the second of (3.27). When $\vartheta \neq 0$, if n decreases with z then the ray curvature is positive, and the ray is bended downward; the opposite happens if n is increasing with z. If n is constant, we obtain that the ray curvature is null and the ray is a straight line, as expected. A ray is rectilinear also for a varying $n(z)$ if $\vartheta = 0$ (i.e., if the ray propagates along the direction of the stratification).

It is also interesting to rewrite the second of (3.27) as

$$\frac{d\vartheta}{ds} = -\frac{1}{n}\frac{dn}{dz}\sin\vartheta = -\frac{1}{n\cos\vartheta}\frac{dn}{ds}\sin\vartheta \Leftrightarrow n\cos\vartheta\frac{d\vartheta}{ds} + \frac{dn}{ds}\sin\vartheta = 0 \Leftrightarrow \frac{d(n\sin\vartheta)}{ds} = 0$$

$$(3.28)$$

which states that $n(z)\sin\vartheta(x, y, z)$ is constant along the ray. This is the generalization of the Snell refraction law [(2.81)] to the case of continuously varying refractive index.

As an example, let us consider a space with $\mu_r = 1$, whose lower homogeneous half-space $z < 0$, with dielectric constant $\varepsilon_r = 4$, is covered by a dielectric stratified slab of thickness $d = 10$ cm with $\varepsilon_r(z) = (2 - z/d)^2$, $0 \leq z \leq d$, as depicted in Figure 3.7; the upper half-space is vacuum, $\varepsilon_r = 1$. Accordingly,

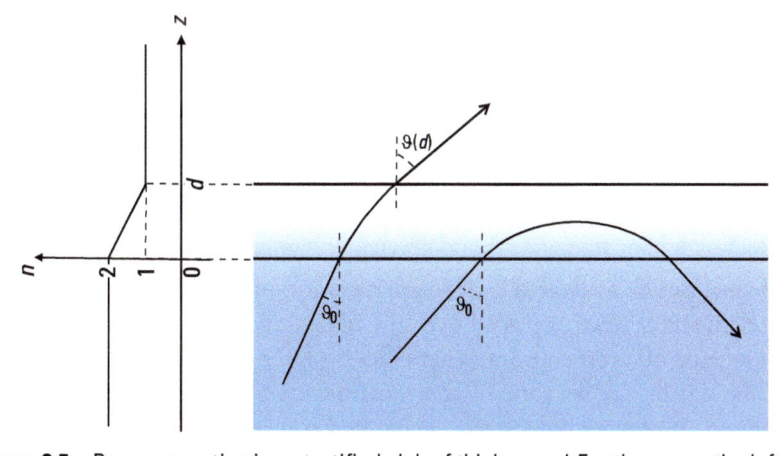

Figure 3.7 Ray propagation in a stratified slab of thickness d. For the ray on the left $\vartheta_0 < \vartheta_c$, whereas for the ray on the right $\vartheta_0 > \vartheta_c$.

we have $n(z) = 2$ for $z \leq 0$, $n(z) = 2 - z/d$ for $0 \leq z \leq d$, and $n(z) = 1$ for $z > d$. Let us assume that a ray coming from the lower half-space impinges on the dielectric slab with an incidence angle $\vartheta(0) = \vartheta_0$. While propagating through the slab, the ray is bended downward, since $dn/dz < 0$; and in view of (3.28), at the height z we have $n(z)\sin\vartheta(z) = n(0)\sin\vartheta_0 = 2\sin\vartheta_0$. Therefore (see Figure 3.7), if $2\sin\vartheta_0 < 1$ [i.e., if the incidence angle ϑ_0 is smaller than the critical angle $\vartheta_c = \arcsin(1/2) = \pi/6 = 30°$], the ray emerges in the vacuum half-space with an angle $\vartheta(d) = \arcsin(2\sin(\vartheta_0))$, where we have used $n(d) = 1$. The ray curvature is given by the second of (3.27); assuming for instance $\vartheta_0 = \pi/10 = 18°$, the ray curvature is 1.545 m^{-1} at $z = 0$ and 6.18 m^{-1} at $z = d = 0.1m$. Conversely, if the incidence angle ϑ_0 is larger than the critical angle ϑ_c, the ray cannot reach the upper vacuum half-space; in fact, it attains a maximum height z_{max} such that $n(z_{max}) = 2\sin\vartheta_0$, so that $\sin\vartheta(z_{max}) = 1$ and $\vartheta(z_{max}) = \pi/2$, and then it is deflected back downward; see Figure 3.7. The ray curvature at $z = z_{max}$, according to the second of (3.27), is $(5\ m^{-1})/\sin\vartheta_0$. Note that the critical angle for this problem is coincident with the critical angle defined Section 2.3.5 and corresponding to a hard transition from the lower to the upper homogeneous half-spaces.

3.2 Fresnel Ellipsoids

According to GO (i.e., for $f \to \infty$), propagation is along the rays, so that if the LOS is free from obstacles, the LOS field at the receiver can be computed by using the free-space expression (2.150). However, in practice (i.e., at finite frequency), this expression, and hence the Friis formula, can only be used if the obstacles are sufficiently far from the LOS connecting the two antennas. To see this, let us consider a transmitting antenna T and a receiving antenna R spaced by a distance r in a homogeneous medium; see Figure 3.8(a). In addition, let us consider a plane surface, orthogonal to TR (i.e., to the LOS), placed at distance r_1 from T and r_2 from R, so that $r_1 + r_2 = r$, and coincident with the x-y plane of a Cartesian reference system whose z-axis is the LOS; see Figure 3.8(a). By using the equivalence theorem (see Section 2.4.4), the field at R can be evaluated as the superposition of fields radiated by equivalent elementary sources placed over the plane, and related to the tangential components of electric and magnetic fields produced by T over the plane. If the plane is in the far zone of the transmitting antenna, and the receiving antenna is in the far zone of the elementary sources over the plane, then the field received at R can be written as

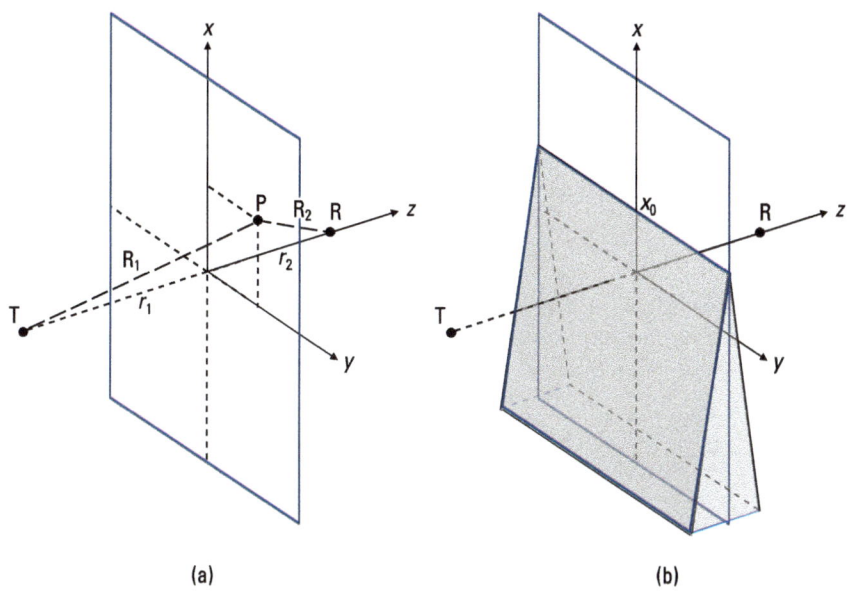

(a)   (b)

Figure 3.8 Geometry for the definition of Fresnel ellipsoids: (a) free-space case, P ≡ (x, y, 0) and $r_1 + r_2 = r$; and (b) perfectly absorbing obstacle, obstructing the x-y plane for $x < x_0$.

$$\mathbf{E}(\mathrm{R}) = \int_{-\infty}^{\infty} \int_{-\infty}^{\infty} \mathbf{f}(x,y) \frac{\exp\left\{-jk\left[R_1(x,y) + R_2(x,y)\right]\right\}}{R_1(x,y)R_2(x,y)} dxdy \qquad (3.29)$$

where $\mathbf{f}(x, y)$ is a vector function related to the equivalent elementary sources over the plane at the generic point P ≡ (x, y, 0), whose detailed expression is of no concern here, and

$$R_1(x,y) = \sqrt{r_1^2 + x^2 + y^2}, \quad R_2(x,y) = \sqrt{r_2^2 + x^2 + y^2} \qquad (3.30)$$

See Figure 3.8(a). If x and y are much smaller than r_1 and r_2, then

$$R_1(x,y) \cong r_1 + \frac{x^2 + y^2}{2r_1}, \quad R_2(x,y) \cong r_2 + \frac{x^2 + y^2}{2r_2} \qquad (3.31)$$

so that

$$R_1(x,y) + R_2(x,y) \cong r + \frac{r}{2r_1r_2}\left(x^2 + y^2\right) \qquad (3.32)$$

and the integrand of (3.29) can be approximated as

$$
\mathbf{f}(x,y)\frac{\exp\left\{-jk\left[R_1(x,y)+R_2(x,y)\right]\right\}}{R_1(x,y)R_2(x,y)}
$$

$$
\cong \frac{\mathbf{f}(0,0)}{r_1 r_2}\exp(-jkr)\exp\left[-j\frac{\pi r}{\lambda r_1 r_2}\left(x^2+y^2\right)\right]
$$

(3.33)

Accordingly, on the *x-y* plane we can define *Fresnel circles*, whose contours are the circumferences on which the last phase term of the integrand is equal to $m\pi$, with m integer:

$$
\frac{\pi r}{\lambda r_1 r_2}\rho^2 = m\pi \quad \text{with} \quad \rho = \sqrt{x^2 + y^2}
$$

Accordingly, the radius of the *m*th Fresnel circle is

$$
\rho_m = \sqrt{m\frac{\lambda r_1 r_2}{r}}
$$

(3.34)

so that these circumferences are very close if λ is small, and they become closer and closer as m increases (i.e., as the distance from the origin increases); accordingly, real and imaginary parts of the integrand oscillate faster and faster (see Figure 3.9), and outside the first few Fresnel circles positive and negative areas cancel out in the integration process of (3.29).

Therefore, the main contribution to the integral comes from the first few Fresnel circles.

If the plane is shifted along the *z*-axis, each Fresnel circumference describes a surface whose points are characterized by $k(R_1 + R_2) = kr + m\pi$ (i.e., $R_1 + R_2 = r + m\lambda/2$); such a surface is an ellipsoid whose foci are T and

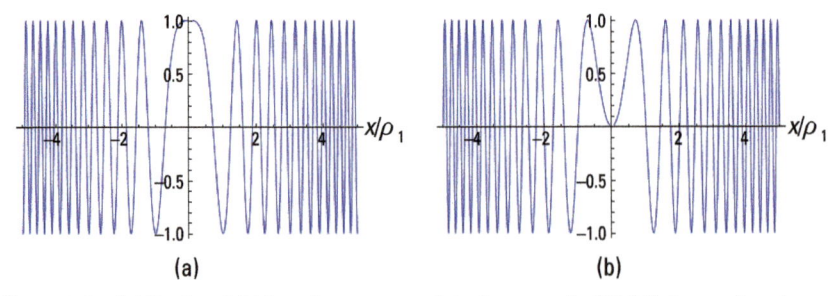

(a)   (b)

Figure 3.9   (a) Real and (b) imaginary parts of the integrand of (3.29), for $y = 0$ and x varying.

R (*Fresnel ellipsoid*). The minor semiaxis b_m of the mth Fresnel ellipsoid is obtained by setting $r_1 = r_2 = r/2$ in (3.34):

$$b_m = \frac{\sqrt{m\lambda r}}{2} \tag{3.35}$$

Therefore, b_m vanishes as the frequency tends to infinity ($\lambda \to 0$); and the minor axis of the first Fresnel ellipsoid ($m = 1$) is the geometric mean of wavelength and transmitter-to-receiver distance: $2b_1 = \sqrt{\lambda r}$. An example of Fresnel ellipsoid is depicted in Figure 3.10.

Based on the considerations above, we can expect that if obstacles do not intersect the first few Fresnel ellipsoids, they do not affect the LOS field at the receiver, and the free-space formula is accurate. Conversely, if perfectly absorbing or reflecting obstacles completely obstruct the first few Fresnel ellipsoids, the field at the receiver is negligible with respect to the LOS one. In intermediate situations, in which the first few Fresnel ellipsoids are only partially obstructed by obstacles (regardless of whether the LOS is obstructed or not), the free-space formula is not accurate, but the received field is still on the same order of magnitude of the LOS one. This qualitative behavior can be quantitatively confirmed by considering a perfectly absorbing obstacle obstructing the half plane $x \leq x_0$; see Figure 3.8(b). In this case, we can assume that the field on the x-y plane is zero for $x \leq x_0$ and that it is equal to the free-space field for $x > x_0$ (*Kirchhoff*, or *physical optics, approximation*), so that the field at the receiver is given by

$$\mathbf{E}(\mathrm{R}) = \int_{-\infty}^{\infty} \int_{x_0}^{\infty} \mathbf{f}(x,y) \frac{\exp\left\{-jk\left[R_1(x,y) + R_2(x,y)\right]\right\}}{R_1(x,y)R_2(x,y)} dx dy \tag{3.36}$$

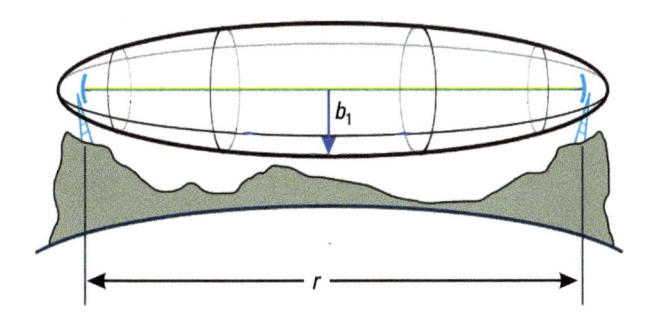

Figure 3.10 Radio link between two directive antennas placed at elevated sites: the first Fresnel ellipsoid.

This integral can be numerically evaluated, and its modulus (normalized to the modulus of the free-space received field) as a function of x_0 is plotted in Figure 3.11. It turns out that for $x_0 < -\rho_1$ the normalized modulus of the received field is always higher than 0.875 and smaller than 1.12; and for $x_0 > \rho_1$ the normalized modulus of the received field is always smaller than $1/(2\pi) = 0.159$, and it is well approximated by $\rho_1/(2\pi x_0)$.

Accordingly, when designing a radio link between two directive antennas placed at elevated sites, it is usually necessary that the first Fresnel ellipsoid be completely free from obstacles; see Figure 3.10.

In conclusion, we can state that the region of space effectively involved in the propagation from the transmitting to the receiving antenna is the volume enclosed in the first few Fresnel ellipsoids, and this volume collapses to the straight-line optic ray as the frequency tends to infinity.

3.3 Stationary Phase Method

The same rationale leading to the definition of the Fresnel circles is at the basis of the stationary phase method for the asymptotic evaluation of oscillatory integrals. In fact, let us consider the following integral:

$$I(\Omega) = \int_{-\infty}^{\infty} f(x)\exp\left[j\Omega q(x)\right]dx \qquad (3.37)$$

where $f(x)$ is a continuous real or complex function and $q(x)$ is a real function assumed continuous with its derivatives. The solution of electromagnetic

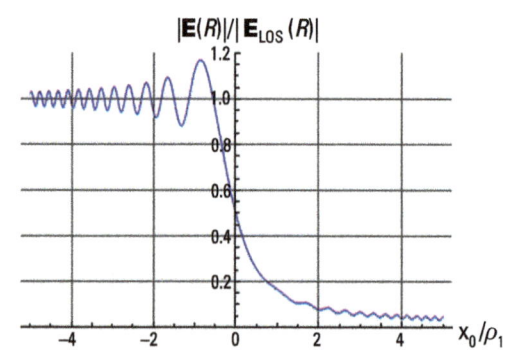

Figure 3.11   Normalized modulus of the received field (3.36) versus the position x_0 of the obstacle of Figure 3.8(b).

propagation and scattering problems can be often expressed by integrals of this kind, see Sections 3.2 and 3.4–3.6.

For large values of Ω, only the region close to the stationary phase point $x = x_s$,

$$\left.\frac{dq}{dx}\right|_{x=x_s} = q'(x_s) = 0 \qquad (3.38)$$

provides a significant contribution to the integral (3.37); see Figure 3.12. In this region, we can set

$$f(x) \cong f(x_s) \quad \text{and} \quad q(x) \cong q(x_s) + \frac{1}{2}q''(x_s)(x - x_s)^2$$

$$\text{with} \quad q''(x_s) = \left.\frac{d^2q}{dx^2}\right|_{x=x_s} \qquad (3.39)$$

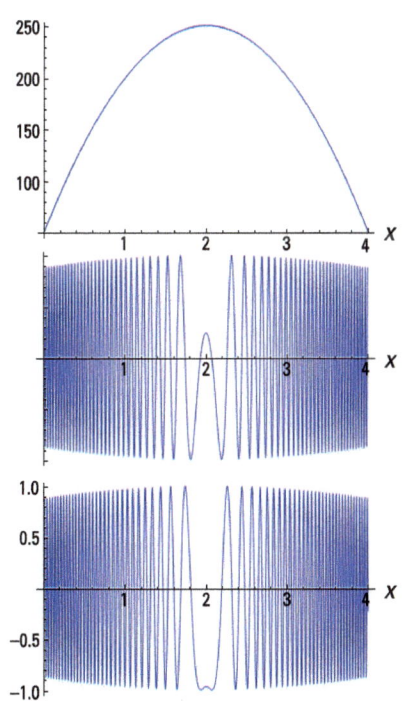

Figure 3.12 Stationary phase method (3.37): plots of the phase function $\Omega q(x)$ (a), and of the real (b) and imaginary (c) parts of the integrand $f(x)\exp[j\Omega q(x)]$. $f(x) = \cos(x/4 - 1/2)$, $\Omega = 50$, $q(x) = -(x-1)^2 + 2x + 2$.

so that

$$I(\Omega) \cong f(x_s)\exp\left[j\Omega q(x_s)\right]\int_{-\infty}^{\infty}\exp\left[j\Omega\frac{q''(x_s)}{2}(x-x_s)^2\right]dx$$

$$= f(x_s)\exp\left[j\Omega q(x_s)\right]\sqrt{\frac{2\pi j}{\Omega q''(x_s)}}$$

(3.40)

where use has been made of (C.2). Actually, it can be demonstrated that for $\Omega \to \infty$ the difference between the exact value of the integral $I(\Omega)$ and its approximation given by (3.40) is $O(1/\Omega^{3/2})$ (i.e., it is of order $1/\Omega^{3/2}$) [6].

In practice, for large but finite Ω, (3.40) accurately approximates $I(\Omega)$ if $f(x)$ can be considered constant over a small interval around x_s such that phase variation is on the order of π. By using (3.39), this interval is evaluated as $x_s - \Delta x < x < x_s + \Delta x$, with

$$\Delta x \approx \sqrt{\frac{2\pi}{\Omega|q''(x_s)|}}$$

(3.41)

If more than one stationary phase point is present, the integral is asymptotically evaluated as the sum of contributions from each stationary point, computed via (3.40).

3.3.1 Finite Integration Interval

Let us consider an integral of the kind of (3.37), but with finite limits of integration:

$$I_{ab}(\Omega) = \int_a^b f(x)\exp\left[j\Omega q(x)\right]dx$$

(3.42)

Let us first assume that no stationary phase point is included in the integration interval. In this case, $q'(x) \neq 0$ in $[a, b]$ and $I_{ab}(\Omega)$ can be integrated by parts:

$$I_{ab}(\Omega) = \int_a^b \frac{f(x)}{j\Omega q'(x)}\frac{d}{dx}\exp\left[j\Omega q(x)\right]dx$$

$$= \left[\frac{f(x)}{j\Omega q'(x)}\exp\left[j\Omega q(x)\right]\right]_{x=a}^{x=b} - \frac{1}{j\Omega}\int_a^b f_1(x)\exp\left[j\Omega q(x)\right]dx$$

(3.43)

where $f_1(x) = \dfrac{d}{dx} \dfrac{f(x)}{q'(x)} = \dfrac{f'(x)}{q'(x)} - \dfrac{f(x)q''(x)}{q'^2(x)}$.

The procedure can be iterated for the integral at the second term of the right-hand side of (3.43), and at each iteration step a further multiplication by $1/\Omega$ is obtained. Accordingly, we have:

$$I_{ab}(\Omega) = \frac{f(b)}{j\Omega q'(b)} \exp\left[\, j\Omega q(b)\right] - \frac{f(a)}{j\Omega q'(a)} \exp\left[\, j\Omega q(a)\right] + O\!\left(\frac{1}{\Omega^2}\right) \quad (3.44)$$

The leading terms of (3.44) only depend on the values assumed by the integrand near a and b; therefore, in this case, the main contribution to the integral comes from the end-point regions of the integration interval.

Finally, if one of the integration limits goes to infinity, the corresponding end-point contribution vanishes if, for $x \to \pm\infty$, $f(x) \to 0$ or $q'(x) \to \infty$.

Let us now assume that a stationary phase point x_s is included in the integration interval: that is, $q'(x_s) = 0$ with $a < x_s < b$. We can then write

$$
\begin{aligned}
I_{ab}(\Omega) &= \int_a^b f(x)\exp\left[\, j\Omega q(x)\right]dx = \int_{-\infty}^{+\infty} f(x)\exp\left[\, j\Omega q(x)\right]dx \\
&\quad - \int_b^{+\infty} f(x)\exp\left[\, j\Omega q(x)\right]dx - \int_{-\infty}^{a} f(x)\exp\left[\, j\Omega q(x)\right]dx
\end{aligned}
\quad (3.45)
$$

where $f(x)$ and $q(x)$ are prolonged outside the interval $[a, b]$ by ensuring that they are continuous with their derivatives, that $\lim\limits_{x\to\pm\infty} f(x) = 0$, and that $q'(x) \neq 0$ for $x \notin [a, b]$. The first term at the right-hand side of (3.45) can be evaluated by using (3.40), whereas the second and third terms can be obtained by resorting to (3.44), so that we have

$$
\begin{aligned}
I_{ab}(\Omega) &= f(x_s)\exp\left[\, j\Omega q(x_s)\right]\sqrt{\frac{2\pi j}{\Omega q''(x_s)}} \\
&\quad + \frac{f(b)}{j\Omega q'(b)}\exp\left[\, j\Omega q(b)\right] - \frac{f(a)}{j\Omega q'(a)}\exp\left[\, j\Omega q(a)\right] + O\!\left(\frac{1}{\Omega^{3/2}}\right)
\end{aligned}
$$

$$(3.46)$$

Accordingly, for $\Omega \to \infty$ the leading term is the stationary phase contribution, which is infinitesimal of order $1/2$, whereas the end-point contributions are infinitesimal of order one. The physical meaning of these terms is discussed in Section 3.4.

In practice, for large but finite Ω, the leading terms of (3.46) accurately approximate $I_{ab}(\Omega)$ if, in addition to the already mentioned conditions on the behavior of $f(x)$ around x_s, the stationary point x_s is not too close to a or b. In fact, if the distance of x_s from a (or b) is very small, we have $q'(a) \cong 0$ (or $q'(b) \cong 0$), and the asymptotic evaluation of the corresponding end-point contribution is not a good approximation of its exact value.

However, if the stationary point is exactly coincident with one of the integration limits (for instance, $x_s = a$, but a similar result is obtained for $x_s = b$), we can exploit the following result, obtained via a procedure similar to the one used to obtain (3.40):

$$
\begin{aligned}
I_{x_s}(\Omega) &= \int_{x_s=a}^{\infty} f(x)\exp\left[j\Omega q(x)\right]dx \cong f\left(x_s\right)\exp\left[j\Omega q\left(x_s\right)\right]\int_{x_s=a}^{\infty}\exp\left[j\Omega\frac{q''\left(x_s\right)}{2}\left(x-x_s\right)^2\right]dx \\
&= f\left(x_s\right)\exp\left[j\Omega q\left(x_s\right)\right]\int_{0}^{\infty}\exp\left[j\Omega\frac{q''\left(x_s\right)}{2}x'^2\right]dx' \\
&= f\left(x_s\right)\exp\left[j\Omega q\left(x_s\right)\right]\frac{1}{2}\int_{-\infty}^{\infty}\exp\left[j\Omega\frac{q''\left(x_s\right)}{2}x'^2\right]dx' \\
&= \frac{1}{2}f\left(x_s\right)\exp\left[j\Omega q\left(x_s\right)\right]\sqrt{\frac{2\pi j}{\Omega q''\left(x_s\right)}}
\end{aligned}
$$

$$\text{(3.47)}$$

which is exactly one half of the value obtained for the integral spanning the entire real axis, (3.40).

3.3.2 Transition Function

When the stationary phase point is close to one of the limits of integration, say $x_s \cong a$, then, as shown in Appendix D, an accurate approximation of the corresponding end-point contribution can be obtained by multiplying the corresponding term of (3.44) or (3.46) by the *transition function*, given by, if $q''(a) < 0$,

$$
F(X) = 2j\sqrt{X}\exp\{jX\}\int_{\sqrt{X}}^{+\infty}\exp\{-jt^2\}dt \qquad \text{(3.48)}
$$

where

$$X = \Omega \frac{|q'(a)|^2}{2|q''(a)|} \tag{3.49}$$

Accordingly,

$$I_a(\Omega) = \int_a^\infty f(x)\exp\left[j\Omega q(x)\right]dx$$

$$\cong \begin{cases} -\dfrac{f(a)}{j\Omega q'(a)}\exp\left[j\Omega q(a)\right]F(X) & \text{if } x_s < a \\[3mm] f(x_s)\exp\left[j\Omega q(x_s)\right]\sqrt{\dfrac{2\pi j}{\Omega q''(x_s)}} - \dfrac{f(a)}{j\Omega q'(a)}\exp\left[j\Omega q(a)\right]F(X) & \text{if } x_s > a \end{cases} \tag{3.50}$$

If $q''(a) > 0$, $F(X)$ must be replaced by $F^*(X)$.

The transition function $F(X)$ can be evaluated numerically and is usually available in the form of a look-up table. Its modulus and phase are plotted in Figure 3.13. By iteratively integrating by parts the integral in (3.48) according to the same procedure as in (3.43), we obtain that $F(X) \to 1$ for $X \to \infty$, so that for x_s far from a and $\Omega \to \infty$ we recover the end-point asymptotic evaluation of (3.44) or (3.46). Conversely, for $X \to 0$ (i.e., for $x_s \to a$), (3.50) reduces to (3.47), as shown in Appendix D. Finally, in the intermediate case of x_s close to a, $F(X)$ ensures a smooth transition of the end-point contribution (3.50) from the $1/\Omega$ behavior of (3.44) or (3.46) to the $1/\sqrt{\Omega}$ behavior of (3.47).

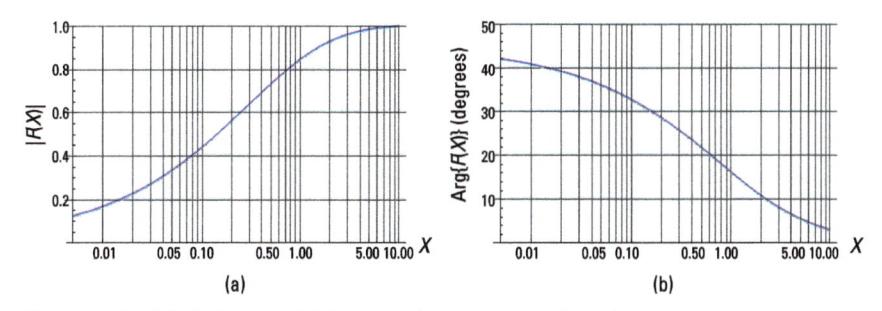

Figure 3.13　(a) Modulus and (b) phase of the transition function.

3.4 Diffraction

Phenomena that occur when a wave encounters an obstacle, and in particular the bending of the wave around the obstacle itself, are generally referred to as *diffraction*. Figure 3.8(b) provides an example of diffraction.

GO is not able to describe diffraction, so it needs to be generalized by resorting to more refined asymptotic techniques. A fruitful approach is to consider some canonical problems for which the analytical solution of Maxwell's equations is available in integral form, and then to asymptotically evaluate the obtained integral by using the approach presented in Section 3.3. It can be expected that the leading term (i.e., the stationary phase point contribution), will be coincident with the GO solution, whereas higher order terms (i.e., the end-point contributions), will be able to account for diffraction. As discussed in Section 3.5, results obtained for canonical problems can be easily exploited to deal with more complex scenarios.

To illustrate the approach outlined in the preceding paragraph, let us consider a plane wave, propagating in a linear, local, isotropic, stationary, lossless, and homogeneous medium (for instance, vacuum), impinging on a PEC half-plane, with its direction of propagation being perpendicular to the half-plane edge and forming an incidence angle $\pi/2 - \varphi_0$ with the half-plane normal unit vector; see Figure 3.14. Let us use a cylindrical reference system, with the z-axis coincident with the half-plane edge and the x positive semi-axis belonging to the half-plane (Figure 3.14). Note that the problem geometry has a cylindrical symmetry; in fact, it is invariant with z. Accordingly, the field solution will also be invariant with z and, with no loss of generality, we can assume that the observation point R, at which we will evaluate the total field, lies in the plane $z = 0$. According to GO, a reflected ray congruence arises, so that the entire space can be divided into three regions. For $0 < \varphi < \pi - \varphi_0$, incident and reflected ray congruences are present, and the total field is the sum of incident and reflected field; for $\pi - \varphi_0 < \varphi < \pi + \varphi_0$, only the incident ray congruence is present, and the total field is equal to the incident field; finally, for $\pi + \varphi_0 < \varphi < 2\pi$, no ray congruence is present, and the total field is null. Therefore, at the critical directions $\varphi = \pi - \varphi_0$ [*reflection boundary* (RB)] and $\varphi = \pi + \varphi_0$ [*lit-shadow boundary* (LSB)] [see Figure 3.14(a)], the GO field is discontinuous, and it is not an accurate approximation of the actual field; in addition, in the shadow region, $\pi + \varphi_0 < \varphi < 2\pi$, the GO field is zero, so that GO does not help evaluating the field in this region.

For the problem at hand, an exact solution of the Maxwell's equations is available [3]. However, its derivation is not straightforward, so here we use a simpler, approximate procedure that is sufficient to illustrate the approach

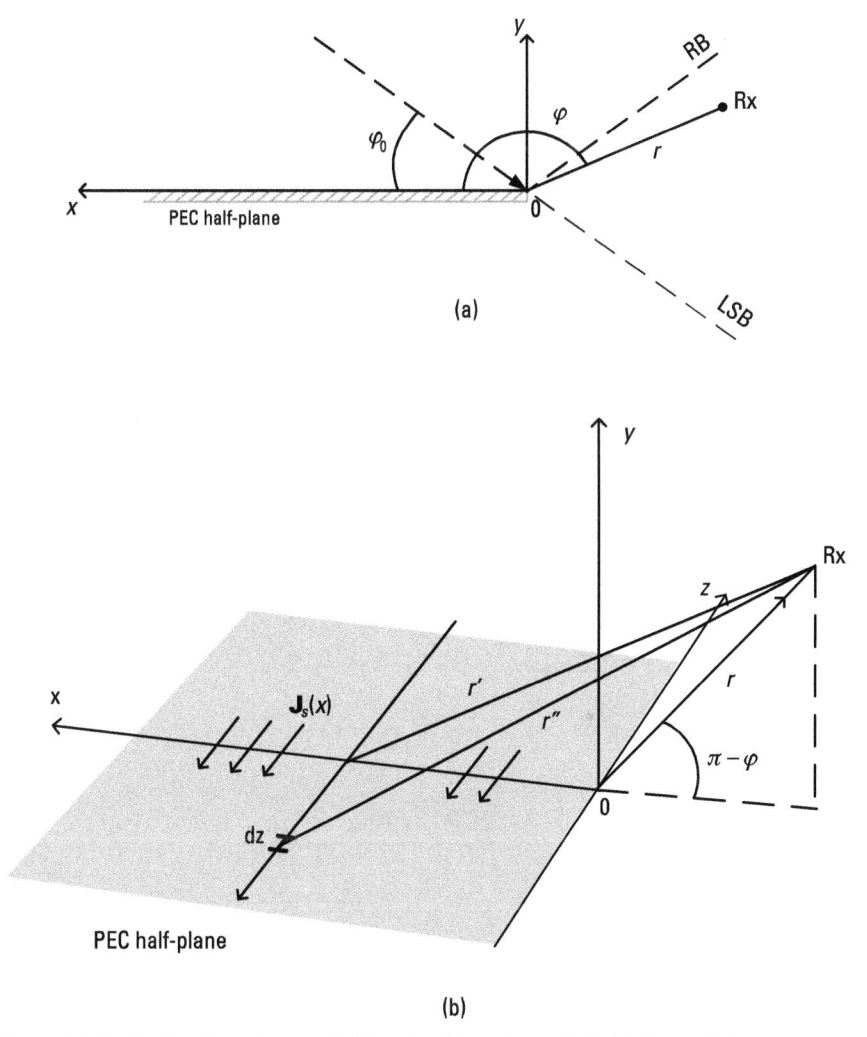

Figure 3.14 Evaluation of diffraction by a PEC half-plane: (a) 2-D view with the definition of critical directions RB and LSB and (b) 3-D view.

and that can easily be generalized to noncanonical geometries (e.g., randomly rough edge [7, 8]). According to this procedure, first the current density induced over the PEC half-plane is evaluated by using the physical optics (PO) approximation (i.e., by assuming that the tangential magnetic field over the PEC half-plane is equal to the one over a PEC plane). Then, the field radiated by such a current density distribution, which is the field scattered by the PEC half-plane, is computed.

By assuming perpendicular (i.e., TE) polarization, the electric field of the incident wave is

$$\mathbf{E}_{\text{inc}}(r) = -\hat{z}E_0 \exp\left\{ jk\left(\cos\varphi_0 x + \sin\varphi_0 y \right) \right\} \tag{3.51}$$

By using the PO approximation, the surface current density on the half-plane $y = 0$, $x > 0$ can be expressed as

$$\mathbf{J}_s(x) = \hat{y} \times \mathbf{H}\Big|_{\substack{y=0 \\ x>0}} \cong 2\hat{y} \times \mathbf{H}_{\text{inc}}\Big|_{\substack{y=0 \\ x>0}} = -\hat{z}2\frac{E_0}{\zeta}\sin\varphi_0 \exp\left(jkx\cos\varphi_0 \right) \tag{3.52}$$

The electric field radiated by this current density distribution (i.e., the field scattered by the PEC half-plane) can be computed according to (2.137):

$$\begin{aligned}
\mathbf{E}_{sc}(r,\varphi) &= -\frac{j\omega\mu}{4\pi}\left(I + \frac{\nabla\nabla\cdot}{k^2} \right) \int\limits_{0}^{+\infty}\int\limits_{-\infty}^{+\infty} \frac{\exp(-jkr'')}{r''}\mathbf{J}_s(x)\,dz\,dx \\
&= \frac{jkE_0\sin\varphi_0}{2\pi}\left(I + \frac{\nabla\nabla\cdot}{k^2} \right)\hat{z}I(r,\varphi)
\end{aligned} \tag{3.53}$$

where

$$I(r,\varphi) = \int\limits_{0}^{+\infty}\int\limits_{-\infty}^{+\infty} \frac{\exp\left\{ jk\left(x\cos\varphi_0 - r'' \right) \right\}}{r''}\,dz\,dx \tag{3.54}$$

$$r'' = \sqrt{z^2 + r'^2} \quad \text{with} \quad r' = \sqrt{x^2 + r^2 - 2rx\cos\varphi}$$

See Figure 3.14(b). Note that $\nabla \cdot \hat{z}I(r,\varphi) = \nabla I(r,\varphi) \cdot \hat{z} = 0$, since I is independent of z, so that its gradient has no component along z. Therefore,

$$\mathbf{E}_{sc}(r,\varphi) = \hat{z}\frac{jkE_0\sin\varphi_0}{2\pi}I(r,\varphi) \tag{3.55}$$

The integral over the entire z-axis can be computed via the stationary phase method, (3.40), by taking $k \to \infty$. The derivative of the phase function is

$$\frac{d\left(x\cos\varphi_0 - r'' \right)}{dz} = -\frac{dr''}{dz} = -\frac{z}{\sqrt{z^2 + r'^2}} \tag{3.56}$$

so that $z_s = 0$ and

$$\frac{d^2\left(x\cos\varphi_0 - r''\right)}{dz^2}\Bigg|_{z=0} = \frac{d}{dz}\left(-\frac{z}{\sqrt{z^2+r'^2}}\right)\Bigg|_{z=0} = -\frac{1}{r'} \qquad (3.57)$$

We then have

$$\mathbf{E}_{sc}\left(r,\varphi\right) = \hat{\mathbf{z}}\sqrt{\frac{jk}{2\pi}}E_0\sin\varphi_0 \int_0^{+\infty} \frac{\exp\left\{jk\left(x\cos\varphi_0 - r'\right)\right\}}{\sqrt{r'}}\,dx \qquad (3.58)$$

The integral in (3.58) is of the kind of (3.42), with $\Omega = k$, $f(x) = 1/\sqrt{r'}$, $q(x) = x\cos\varphi_0 - r'$, $a = 0$, $b \to \infty$. Therefore, it can be asymptotically evaluated according to the approach described in Section 3.3: Its leading term is the stationary phase point contribution, and the following term is the endpoint contribution.

3.4.1 Stationary Phase Point Contribution: The GO Field

The derivative of the phase function in (3.58) is

$$\frac{dq(x)}{dx} = \frac{d\left(x\cos\varphi_0 - r'\right)}{dx} = \cos\varphi_0 - \frac{x - r\cos\varphi}{r'} \qquad (3.59)$$

so that the stationary point x_s satisfies

$$x_s = r_s'\cos\varphi_0 + r\cos\varphi \qquad (3.60)$$

where $r_s' = r'(x_s)$. Before finding a more explicit expression for x_s, it is useful to notice that the abscissa x of a generic point P over the half-plane (see Figure 3.15), can be written as

$$x = r'\cos\varphi' + r\cos\varphi \qquad (3.61)$$

The definition of φ' is provided in Figure 3.15. By comparing (3.60) with (3.61), it can be seen that the stationary point x_s corresponds to the point P_s of the half-plane such that $\varphi' = \varphi_0$. For $\varphi < \pi$ (i.e., if the observation point r,φ is in the upper half-space), this means that the scattering angle is equal to the incidence one, so that P_s is the GO specular reflection point; see Figure 3.15(a). For $\varphi > \pi$ (i.e., if the observation point r,φ is in the lower half-space), this means that the transmission angle is equal to the incidence one, so that P_s is the intersection of the half-plane with the GO ray reaching the observation point; see Figure 3.15(b).

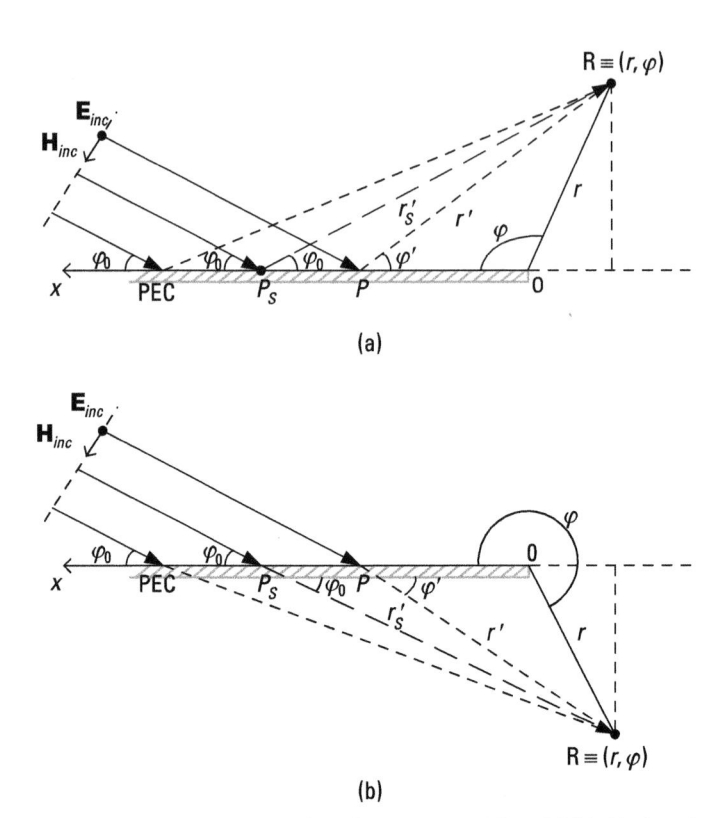

(a)

(b)

Figure 3.15 Asymptotic evaluation of the field scattered by a PEC half-plane in the (a) upper half-space (i.e., $\varphi < \pi$) and in the (b) lower half-space (i.e., $\varphi > \pi$).

To explicitly evaluate x_s, we use the equality $r'_s \sin\varphi_0 = \pm r \sin\varphi$; see Figure 3.15, where the plus and minus signs hold for the upper ($\varphi < \pi$) and lower ($\varphi > \pi$) half-spaces, respectively. We then have

$$r'_s = \pm \frac{r\sin\varphi}{\sin\varphi_0} \tag{3.62}$$

By replacing (3.62) in (3.60) we get

$$x_s = \pm \frac{r\sin\varphi}{\sin\varphi_0}\cos\varphi_0 + r\cos\varphi = r\frac{\sin(\varphi_0 \pm \varphi)}{\sin\varphi_0} \tag{3.63}$$

This result shows that $x_s > 0$, and the stationary point belongs to the integration interval, if $\varphi < \pi - \varphi_0$ or if $\varphi > \pi + \varphi_0$.

To obtain the stationary phase evaluation of (3.58), $\mathbf{E}_{scSP}$, we need to evaluate $q''(x_s)$:

$$q''(x_s) = \frac{d^2 q(x)}{dx^2}\bigg|_{x=x_s} = \frac{d^2\left(x\cos\varphi_0 - r'\right)}{dx^2}\bigg|_{\substack{x=x_s \\ r'=r'_s}} = -\frac{\sin^2\varphi_0}{r'_s} \qquad (3.64)$$

Therefore, for $\varphi < \pi - \varphi_0$ and $\varphi > \pi + \varphi_0$ we have:

$$\begin{aligned}
\mathbf{E}_{scSP}(r,\varphi) &= \hat{\mathbf{z}}\sqrt{\frac{jk}{2\pi}}E_0\sin\varphi_0\frac{\exp\left\{jk\left(x_s\cos\varphi_0 - r'_s\right)\right\}}{\sqrt{r'_s}}\sqrt{\frac{-2\pi jr'_s}{k\sin^2\varphi_0}} \\
&= \hat{\mathbf{z}}E_0\exp\left(jkx_s\cos\varphi_0\right)\exp\left(-jkr'_s\right) = -\mathbf{E}_{inc}\left(P_s\right)\exp\left(-jkr'_s\right)
\end{aligned} \qquad (3.65)$$

whereas for $\pi - \varphi_0 < \varphi < \pi + \varphi_0$ $\mathbf{E}_{scSP}(r, \varphi) = 0$.

In the upper half-space, with $\varphi < \pi - \varphi_0$, the scattered field of (3.65) is coincident with the GO reflected field, and the total field is the sum of incident and reflected field. Conversely, in the lower half-space, with $\varphi > \pi + \varphi_0$, the scattered field of (3.65) is the opposite of the incident field, and the total field is zero, as predicted by GO. Finally, for $\pi - \varphi_0 < \varphi < \pi + \varphi_0$ the stationary phase method leads to a null scattered field, so that the total field is coincident with the incident field. In conclusion, the stationary phase point contribution to the integral of (3.58) leads to the GO field.

3.4.2 End-Point Contribution: The Edge Diffracted Field

Let us now analyze the end-point contribution, given by (3.44) with [see (3.58)] $f(0) = 1/\sqrt{r}$, $f(\infty) = 0$ (so that only the end-point $x = 0$ provides a contribution), $q(0) = -r$, and [see (3.59)] $q'(0) = \cos\varphi_0 + \cos\varphi$. Accordingly, the end-point contribution $\mathbf{E}_d$ to the scattered field of (3.58) is

$$\begin{aligned}
\mathbf{E}_d(r,\varphi) &= -\hat{\mathbf{z}}\sqrt{\frac{jk}{2\pi}}E_0\sin\varphi_0\frac{\exp\left(-jkr\right)}{jk\left(\cos\varphi_0 + \cos\varphi\right)\sqrt{r}} \\
&= -\hat{\mathbf{z}}E_0\frac{\exp\left(-j\frac{\pi}{4}\right)}{\sqrt{2\pi k}}\frac{\sin\varphi_0}{\cos\varphi_0 + \cos\varphi}\frac{\exp\left(-jkr\right)}{\sqrt{r}} = \mathbf{E}_{inc}(O)D(\varphi_0,\varphi)\frac{\exp\left(-jkr\right)}{\sqrt{r}}
\end{aligned}$$

$$(3.66)$$

where
$$D(\varphi_0,\varphi) = \frac{\exp\left(-j\frac{\pi}{4}\right)}{\sqrt{2\pi k}} \frac{2\sin\frac{\varphi_0}{2}\cos\frac{\varphi_0}{2}}{\cos\varphi_0 + \cos\varphi} \tag{3.67}$$

is the *edge diffraction coefficient*. The field of (3.66) and (3.67) is termed *asymptotic physical optics* (APO) solution for the diffraction from a half-plane edge, and [see the last factor of (3.66)], it can be described as a cylindrical ray congruence. The half-plane edge behaves as a linear source that generates *diffracted rays*; see Figure 3.16(a). This new ray species is not predicted by GO and represents a useful generalization, since it makes it possible to compute the field even in the shadow region, without abandoning the convenient description of propagation as occurring along (rectilinear, in homogeneous media) rays. The field intensity of diffracted rays depends on the field intensity of the incident ray that hits the edge, on incidence and observation directions φ_0 and φ, and on the distance from the edge r. In addition, it vanishes as the square root of k, so that the diffracted field becomes increasingly smaller than the GO one as the frequency increases.

Finally, it must be noted that the diffraction coefficient of (3.67) diverges at the critical directions RB and LSB, so that it cannot be used near to these directions. This is due to the fact that in this case the stationary point is close to the end point $x = 0$; see (3.63). By using the approach of Section 3.3.2, this problem is solved by multiplying the diffraction coefficient (3.67) by the transition function $F(X)$ of (3.48), with

$$X = \frac{kr}{2} \frac{\left(\cos\varphi_0 + \cos\varphi\right)^2}{\sin^2\varphi} \tag{3.68}$$

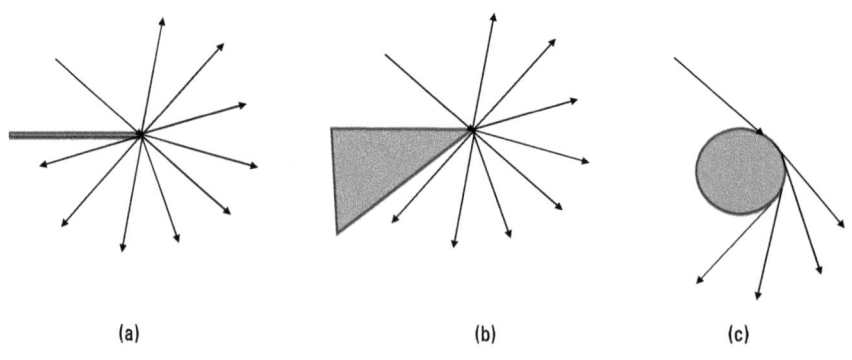

(a)   (b)   (c)

Figure 3.16 Diffracted ray species: (a) edge diffracted rays from a half-plane, (b) edge diffracted rays from a wedge, and (c) creeping rays on a circular cylinder.

as obtained from (3.49) by considering that in the case at hand $q'(0) = \cos\varphi_0 + \cos\varphi$ and $q''(0) = -\sin^2\varphi/r$.

In this way, we obtain the *uniform APO* (UAPO) solution: In correspondence with the critical directions RB and LSB, the UAPO diffracted field discontinuity exactly compensates for the GO field discontinuity (see Section 3.3.2 and Appendix D), so that the total field is continuous. Its value at the LSB is

$$\mathbf{E}\big(r,\pi+\varphi_0\big) = \mathbf{E}_{sc}\big(r,\pi+\varphi_0\big) = \frac{1}{2}\mathbf{E}_{inc}(O)\exp\big(-jkr\big) \qquad (3.69)$$

and its value at the RB is

$$\mathbf{E}\big(r,\pi-\varphi_0\big) = \mathbf{E}_{inc}\big(r,\pi-\varphi_0\big) + \mathbf{E}_{sc}\big(r,\pi-\varphi_0\big)$$
$$= \mathbf{E}_{inc}\big(r,\pi-\varphi_0\big) + \frac{1}{2}\mathbf{E}_{inc}(O)\exp\big(-jkr\big) \qquad (3.70)$$

Equations (3.69) and (3.70) can be also obtained by using (3.47) to evaluate the integral in (3.58).

Close to the critical directions, use of the transition function $F(X)$ ensures a smooth transition from a cylindrical, (3.66), to a plane, (3.69) and (3.70), wavefront.

Far from the critical directions, $X \gg 1$ and $F(X) \cong 1$. Actually, it is sufficient that X is larger than 3 to have a modulus of $F(X)$ larger than 0.95 [see Figure 3.13(a)], and a phase of $F(X)$ smaller than 8 degrees; see Figure 3.13(b). Accordingly, the width δ of the angular region around the critical directions, in which $F(X)$ cannot be considered unitary and must be explicitly evaluated, can be obtained by letting $X = 3$ in (3.68), and it is very small if $kr \gg 1$. In fact, for $\varphi = \pi \pm \varphi_0 + \delta$, with small δ, we have

$$X = \frac{kr}{2}\frac{\big(\cos\varphi_0 + \cos\varphi\big)^2}{\sin^2\varphi} \cong \frac{kr}{2}\frac{\big(\cos\varphi_0 + \cos(\pi\pm\varphi_0) - \delta\sin(\pi\pm\varphi_0)\big)^2}{\sin^2\big(\pi\pm\varphi_0\big)} = \frac{kr}{2}\delta^2$$

$$(3.71)$$

so that $X = 3$ leads to $\delta = \sqrt{6/kr} \approx \sqrt{\lambda/r}$.

This section's results have been obtained by using the PO approximation. As already noted, this approximation is not necessary for the problem at hand, for which an exact solution is available. The asymptotic evaluation of the latter leads substantially to the same results reported above: The stationary phase evaluation still leads to the GO field, and the higher-order contributions still lead to a cylindrical congruence of diffracted rays; however, the diffraction coefficient of (3.67) must be replaced by

$$D(\varphi_0,\varphi) = -\frac{\exp\left(-j\frac{\pi}{4}\right)}{2\sqrt{2\pi k}}\left[\frac{1}{\cos\left(\dfrac{\varphi-\varphi_0}{2}\right)} - \frac{1}{\cos\left(\dfrac{\varphi+\varphi_0}{2}\right)}\right]$$

$$= \frac{\exp\left(-j\frac{\pi}{4}\right)}{\sqrt{2\pi k}}\frac{2\sin\dfrac{\varphi_0}{2}\cos\dfrac{\varphi}{2}}{\cos\varphi_0 + \cos\varphi}$$

(3.72)

This is the *geometrical theory of diffraction* (GTD) solution to the problem of diffraction by a PEC half-plane. The APO solution is a very good approximation of the GTD one for points far from the half-plane, but, at variance of the latter, it does not satisfy the boundary condition on the PEC half-plane. The diffraction coefficient of (3.72), too, similar to the one of (3.67), diverges at the critical directions RB and LSB and cannot be used in their vicinity. Again, by using the approach of Section 3.3.2, this problem is solved by resorting to the transition function $F(X)$ of (3.52):

$$D(\varphi_0,\varphi) = \frac{\exp\left(-j\frac{\pi}{4}\right)}{2\sqrt{2\pi k}}\left[\frac{F(X^-)}{\cos\left(\frac{\varphi-\varphi_0}{2}\right)} - \frac{F(X^+)}{\cos\left(\frac{\varphi+\varphi_0}{2}\right)}\right]$$

(3.73)

with

$$X^\pm = 2kr\cos^2\left(\frac{\varphi\pm\varphi_0}{2}\right)$$

(3.74)

This is the *uniform geometrical theory of diffraction* (UGTD)—sometimes shortened to UTD—solution to the problem of diffraction by a PEC half-plane. Similar to the UAPO case, in correspondence with the critical directions RB and LSB, the UTD diffracted field discontinuity exactly compensates for the GO field discontinuity, so that the total field is continuous. It is worth noting that for $\varphi = \pi \pm \varphi_0 + \delta$, with small δ, we have $X^\pm \cong (kr/2)\delta^2$, in agreement with (3.71), so that the width of the angular region around the critical directions in which $F(X)$ cannot be considered unitary and must be explicitly evaluated is again of the order of $\sqrt{\lambda/r}$.

3.5 Geometrical Theory of Diffraction and Its Uniform Extension

Section 3.4 demonstrated that the asymptotic evaluation of the diffraction integral in the case of a scattering object with a sharp edge (the PEC half-plane) leads to a new ray species, the edge diffracted rays. This is only one of the possible situations when new ray species, in addition to the GO rays, can be predicted. These rays are obtained by considering canonical cases (i.e., geometrical objects that *locally* coincide with the actual scattering objects, and for which the analytical solution of Maxwell's equations is available in integral form). The resulting diffraction integral is evaluated asymptotically in the high-frequency limit, $k \to \infty$, and the result is interpreted in terms of new ray congruences. Some other examples, in addition to the edge diffracted rays from a half-plane [Figure 3.16(a)] are the edge diffracted rays from a PEC wedge [Figure 3.16(b)] and the *creeping ray*, propagating and attenuating along the lateral surface of a PEC circular cylinder [Figure 3.16(c)]. These rays are consistent with a generalized form of the Fermat's principle: The ray is the shortest path connecting the source with the observation point via, but not crossing, the scattering body.

This extension of GO—named GTD by Keller [9], one of its founders—as already noted, has the advantage of providing a ray description of propagation. The latter may be invalid in the close vicinity of critical directions, where smoothing transition functions (see Sections 3.3.2 and 3.4.2) must be used. This further extension was named UGTD by Kouyoumjian and Pathak [10], who first applied this procedure to diffraction problems.

In the following, the results for the case of diffraction from a wedge are reported, since this canonical case is of great relevance for the description of diffraction from buildings' edges.

3.5.1 Diffraction from a Perfectly Conducting Wedge: GTD and UTD Solutions

Let us consider a PEC wedge, and let us use a reference system whose z-axis coincides with the edge of the wedge, and whose x-axis belongs to one of the wedge's faces; see Figure 3.17. Accordingly, the points of this wedge's face (often referred to as *0-face*) have a φ coordinate equal to zero. The external angle of the wedge is $n\pi$, where n is a real number with $1 \leq n \leq 2$, so that the wedge internal angle is $(2 - n)\pi$, and the points of the other wedge's face (often referred to as *$n\pi$-face* or *n-face*) have a φ coordinate equal to $n\pi$. Note that for $n = 2$ the PEC wedge reduces to the PEC half-plane of Section 3.4, and for $n = 1$ the wedge becomes a PEC half-space with a plane surface.

According to GTD, if a ray impinges on the edge at a point O from a direction forming an angle β_0 with the edge, then diffracted rays emanate from O forming a cone whose aperture half-angle is β_0 and whose axis is the edge (*Keller cone*); see Figure 3.17. The incident ray direction $\hat{\mathbf{k}}$ is individuated by the incidence angles φ_0 and β_0, whereas the direction $\hat{\mathbf{k}}_s$ of each diffracted ray is individuated by the diffraction angles φ and $\beta = \beta_0$.

It is now convenient to define the *edge-fixed incidence plane* as the plane containing the incident ray and the edge; its normal unit vector is $\hat{\boldsymbol{\varphi}}_0 = \hat{\mathbf{k}} \times \hat{\mathbf{z}} / |\hat{\mathbf{k}} \times \hat{\mathbf{z}}|$, whereas the unit vector parallel to this plane and normal to the incident ray is $\hat{\boldsymbol{\beta}}_0 = \hat{\boldsymbol{\varphi}}_0 \times \hat{\mathbf{k}}$. The incident electric field $\mathbf{E}_{\text{inc}}$ can be then expressed as

$$\mathbf{E}_{\text{inc}} = E_{\text{inc}}^s \hat{\boldsymbol{\beta}}_0 + E_{\text{inc}}^h \hat{\boldsymbol{\varphi}}_0 \tag{3.75}$$

where E_{inc}^s and E_{inc}^h are referred to as *soft-polarized* and *hard-polarized* components of the incident field, respectively.

Similarly, the *edge-fixed diffraction plane* is defined as the plane containing the diffracted ray and the edge; its normal unit vector is $\hat{\boldsymbol{\varphi}} = \hat{\mathbf{z}} \times \hat{\mathbf{k}}_s / |\hat{\mathbf{z}} \times \hat{\mathbf{k}}_s|$,

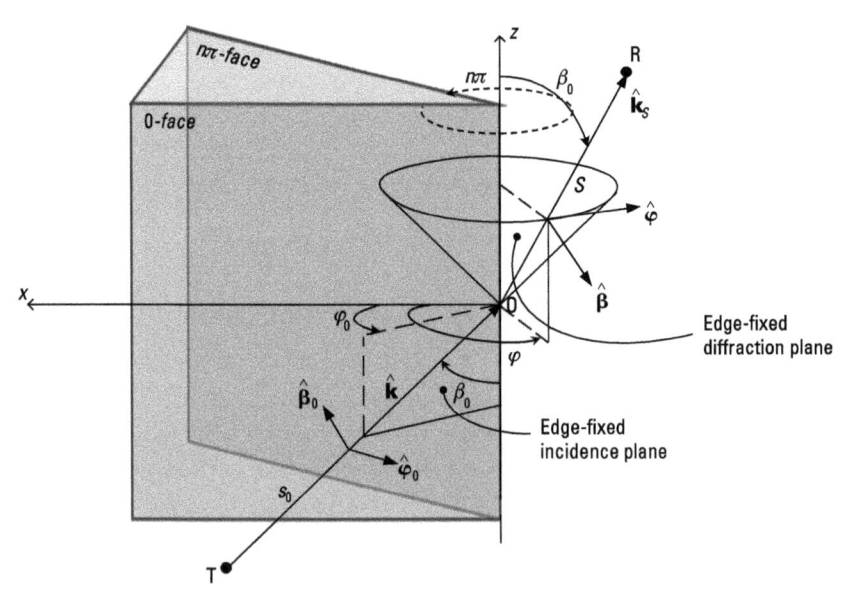

Figure 3.17   Diffraction from a wedge: obliquus incidence.

whereas the unit vector parallel to this plane and normal to the diffracted ray is $\hat{\beta}_0 = \hat{\varphi} \times \hat{k}_s$. The diffracted electric field E_d can be then expressed as

$$\mathbf{E}_d = E_d^s \hat{\beta} + E_d^h \hat{\varphi} \tag{3.76}$$

where E_d^s and E_d^h are the soft- and hard-polarized components of the diffracted field, respectively.

Note that for $\beta_0 = \pi/2$ (i.e., for normal-to-the-edge incidence), the edge-fixed incidence and diffraction planes are perpendicular to the usual incidence plane, so that in this case, soft and hard polarizations coincide with usual perpendicular and parallel polarizations, respectively.

Assuming that the incident ray belongs to a spherical ray congruence radiated by a point-like source placed in the point T of spherical coordinates s_0, $\pi - \beta_0$, φ_0 (see Figure 3.17), the diffracted field at the point R of spherical coordinates s, β_0, φ is:

$$\mathbf{E}_d = \left(s,\beta_0,\varphi\right) = D\left(\beta_0,\varphi_0,\varphi\right)\mathbf{E}_{\mathrm{inc}}(O)A\left(s_0,s\right)\exp\left(-jks\right) \tag{3.77}$$

where

$$A\left(s_0,s\right) = \sqrt{\frac{s_0}{\left(s_0 + s\right)s}} \tag{3.78}$$

describes how the amplitude of the field varies along the diffracted ray, and

$$D = \begin{pmatrix} D_s & 0 \\ 0 & D_h \end{pmatrix} \tag{3.79}$$

is the *dyadic diffraction coefficient*, with D_s and D_h being the scalar diffraction coefficients at soft and hard polarization, respectively, expressed as

$$D_{s,h}\left(\beta_0,\varphi_0,\varphi\right) = \frac{\exp\left(-j\frac{\pi}{4}\right)\sin\left(\frac{\pi}{n}\right)}{n\sqrt{2\pi k}\,\sin\beta_0}\left[\frac{1}{\cos\left(\frac{\pi}{n}\right) - \cos\left(\frac{\varphi-\varphi_0}{n}\right)} \mp \frac{1}{\cos\left(\frac{\pi}{n}\right) - \cos\left(\frac{\varphi+\varphi_0}{n}\right)}\right] \tag{3.80}$$

where the minus and plus sign hold for soft and hard polarizations, respectively. Note that for $n = 1$ (i.e., plane surface), the diffraction coefficients are zero as expected, since in this case there is no edge and hence no diffracted field. In addition, for $n = 2$ (i.e., PEC half-plane) and $\beta_0 = \pi/2$ (i.e., normal incidence) D_s reduces to (3.72).

It can be verified that the first term in the square brackets of (3.80) diverges at the LSB of the 0-face (SB$_0$), $\varphi = \pi + \varphi_0$, and of the $n\pi$-face (SB$_n$), $\varphi = \varphi_0 - \pi$; and the second term in the square brackets of (3.80) diverges at the RB of the 0-face (RB$_0$), $\varphi = \pi - \varphi_0$, and of the $n\pi$-face (RB$_n$), $\varphi = (2n - 1)\pi - \varphi_0$. Therefore, the GTD diffraction coefficients (3.80) cannot be used near these critical directions. In order to use proper transition functions, as dictated by the UTD, (3.80) must be reformulated as follows:

$$D_{s,h}(\beta_0,\varphi_0,\varphi) = -\frac{\exp\left(-j\frac{\pi}{4}\right)\sin\left(\frac{\pi}{n}\right)}{2n\sqrt{2\pi k}\sin\beta_0}\left\{\left[\cot\left(\frac{\pi+(\varphi-\varphi_0)}{2n}\right)+\cot\left(\frac{\pi-(\varphi-\varphi_0)}{2n}\right)\right]\right.$$
$$\left.\mp\left[\cot\left(\frac{\pi+(\varphi+\varphi_0)}{2n}\right)+\cot\left(\frac{\pi-(\varphi+\varphi_0)}{2n}\right)\right]\right\}$$

(3.81)

Equation (3.81) has the advantage that each cotangent term diverges at one of the four critical directions: SB$_n$, SB$_0$, RB$_n$, and RB$_0$. The UTD expression of the wedge diffraction coefficients is then obtained by multiplying each cotangent term by the proper transition function:

$$D_{s,h}(\beta_0,\varphi_0,\varphi) = \frac{\exp\left(-j\frac{\pi}{4}\right)\sin\left(\frac{\pi}{n}\right)}{2n\sqrt{2\pi k}\sin\beta_0}\left\{\left[\cot\left(\frac{\pi+(\varphi-\varphi_0)}{2n}\right)F(X_{SBn})+\cot\left(\frac{\pi-(\varphi-\varphi_0)}{2n}\right)F(X_{SBo})\right]\right.$$
$$\left.\mp\left[\cot\left(\frac{\pi+(\varphi+\varphi_0)}{2n}\right)F(X_{RBn})+\cot\left(\frac{\pi-(\varphi+\varphi_0)}{2n}\right)F(X_{RBo})\right]\right\}$$

(3.82)

in which $F(X)$ is still given by (3.48) and:

$$\begin{cases} X_{SBn} = 2kL\cos^2\left(\frac{2n\pi N^+ - (\varphi-\varphi_0)}{2}\right) \\ X_{SBo} = 2kL\cos^2\left(\frac{2n\pi N^- - (\varphi-\varphi_0)}{2}\right) \\ X_{RBn} = 2kL\cos^2\left(\frac{2n\pi N^+ - (\varphi+\varphi_0)}{2}\right) \\ X_{RBo} = 2kL\cos^2\left(\frac{2n\pi N^- - (\varphi+\varphi_0)}{2}\right) \end{cases}$$

(3.83)

$$\text{where } L = \frac{s_0 s}{s_0 + s} \sin^2 \beta_0 \tag{3.84}$$

and N^+ and N^- are the integers which most nearly satisfy the equations

$$2n\pi N^+ - \left(\varphi \pm \varphi_0\right) = \pi$$
$$2n\pi N^- - \left(\varphi \pm \varphi_0\right) = -\pi \tag{3.85}$$

Note however that $N^\pm$ only need to be evaluated near the critical directions, where they are all equal to zero, except for N^+ near the RBn, which is equal to 1.

For scattering by a PEC wedge, too, the UTD diffracted field has the noteworthy property that its discontinuity at the critical directions exactly compensates for the discontinuity of the GO field, so that the total field (GO + UTD) is continuous.

Finally, it can be noted that if $s_0 \to \infty$ (plane wave incidence) and $\beta_0 = \pi/2$ (normal incidence), then $L = s \equiv r$. If in addition $n = 2$, then $N^\pm$ have no role in (3.83), so that (3.83) coincides with (3.74), and the UTD diffraction coefficient (3.82) for soft polarization reduces to (3.73).

3.5.2 Lossy Dielectric Wedge

Wedges are often used to model diffraction by the straight edges of buildings in urban areas. Therefore, it is of interest to consider a wedge made of a lossy dielectric material. Unfortunately, at variance with the PEC case, for a dielectric wedge an exact analytical solution to be asymptotically evaluated is not available. However, in the case of small losses it is expected that the GO field transmitted through the wedge is dominant with respect to the diffracted field, so that the latter can be neglected. Conversely, if the material losses are significant, and the internal angle of the wedge is not small, the transmitted field is negligible and the diffracted field must be evaluated. In this case, a fairly accurate heuristic solution is obtained by modifying the UTD solution for the PEC case in such a way as to preserve the property that the UTD field discontinuity at the critical directions exactly compensates for the discontinuity of the GO field. This is achieved by multiplying the terms of (3.82) that are discontinuous at the reflection boundaries (i.e., the third and fourth terms) by a proper reflection coefficient [11]. In the case of normal incidence ($\beta_0 = \pi/2$), when soft and hard polarizations coincide with usual perpendicular and parallel polarizations, respectively, the Fresnel reflection coefficient for the proper polarization can be used, and (3.82) is slightly modified as follows:

$$
D_{s,b}\left(\frac{\pi}{2},\varphi_0,\varphi\right) = -\frac{\exp\left(-j\frac{\pi}{4}\right)\sin\left(\frac{\pi}{n}\right)}{2n\sqrt{2\pi k}}\left\{\left[\cot\left(\frac{\pi+(\varphi-\varphi_0)}{2n}\right)F\left(X_{SBn}\right)+\cot\left(\frac{\pi-(\varphi-\varphi_0)}{2n}\right)F\left(X_{SBo}\right)\right]\right.
$$
$$
\left.\pm\Gamma_{\perp,\parallel}\left[\cot\left(\frac{\pi+(\varphi+\varphi_0)}{2n}\right)F\left(X_{RBn}\right)+\cot\left(\frac{\pi-(\varphi+\varphi_0)}{2n}\right)F\left(X_{RBo}\right)\right]\right\}
$$

(3.86)

In the more general case, the rotation from the reference system based on the edge-fixed incidence plane to the one based on the usual incidence plane should be accounted for.

Results obtained via the heuristic diffraction coefficient of (3.86), with possible small corrections to account for the presence of a transmitted field [12, 13], turn out to be in good agreement with numerical simulations [12] and measurements [13].

3.6   Rough-Surface Scattering

The canonical problem of reflection from the smooth boundary between two different homogeneous media, discussed in Section 3.1.3, is often employed to model reflection on soil or sea surfaces, or on building walls. However, such surfaces may have small irregularities or undulations, on the order of the electromagnetic wavelength or a few times larger. Therefore, it is sometimes useful to modify the geometry of this problem by assuming that the boundary between the two media is rough (i.e., that it is a surface with small height deviations with respect to a mean, usually plane, smooth surface). If this mean plane coincides with the x-y plane, surface height deviations are described by the function $z(x, y)$. In most practical situations, surface roughness is not deterministically known (and, in the case of the sea surface, it is also time-varying), so that the height function $z(x, y)$ must be modeled as a *random process*, described by its statistical properties. For the purposes of this book, it is sufficient to specify the *probability density function* (pdf) $p_z(z)$, being $p_z(z)dz$ the probability that the surface height z lies between z and $z + dz$.

The pdf provides an estimate of the amplitude of the surface roughness, and it can be synthetically characterized by mean value $< z >$ and variance σ_z^2 of z:

$$
< z > = \int_{-\infty}^{+\infty} z p_z(z)dz,
$$

(3.87)

$$
\sigma_z^2 = <\left(z - <z>\right)^2> = \int_{-\infty}^{+\infty}\left(z - <z>\right)^2 p_z(z)dz = <z^2> - \left(<z>\right)^2
$$

Since $z(x, y)$ represents the height deviation with respect to the mean plane, it is set $<z> = 0$, so that $\sigma_z^2 = <z^2>$; in addition, it is usually assumed that the pdf is a Gaussian function:

$$p_z(z) = \frac{1}{\sqrt{2\pi}\sigma_z} \exp\left(-\frac{z^2}{2\sigma_z^2}\right) \tag{3.88}$$

Via the numerical integration of (3.88) it is possible to verify that the probability that $-2\sigma_z < z < 2\sigma_z$ is about 0.95, so that the square root of the height variance (i.e., the height *standard deviation*) σ_z can be considered as the surface roughness effective amplitude.

Let us now consider a plane boundary between two different homogeneous media, medium 1 and medium 2, with refractive indexes n_1 and n_2, respectively, and let us consider a source placed in medium 1 at the point T at distance h_1 from the plane; see Figure 3.18(a). Let us first assume that the boundary is smooth; in this case, according to GO, the total field at a point R of medium 1, at distance h_2 from the plane and at horizontal distance d from the source, will be the (coherent) sum of the direct (LOS) contribution $\mathbf{E}_i(R)$, computed via (2.150) and of the reflected field $\mathbf{E}_r(R)$, associated to the ray reflected by the plane at the specular point O, such that incidence and reflection angles are equal; see Figure 3.18(a).

The reflected field $\mathbf{E}_r(R)$ can be computed via (3.20). However, in order to generalize this result to the rough boundary case, it is useful to recover (3.20) by using the equivalence theorem (see Section 2.4.4) and writing a scattering integral to be evaluated asymptotically via the stationary phase method. Accordingly, let us consider a reference system with the x-y plane coincident with the boundary, the x-axis defined by the projections T′ and R′ of T and R over the boundary plane, and the origin coincident with the specular point O; see Figure 3.18(a). The reflected field received at R can be then written as

$$\mathbf{E}_r(R) = \int\limits_{-\infty}^{\infty}\int\limits_{-\infty}^{\infty} \mathbf{f}(x,y)\frac{\exp\left\{-jk\left[R_1(x,y)+R_2(x,y)\right]\right\}}{R_1(x,y)R_2(x,y)}\,dx\,dy \tag{3.89}$$

where $\mathbf{f}(x, y)$ is a vector function related to the equivalent elementary sources over the boundary at the generic point $P \equiv (x, y, 0)$, whose detailed expression is of no concern here,

$$R_1(x,y) = \sqrt{h_1^2 + (x+d_1)^2 + y^2}, \quad R_2(x,y) = \sqrt{h_2^2 + (x-d_2)^2 + y^2} \tag{3.90}$$

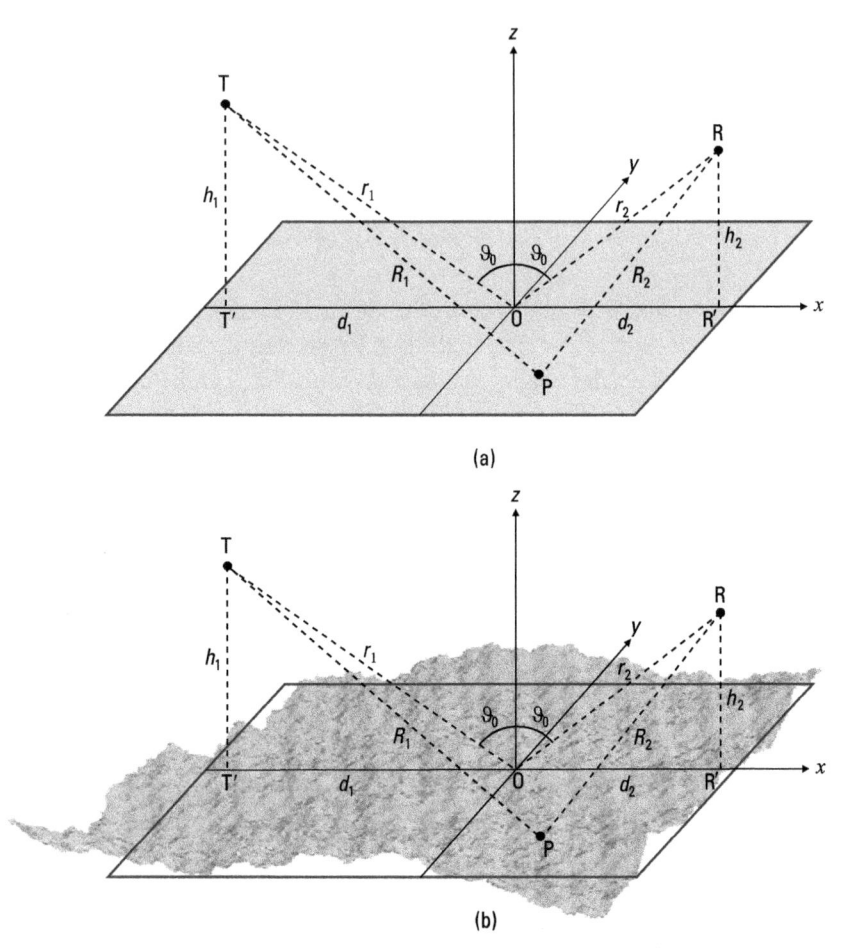

Figure 3.18 Scattering by: (a) a smooth interface, $P \equiv (x, y, 0)$, and (b) a rough interface, $P \equiv (x, y, z(x, y))$.

and we have set $T \equiv (-d_1, 0, h_1)$ and $R \equiv (d_2, 0, h_2)$, with $d_1 + d_2 = d$ and $d_1/d_2 = h_1/h_2$; see Figure 3.18(a).

The derivatives of the phase function are:

$$u_x(x,y) = \frac{\partial(R_1 + R_2)}{\partial x} = \frac{x + d_1}{R_1(x,y)} + \frac{x - d_2}{R_2(x,y)}$$

$$u_y(x,y) = \frac{\partial(R_1 + R_2)}{\partial y} = y\left(\frac{1}{R_1(x,y)} + \frac{1}{R_2(x,y)}\right)$$

$$(3.91)$$

Noting that $d_1/r_1 = d_2/r_2$, it is easy to verify that u_x and u_y are both zero for $x = 0$ and $y = 0$, so that, as expected, the stationary phase point is the origin O. The second derivatives of the phase function, evaluated at $x = 0$, $y = 0$ are:

$$\left.\frac{\partial^2(R_1 + R_2)}{\partial x^2}\right|_{\substack{x=0 \\ y=0}} = \left.\frac{\partial u_x}{\partial x}\right|_{\substack{x=0 \\ y=0}} = \frac{1}{r_1}\left(1 - \frac{d_1^2}{r_1^2}\right) + \frac{1}{r_2}\left(1 - \frac{d_2^2}{r_2^2}\right) = \frac{r_1 + r_2}{r_1 r_2}\cos^2\vartheta_0$$

$$\left.\frac{\partial^2(R_1 + R_2)}{\partial y^2}\right|_{\substack{x=0 \\ y=0}} = \left.\frac{\partial u_y}{\partial y}\right|_{\substack{x=0 \\ y=0}} = \frac{1}{r_1} + \frac{1}{r_2} = \frac{r_1 + r_2}{r_1 r_2}$$

$$\left.\frac{\partial^2(R_1 + R_2)}{\partial x\,\partial y}\right|_{\substack{x=0 \\ y=0}} = \left.\frac{\partial u_y}{\partial x}\right|_{\substack{x=0 \\ y=0}} = 0$$

$$(3.92)$$

Accordingly, the stationary phase evaluation of the integrals in (3.89) leads to

$$E_r(R) = f(0,0)\frac{\exp\{-jk[r_1 + r_2]\}}{r_1 r_2}\sqrt{\frac{-2\pi j}{k}\frac{r_1 r_2}{(r_1 + r_2)\cos^2\vartheta_0}}\sqrt{\frac{-2\pi j}{k}\frac{r_1 r_2}{(r_1 + r_2)}}$$

$$= \frac{-2\pi j}{k\cos\vartheta_0}f(0,0)\frac{\exp\{-jk[r_1 + r_2]\}}{r_1 + r_2}$$

$$(3.93)$$

which is in agreement with (3.20) when $f(0,0)$ is properly computed. Note that to obtain the reflected field of (3.93), it is not necessary that the whole plane is illuminated by the incident field; based on the discussion of Sections 3.2 and 3.3, it is sufficient to illuminate an area centered in O and with linear sizes $\sqrt{\lambda r_1 r_2 / r_1 + r_2}(1/\cos\vartheta_0)$ along x and $\sqrt{\lambda r_1 r_2 / r_1 + r_2}$ along y (the first Fresnel zone).

If the boundary surface is rough, then the generic point on the boundary is $P \equiv (x, y, z(x, y))$, and the field scattered by the rough surface and received at R is

$$\tilde{E}_r(R) = \int_{-\infty}^{\infty}\int_{-\infty}^{\infty}\tilde{f}(x,y)\frac{\exp\{-jk[\tilde{R}_1(x,y) + \tilde{R}_2(x,y)]\}}{\tilde{R}_1(x,y)\tilde{R}_2(x,y)}dx\,dy \quad (3.94)$$

where $\tilde{\mathbf{f}}(x,y)$ is a vector function related to the equivalent elementary sources over the boundary, which differs from $\mathbf{f}(x, y)$ mainly because the surface local slopes are now nonnull (as well as being zero-mean random variables), and

$$\tilde{R}_1(x,y) = \sqrt{(h_1 - z)^2 + (x + d_1)^2 + y^2}$$
$$= R_1(x,y)\sqrt{1 + \frac{z^2 - 2h_1 z}{R_1^2(x,y)}} \cong R_1(x,y) - \frac{h_1 z(x,y)}{R_1(x,y)}$$

$$\tilde{R}_2(x,y) = \sqrt{(h_2 - z)^2 + (x - d_2)^2 + y^2}$$
$$= R_2(x,y)\sqrt{1 + \frac{z^2 - 2h_2 z}{R_2^2(x,y)}} \cong R_2(x,y) - \frac{h_2 z(x,y)}{R_2(x,y)}$$

(3.95)

In (3.95) R_1 and R_2 are given by (3.90), and it is assumed that $\sigma_z \ll h_1$, $\sigma_z \ll h_2$. We can then write

$$\tilde{\mathbf{E}}_r(R) = \int_{-\infty}^{\infty}\int_{-\infty}^{\infty} \tilde{\mathbf{f}}(x,y)\frac{\exp\{-jk[R_1(x,y) + R_2(x,y) - u_z(x,y)z(x,y)]\}}{R_1(x,y)R_2(x,y)}\, dx\, dy$$

(3.96)

where

$$u_z(x,y) = \frac{h_1}{R_1(x,y)} + \frac{h_2}{R_2(x,y)}$$

(3.97)

Note that the scattered field of (3.96) is a random vector, due to the randomness of the surface roughness $z(x, y)$, which also implies the randomness of $\tilde{\mathbf{f}}(x,y)$. The scattered field can be then characterized by computing its mean value and its variance. The former represents the deterministic, *coherent* component of the scattered field, whereas the latter is the intensity of its random, *incoherent* component.

3.6.1 Mean Value of the Scattered Field

The mean value of the scattered field can be computed by recalling that the statistical mean is linear [see (3.87)], so that mean and integral operators can be exchanged, and that in the integrand of (3.96) only $\tilde{\mathbf{f}}(x,y)$ and $z(x, y)$ are random. Therefore,

$$< \tilde{\mathbf{E}}_r(R) > \cong \int\limits_{-\infty}^{\infty} \int\limits_{-\infty}^{\infty} \frac{\exp\{-jk[R_1(x,y)+R_2(x,y)]\}}{R_1(x,y)R_2(x,y)}$$
$$< \tilde{\mathbf{f}}(x,y)\exp[jku_z(x,y)z(x,y)] > dx\,dy \tag{3.98}$$

Considering that $\tilde{\mathbf{f}}(x,y)$ is dependent on the local surface slopes, rather than on the local surface height (recall that it is related to surface tangential field components), and if it is reasonably assumed that local slopes are independent of local height, then we can let $< \tilde{\mathbf{f}}(x,y)\exp[jku_z(x,y)x(x,y)] > = < \tilde{\mathbf{f}}(x,y) > < \exp[jku_z(x,y)z(x,y)] >$. In addition, if the dependence of $\mathbf{f}(x,y)$ on the slopes is approximated with a linear function, then the mean value of $\tilde{\mathbf{f}}(x,y)$ is equal to its value at zero slopes, (i.e., $< \tilde{\mathbf{f}}(x,y) > \cong \mathbf{f}(x,y)$). Finally,

$$< \exp[jku_z(x,y)z(x,y)] > = \int\limits_{-\infty}^{\infty} p_z(z)\exp[jku_z(x,y)z(x,y)]dz$$
$$= \int\limits_{-\infty}^{\infty} \frac{1}{\sqrt{2\pi}\sigma_z}\exp\left(-\frac{z^2}{2\sigma_z^2}\right)\exp[jku_z(x,y)z(x,y)]dz = \exp\left[\frac{k^2\sigma_z^2 u_z^2(x,y)}{2}\right] \tag{3.99}$$

where use has been made of (C.3). We then have

$$< \tilde{\mathbf{E}}_r(R) > \cong \int\limits_{-\infty}^{\infty} \int\limits_{-\infty}^{\infty} \mathbf{f}(x,y)\exp\left[-\frac{k^2\sigma_z^2 u_z^2(x,y)}{2}\right]\frac{\exp\{-jk[R_1(x,y)+R_2(x,y)]\}}{R_1(x,y)R_2(x,y)}dx\,dy \tag{3.100}$$

which is of the same form of the reflected field of the smooth surface case, (3.89), and can be evaluated via the stationary phase method. Since the integrand phase function is the same as in (3.89), then (3.91) and (3.92) also hold in this case, and (3.100) is computed as

$$< \tilde{\mathbf{E}}_r(R) > \cong \frac{-2\pi j}{k\cos\vartheta_0}\mathbf{f}(0,0)\exp\left[-\frac{k^2\sigma_z^2 u_z^2(0,0)}{2}\right]\frac{\exp\{-jk[r_1+r_2]\}}{r_1+r_2} \tag{3.101}$$
$$= \mathbf{E}_r(R)\exp\left(-2k^2\sigma_z^2\cos^2\vartheta_0\right)$$

where use has been made of $u_z(0,0) = h_1/r_1 + h_2/r_2 = 2\cos\vartheta_0$. Therefore, the mean scattered field from the rough surface is equal to the reflected field from the smooth, mean plane, attenuated by a factor depending on the roughness standard deviation: The higher the roughness standard deviation, the lower the mean field. When $k\sigma_z\cos\vartheta_0$ (sometimes called the *Rayleigh parameter*) is much larger than unity, the mean value of the scattered field is negligible, so that the scattered field is completely random.

3.6.2 Variance of the Scattered Field

It can be demonstrated (see, e.g., [4], Section 9.3.2) that

$$<\left|\tilde{\mathbf{E}}_r(R)\right|^2 >\cong \frac{4\pi^2}{k^2}\frac{\left|\mathbf{f}(0,0)\right|^2}{\cos^2\vartheta_0\left(r_1+r_2\right)^2}=\left|\mathbf{E}_r(R)\right|^2 \tag{3.102}$$

(i.e., the mean value of the square modulus of the field scattered by the rough plane is equal to the square modulus of the field reflected by the smooth plane). This is physically explained by considering that when a wave impinges on a rough surface, it is randomly scattered in all directions; this weakens the field at the receiver coming from the specular region, but it also allows for surface regions far from the specular point to contribute to the field at the receiver. The latter effect compensates for the former, and the average power density remains substantially the same as in the smooth surface case.

The scattered field variance can be then evaluated as

$$\mathrm{VAR}\left(\tilde{\mathbf{E}}_r(R)\right)=<\left|\tilde{\mathbf{E}}_r(R)\right|^2 >-\left|<\tilde{\mathbf{E}}_r(R)>\right|^2$$
$$\cong\left|\tilde{\mathbf{E}}_r(R)\right|^2\left[1-\exp\left(-4k^2\sigma_z^2\cos^2\vartheta_0\right)\right] \tag{3.103}$$

If $(2k\sigma_z\cos\vartheta_0)^2 << 1$ (i.e., σ_z is much smaller than wavelength), then the field variance (3.103) is negligible, the scattered field at the receiver is nearly deterministic and coincident with the reflected field, and the roughness can be ignored.

For intermediate roughness (σ_z slightly smaller than wavelength), as the roughness standard deviation increases, the mean of the field at the receiver (i.e., the coherent reflected field) decreases, whereas its variance (i.e., the intensity of the incoherent random component) increases.

Finally, if $2(k\sigma_z\cos\vartheta_0)^2 >> 1$ (i.e., σ_z is much larger than wavelength, or even approximately equal to wavelength), then the mean value of the field is negligible, and its variance coincides with its mean square value. This means that the scattered field is totally random and that only the incoherent component is present at the receiver.

References

[1]   Sadiku, M. N. O., *Numerical Techniques in Electromagnetics*, Boca Raton: CRC Press, 1992.

[2]   Born, M., and E. Wolf, *Principles of Optics* (Sixth Edition), Oxford: Pergamon Press, 1980.

[3]   Balanis, C. A., *Advanced Engineering Electromagnetics*, New York: John Wiley and Sons, 1989.

[4]   Franceschetti, G., *Electromagnetics: Theory, Techniques, and Engineering Paradigms*, New York: Plenum Press, 1997.

[5]   Collin, E. R., *Antennas and Radiowave Propagation*, New York: McGraw-Hill, 1985.

[6]   Dingle, R. B., *Asymptotic Expansions. Their Derivation and Interpretation*, London: Academic Press, 1973.

[7]   Franceschetti, G., et al., "Stochastic Theory of Edge Diffraction," *IEEE Trans. Antennas Propag.*, Vol. 56, No. 2, Feb. 2008, pp. 437–449.

[8]   Franceschetti, G., et al., "Stochastic Theory of Edge Diffraction: Its Physical Reading," *IEEE Trans. Antennas Propag.*, Vol. 58, No. 12, Dec. 2010, pp. 4078–4081.

[9]   Keller, J. B., "Geometrical Theory of Diffraction," *J. Opt. Soc. Amer.*, Vol. 52, No. 2, 1962, pp. 116–130.

[10] Kouyoumjian, R. G., and P. H. Pathak, "A Uniform Geometrical Theory of Diffraction for a Wedge in a Perfectly Conducting Surface," *Proc. IEEE*, Vol. 62, 1974, pp. 1448–1461.

[11] Luebbers, R. J., "Finite Conductivity Uniform GTD Versus Knife Edge Diffraction in Prediction of Propagation Path Loss," *IEEE Trans. Antennas Propagat.*, Vol. AP-32, No.1, Jan. 1984, pp. 70–76.

[12] Rouviere, J.-F., N. Douchin, and P. F. Combes, "Diffraction by Lossy Dielectric Wedges Using Both Heuristic UTD Formulations and FDTD," *IEEE Trans. Antennas Propagat.*, Vol. AP-47, No. 11, Nov. 1999, pp. 1702–1708.

[13] Bernardi, P., R. Cicchetti, and O. Testa, "A Three-Dimensional UTD Heuristic Diffraction Coefficient for Complex Penetrable Wedges," *IEEE Trans. Antennas Propagat.*, Vol. AP-50, No. 2, Feb. 2002, pp. 217–224.

4

Propagation Over a Flat or Spherical Earth

This chapter analyzes electromagnetic propagation over the Earth's surface and through its atmosphere. As a matter of fact, if employed antennas are not very directive, which is often the case in radio broadcasting and in mobile communication systems, then the presence of the Earth's surface cannot be ignored, even for LOS links; Sections 4.1 and 4.2 provide useful tools to account for the soil (or sea) surface effects on electromagnetic propagation. In addition, although the presence of the atmosphere can be neglected for links over distances of up to few kilometers and at frequencies not exceeding a few gigahertz, atmospheric effects over longer distances and/or higher frequencies must be taken into account, as illustrated in Sections 4.3 and 4.4. Finally, over-the-horizon links at low (on the order of a few megahertz) frequencies are possible thanks to ionospheric reflection, and the presence of the ionosphere also plays a role, via the *Faraday rotation* effect (see Section 4.5), in satellite links at frequencies on the order of the gigahertz. Therefore, Section 4.5 provides a brief analysis of propagation through the ionosphere.

In summary, this chapter offers an overview of the different ways in which electromagnetic waves can propagate, through the atmosphere and in the vicinity of the Earth's surface, from the transmitter to the receiver [1–4]: Waves propagating very close to the Earth's surface are called *ground waves* (see Sections 4.1–4.3); those reaching the receiver thanks to refraction or scattering phenomena occurring in the lower atmosphere (the troposphere) are termed *tropospheric waves* (see Section 4.3.2); and finally, waves reaching the receiver thanks to ionospheric reflection are called *sky waves* (see Section 4.5).

4.1 Ground-Wave Propagation and Two-Ray Model

Let us consider a radio link between a transmitting and a receiving antenna, placed in the vicinity of the Earth's surface, with no obstacles around them. In this section, the soil or sea surface is modeled as a plane interface between a vacuum and a homogeneous lossy dielectric medium, so that the effects of the Earth surface's curvature and of the presence of the atmosphere are neglected. This simple model is progressively refined in the next sections; however, as we will verify later, it can be safely employed for short distances, up to few kilometers, and frequencies up to some gigahertz.

The geometry of the problem is depicted in Figure 4.1, where transmitting and receiving antennas are placed at points T and R, respectively; their distances from the ground are h_1 and h_2; and transmitter-to-receiver distance is R_D. This problem is often referred to as *ground-wave propagation*. Its exact analytical solution is available [5–7] and can be expressed as the sum of three terms; the first two terms correspond to the direct and ground-reflected waves of the GO solution (see Section 3.1), and their sum is sometimes referred to as *space wave* [1, 2]. The remaining term can be interpreted as a wave guided

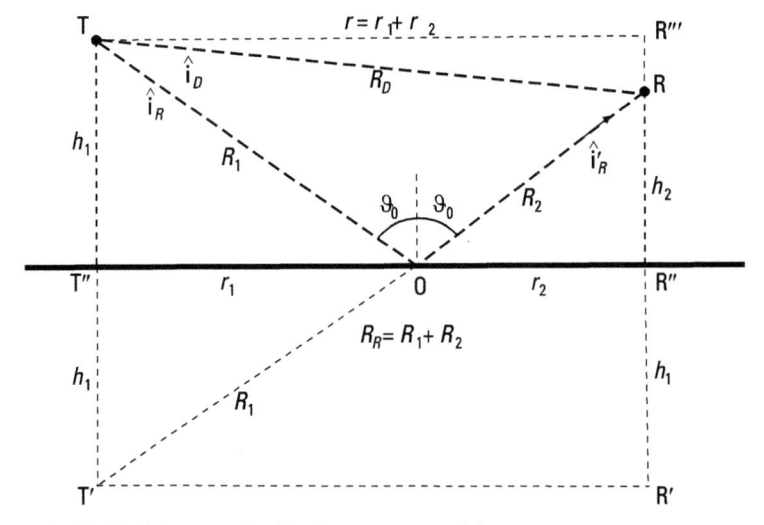

Figure 4.1   Radio link over a flat Earth; two-ray model.

by the ground surface, and it is called *surface wave* (or *Norton surface wave*, or *Norton wave*) [1–4]. The surface wave field is negligible with respect to the GO one if, as it happens in modern wireless communication systems, the frequency is higher than some tens of megahertz, and the antennas are at not less than some wavelengths above the ground. In this case, the GO solution is in very good agreement with the exact one. Accordingly, ground-wave propagation is here treated by using GO: The open-circuit voltage V_0 at the receiving antenna is expressed as the sum of the open-circuit voltage V_{0D} produced by the direct ray joining T and R and the open-circuit voltage V_{0R} produced by the reflected ray, which reaches R from T after a reflection over the specular surface point O. This is the *two-ray model*. If both antennas are either vertically or horizontally polarized, as it is usually the case, then by using (2.150) for the direct ray field, (3.20) for the reflected ray field, and (2.158) for the open-circuit voltages, we can write

$$V_0 = V_{0D} + V_{0R}$$

$$V_{0D} = \frac{jk\zeta I_0}{4\pi R_D}\exp\left(-jkR_D\right)l_T\left(\hat{\mathbf{i}}_D\right)l_R\left(\hat{\mathbf{i}}_D\right) \tag{4.1}$$

$$V_{0R} = \frac{jk\zeta I_0}{4\pi R_R}\exp\left(-jkR_R\right)l_T\left(\hat{\mathbf{i}}_R\right)l_R\left(\hat{\mathbf{i}}'_R\right)\Gamma\left(\vartheta_0\right)$$

where (see Figure 4.1):

$R_R = R_1 + R_2$ is the path length of the reflected ray, with R_1 and R_2 being the lengths of the segments TO and OR, respectively;

$k = 2\pi/\lambda$ is the propagation constant;

I_0 is the input current of the transmitting antenna;

$l_T(\hat{\mathbf{i}}_D)$ and $l_T(\hat{\mathbf{i}}_R)$ are the amplitudes of the transmitting antenna effective lengths for the directions of direct and reflected rays, respectively;

$l_R(\hat{\mathbf{i}}_D)$ and $l_R(\hat{\mathbf{i}}'_R)$ are the amplitudes of the receiving antenna effective lengths for the directions of direct and reflected rays, respectively;

$\Gamma = \Gamma_\perp$ or $\Gamma = -\Gamma_\parallel$ are the Fresnel reflection coefficients for horizontal or vertical polarization,[1] respectively; see (2.92);

ϑ_0 is the incidence angle.

---

1. The minus sign for vertical polarization is due to the signum convention employed in the definition of the corresponding reflection coefficient in Section 2.3.5.

The position of the specular point O is individuated by its horizontal distance to the transmitter r_1 or to the receiver r_2, where, see the similar right triangles TT″O, RR″O, and RR′T′ in Figure 4.1:

$$r_{1,2} = r \frac{h_{1,2}}{h_1 + h_2} \tag{4.2}$$

with $r = r_1 + r_2$ being the transmitter-to-receiver horizontal distance.

The path lengths of direct and reflected rays, R_D and R_R, are expressed as (see the right triangles RR‴T and RR′T′ in Figure 4.1)

$$R_D = \sqrt{r^2 + \left(h_1 - h_2\right)^2}$$
$$R_R = \sqrt{r^2 + \left(h_1 + h_2\right)^2} \tag{4.3}$$

Equation (4.1) can be rewritten as

$$V_0 = V_{0D}\left\{1 + \frac{V_{0R}}{V_{0D}}\right\}$$
$$= V_{0D}\left\{1 + \frac{R_D}{R_R}\frac{l_T\left(\hat{\mathbf{i}}_R\right)l_R\left(\hat{\mathbf{i}}_R'\right)}{l_T\left(\hat{\mathbf{i}}_D\right)l_R\left(\hat{\mathbf{i}}_D\right)}\Gamma\left(\vartheta_0\right)\exp\left[-jk\left(R_R - R_D\right)\right]\right\} \tag{4.4}$$

Note that V_{0D} is the open-circuit voltage that would be obtained in a free-space link.

The modulus of the second term in the curly brackets of (4.4) is not larger than one, since $R_D \le R_R$, $|\Gamma(\vartheta_0)| \le 1$ and, assuming that antennas are properly pointed so that the maxima of their effective lengths are along the direct ray direction, $|l_T(\hat{\mathbf{i}}_R)l_R(\hat{\mathbf{i}}_R')| \le |l_T(\hat{\mathbf{i}}_D)l_R(\hat{\mathbf{i}}_D)|$. If the antennas are very directive, then $|l_T(\hat{\mathbf{i}}_R)l_R(\hat{\mathbf{i}}_R')| << |l_T(\hat{\mathbf{i}}_D)l_R(\hat{\mathbf{i}}_D)|$ and $V_0 \cong V_{0D}$, so that a free-space link is recovered, and the Friis formula (2.167) can be employed. If this is not the case, the more general equation (4.4) must be used.

The received power is directly proportional to the square modulus of V_0 (see Section 2.5.2) and therefore, in view of (4.4), it can be expressed as:

$$P_R = P_{FS}G_{2R} \tag{4.5}$$

where P_{FS} is the free-space received power, given by the Friis formula (2.167), and G_{2R} is the *two-ray gain factor*:

$$G_{2R} = \left| 1 + \frac{R_D}{R_R} \frac{l_T(\hat{\mathbf{i}}_R) l_R(\hat{\mathbf{i}}'_R)}{l_T(\hat{\mathbf{i}}_D) l_R(\hat{\mathbf{i}}_D)} \Gamma(\vartheta_0) \exp\left[-jk(R_R - R_D) \right] \right|^2 \tag{4.6}$$

As the receiver is moved, this factor oscillates from minima smaller than one to maxima larger than one but smaller than four: maxima are obtained when the open-circuit voltages of direct and reflected rays are in phase, so that they interfere constructively, while minima are obtained when they have a 180° phase difference, so that they interfere disruptively. Plots of $|V_0|$ and P_R versus r are presented in Figure 4.2.

If, as it is usually the case, $r \gg h_1$ and $r \gg h_2$, then the vertical dimension in Figure 4.1 should be much shorter than the horizontal one. Accordingly,

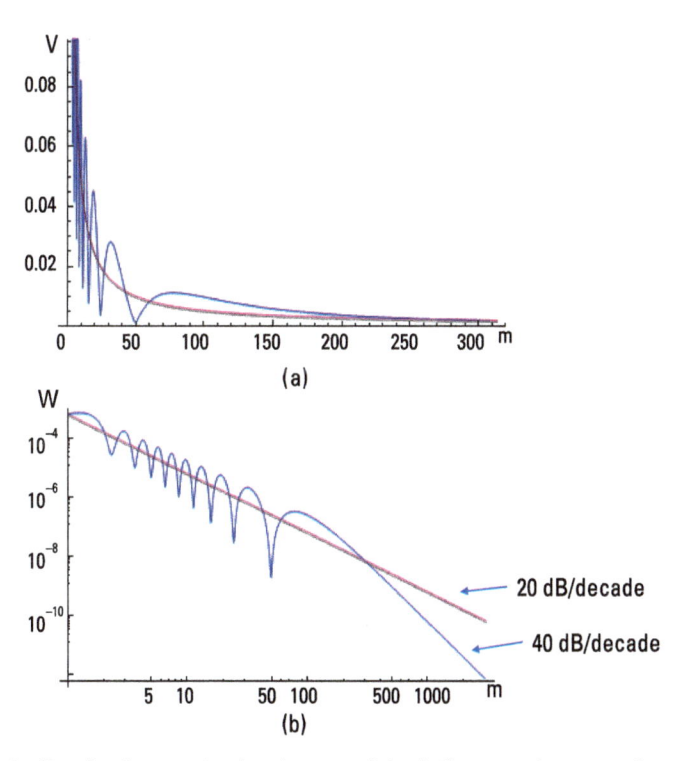

(a)

(b)

Figure 4.2 Received open-circuit voltage modulus in linear scale versus distance (a), and received power in logarithmic scale versus distance (b), for free space (purple line) and two-ray model (blue line) at 1 GHz for 1-W transmitted power and horizontally polarized unitary-gain antennas placed at 1.5m from the ground and with 50-Ω input resistance.

$R_D/R_R \cong 1$; $\hat{\mathbf{i}}_R \cong \hat{\mathbf{i}}'_R \cong \hat{\mathbf{i}}_D$, so that $(l_T(\hat{\mathbf{i}}_R)l_R(\hat{\mathbf{i}}'_R))/(l_T(\hat{\mathbf{i}}_D)l_R(\hat{\mathbf{i}}_D)) \cong 1$; $\vartheta_0 \cong \pi/2$, so that [see (2.92)], $\Gamma(\vartheta_0) \cong -1$.

Therefore, (4.6) simplifies as follows:

$$G_{2R} \cong \left| 1 - \exp\left[-jk\left(R_R - R_D \right) \right] \right|^2 = 2\left\{ 1 - \cos\left[k\left(R_R - R_D \right) \right] \right\} \quad (4.7)$$

We can also approximate (4.3) as

$$R_D = \sqrt{r^2 + \left(h_1 - h_2 \right)^2} \cong r\left[1 + \frac{1}{2}\frac{\left(h_1 - h_2 \right)^2}{r^2} - \frac{1}{8}\frac{\left(h_1 - h_2 \right)^4}{r^4} \right]$$

$$= r + \frac{1}{2}\frac{\left(h_1 - h_2 \right)^2}{r} - \frac{1}{8}\frac{\left(h_1 - h_2 \right)^4}{r^3}$$

$$R_R = \sqrt{r^2 + \left(h_1 + h_2 \right)^2} \cong r\left[1 + \frac{1}{2}\frac{\left(h_1 + h_2 \right)^2}{r^2} - \frac{1}{8}\frac{\left(h_1 + h_2 \right)^4}{r^4} \right] \quad (4.8)$$

$$= r + \frac{1}{2}\frac{\left(h_1 + h_2 \right)^2}{r} - \frac{1}{8}\frac{\left(h_1 + h_2 \right)^4}{r^3}$$

Therefore, if, in addition to the previous hypothesis, we also assume that

$$\frac{1}{8}\frac{\left(h_1 + h_2 \right)^4}{\lambda r^3} \ll 1 \quad \text{(i.e.,} \quad r \gg \frac{h_1 + h_2}{2}\sqrt[3]{\frac{h_1 + h_2}{\lambda}}) \quad (4.9)$$

then we have

$$\cos\left[k\left(R_R - R_D \right) \right] \cong \cos\left[\pi \frac{\left(h_1 + h_2 \right)^2 - \left(h_1 - h_2 \right)^2}{\lambda r} \right] = \cos\left[\frac{4\pi h_1 h_2}{\lambda r} \right] \quad (4.10)$$

and

$$G_{2R} \cong 2\left\{ 1 - \cos\left[\frac{4\pi h_1 h_2}{\lambda r} \right] \right\} \quad (4.11)$$

Maxima of this function, equal to four, are obtained for

$$\frac{4\pi h_1 h_2}{\lambda r} = \left(2n + 1 \right)\pi; \text{ that is, } r_n^{(\max)} = \frac{4 h_1 h_2}{\lambda\left(2n + 1 \right)} \quad (4.12)$$

whereas its minima, equal to zero, are obtained for

$$\frac{4\pi h_1 h_2}{\lambda r} = 2n\pi; \text{ that is, } r_n^{(\min)} = \frac{2h_1 h_2}{\lambda n} \tag{4.13}$$

Finally, for $r \to \infty$ we have

$$\cos\left[\frac{4\pi h_1 h_2}{\lambda r}\right] \cong 1 - \frac{1}{2}\left(\frac{4\pi h_1 h_2}{\lambda r}\right)^2, \text{ so that}$$

$$G_{2R} \cong \left(\frac{4\pi h_1 h_2}{\lambda r}\right)^2 \text{ and}$$

$$P_R = P_{FS} G_{2R} \cong P_T G_T G_R \left(\frac{\lambda}{4\pi r}\right)^2 \left(\frac{4\pi h_1 h_2}{\lambda r}\right)^2 = P_T G_T G_R \left(\frac{h_1 h_2}{r^2}\right)^2 \tag{4.14}$$

where P_T is the transmitted power, and G_T and G_R are the gains of transmitting and receiving antennas, respectively.

By summarizing the results reported above (see also Figure 4.2), we can state that, according to the two-ray model, as the receiver is moved away from the transmitter, the received power oscillates around its free-space value, with maxima and minima placed according to (4.12) and (4.13), respectively, until the last maximum is reached. This happens at distance

$$r_0^{(\max)} = \frac{4h_1 h_2}{\lambda} \tag{4.15}$$

from the transmitter (*break point distance*). From this distance on, the received power monotonically decreases, and for very large distances it attenuates as one over the fourth power of r. This attenuation is significantly faster than the one experienced in free space, when received power asymptotically attenuates as one over r squared. By using logarithmic units, the attenuation is 20 dB/decade for free-space propagation and 40 dB/decade for ground-wave propagation[2] at a distance much larger than the break point one; see Figure 4.2(b).

It must be finally noted that, in the preceding analysis, the ground has been modeled as a lossy dielectric medium, which is a good model at frequencies higher than a few megahertz (see Sections 2.2.2 and 2.2.3). This is the primary topic of interest for modern wireless communication systems in urban areas. At lower frequencies, however, ground is better modeled as a good

---

2. A *decade* is a factor of ten.

conductor (see Section 2.2.3), whose refractive index with respect to air n_{21} is much larger than one: $|n_{21}| \gg 1$. This does not change the above analysis for horizontal polarization. Conversely, for vertical polarization, even at large distances, the reflection coefficient $\Gamma = -\Gamma_{\parallel}$ cannot be approximated as -1; in fact, it is approximately equal to 1 up to incidence angles very close to $\pi/2$, as it can be verified by using $|n_{21}| \gg 1$ in (2.92). Therefore, the minus sign in (4.11) is replaced by a plus sign, so that the roles of maxima and minima in (4.12) and (4.13) are exchanged, and direct and reflected contributions interfere constructively up to large distances, where the received power is four times the free-space one. Although the precise evaluation of the field strength at these low frequencies would require considering the Norton surface wave, this is still a good approximation, as far as the Earth's curvature can be neglected.

4.1.1   Example: Link Between Two Walkie-Talkies

Let us consider a pair of walkie-talkies (i.e., handheld transceivers) whose technical data are listed in Table 4.1. According to the manufacturer, they can be used up to a distance of about 5 km. To check this information, we want to compute their maximum range of operation; that is, the maximum distance from the transmitter at which a sufficient SNR is obtained at the receiver. From Table 4.1, we note that this is obtained for a received power not smaller than -113 dBm (i.e., 0.5×10^{-14}W).

Since the antennas are not directive, and the walkie-talkies are obviously used close to the Earth's surface, we employ the two-ray model. First of all, let us evaluate the distance at which the last maximum of the received power is reached. For the given frequency, we have $\lambda = c/f \cong 0.67$m; in addition, it is reasonable to assume that both transceivers are held at a distance from the ground of about 1.6m: $h_1 \cong h_2 \cong 1.6$m. By using (4.15) we get $r_0^{(max)} \cong 15.28$m. It can be expected that the maximum range of operation r_{max} is much larger than this distance, so that r_{max} can be computed by inverting (4.14):

Table 4.1
Parameters of the Considered Walkie-Talkie System

Parameter	Value
Frequency (f)	446 MHz
Transmitted power (P_T)	500 mW
Antenna gain $G_T = G_R$	$\cong 1.5$
Receiver sensitivity $P_{R min}$ at 12 dB SNR	-113 dBm

$$r_{\max} = \sqrt[4]{\frac{P_T G_T G_R}{P_{R\min}}} \sqrt{h_1 h_2} \cong 6200\,\text{m}.$$

This value is a bit larger than the 5-km maximum range declared by the manufacturer. Section 4.3.1 shows that a better agreement is obtained by accounting for the Earth's curvature. Note that if we use the free-space Friis formula to compute the maximum range, we get $r_{FS\max} = \sqrt{P_T G_T G_R / P_{R\min}}\,(\lambda/4\pi) \cong 7.99 \times 10^5\,\text{m}$ (i.e., a range of almost 800 km), which is completely unrealistic.

4.1.2 Effect of Surface Roughness

The effect of a random surface roughness with height standard deviation σ_z (small with respect to h_1 and h_2) can be accounted for by applying the results of Section 3.6 to the field of the reflected ray. Accordingly, the open-circuit voltage at the receiving antenna can be modeled as a random variable $\tilde{V}_0$: its mean value is obtained by using (3.101) in (4.1) and (4.4), thus having

$$\langle \tilde{V}_0 \rangle = V_{0D}\left\{ 1 + \frac{R_D}{R_R}\frac{l_T(\hat{i}_R)l_R(\hat{i}_R')}{l_T(\hat{i}_D)l_R(\hat{i}_D)}\Gamma(\vartheta_0)\exp\left(-2k^2\sigma_z^2\cos^2\vartheta_0\right)\exp\left[-jk\left(R_R - R_D\right)\right]\right\}$$

$$(4.16)$$

and its variance coincides with the one of the reflected contribution $\tilde{V}_{0R}$, obtained by using (3.103), thus having

$$\text{VAR}\left(\tilde{V}_0\right) = \text{VAR}\left(\tilde{V}_{0R}\right) \cong \left|V_{0R}\right|^2\left[1 - \exp\left(-4k^2\sigma_z^2\cos^2\vartheta_0\right)\right] \quad (4.17)$$

where V_{0R} is the reflected contribution in case of smooth plane. This means that, in the presence of a surface roughness, the fast oscillations of the received signal level around the free-space value, observed at short distance from the transmitter, are, at least in part, random, and they are fully random if $\sigma_z \gg \lambda$. In the latter case, the real exponential function in (4.16) and (4.17) is very close to zero, so that the mean value of the received signal is equal to the free space received signal, and its variance is equal to the square modulus of the smooth-surface reflected contribution. However, at a large distance from the transmitter (i.e., for $r \gg h_1$ and $r \gg h_2$), we have $\vartheta_0 \cong \pi/2$ and $\cos\vartheta_0 \cong 0$, so that the exponential function in (4.17) is very close to one, and the received signal variance is negligible. Accordingly, in this case the received signal is deterministic and [see also (4.16)] coincident with the one in the presence of a smooth surface. In conclusion, the presence of a surface roughness small with respect to h_1 and h_2 does not significantly affect the received power at a large distance from the transmitter.

4.2 Effect of the Earth's Curvature

The Earth is approximately spherical, with a mean radius a of about 6,370 km. The main effect of the Earth surface's curvature on propagation is that, even in the absence of other obstacles, it may obstruct the LOS between a transmitter and a receiver. We recall that the horizon is defined as the locus of points at which straight lines from an observer are tangential to the Earth's surface. It is then easy to realize that two points above the Earth's surface at heights h_1 and h_2 are in LOS if their mutual distance r does not exceed the sum r_h of their distances to the horizon r_{h1} and r_{h2} (see Figure 4.3):

$$r < r_h = r_{h1} + r_{h2} = \sqrt{\left(a + h_1\right)^2 - a^2} + \sqrt{\left(a + h_2\right)^2 - a^2}$$
$$= \sqrt{2ah_1 + h_1^2} + \sqrt{2ah_2 + h_2^2} \cong \sqrt{2ah_1} + \sqrt{2ah_2} \tag{4.18}$$

where we have exploited the fact that $a \gg h_{1,2}$. For instance, for $h_1 = h_2 = 1.6$m we get $r_h \cong 9,030$m $\cong 9$ km, and for $h_1 = h_2 = 10$m we get $r_h \cong 22,574$m

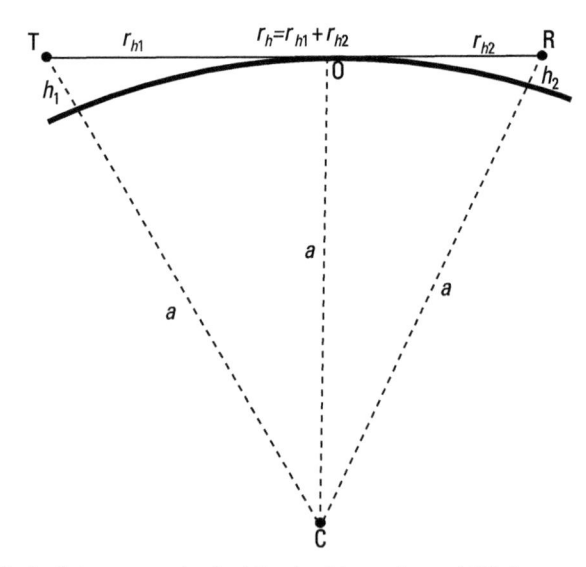

Figure 4.3 Radio link over a spherical Earth with maximum LOS distance r_h and distances to the horizon r_{h1} and r_{h2} of transmitter and receiver.

$\cong 22.5$ km. GO can be used for distances up to about 70% of r_h; for longer distances, diffraction cannot be neglected, and there is no simple method for precisely calculating the field strength.

Assuming that GO can be employed, the same formulation of Section 4.1 can be used, provided that the actual distances from the ground h_1 and h_2 are replaced by the distances h'_1 and h'_2 from the plane tangent to the Earth's surface at the specular reflection point O (see Figure 4.4). Therefore, received power at a large distance is decreased with respect to the flat-Earth case. Considering that h_1, h_2, and r are much smaller than a, so that approximations similar to those in (4.18) can be used, distances h'_1 and h'_2 can be computed by solving the following system of six nonlinear algebraic equations:

$$\begin{cases} h'_1 \cong h_1 - \Delta h_1 \\ h'_2 \cong h_2 - \Delta h_2 \\ r_1 \cong \sqrt{2a\Delta h_1} \Leftrightarrow \Delta h_1 \cong r_1^2/2a \\ r_2 \cong \sqrt{2a\Delta h_2} \Leftrightarrow \Delta h_2 \cong r_2^2/2a \\ r_1 = r\dfrac{h'_1}{h'_1 + h'_2} \\ r_2 = r\dfrac{h'_2}{h'_1 + h'_2} \end{cases} \tag{4.19}$$

The six unknowns are h'_1, h'_2, Δh_1, Δh_2, r_1, and r_2, with Δh_1 and Δh_2 defined in Figure 4.4. This system can be solved by using numerical methods. A simple iterative approach consists of first letting $h'_1 \cong h_1$ and $h'_2 \cong h_2$ in the fifth and sixth of (4.19) to obtain first estimates of r_1 and r_2, used in the third and fourth of (4.19) to compute Δh_1 and Δh_2. The latter can be finally used in the first and second of (4.19) to obtain new estimates of h'_1 and h'_2. This procedure can be iterated by using the new estimates of h'_1 and h'_2 again in the fifth and sixth of (4.19) and proceeding as before. The iterative procedure is arrested when the difference between two subsequent estimates of h'_1 and h'_2

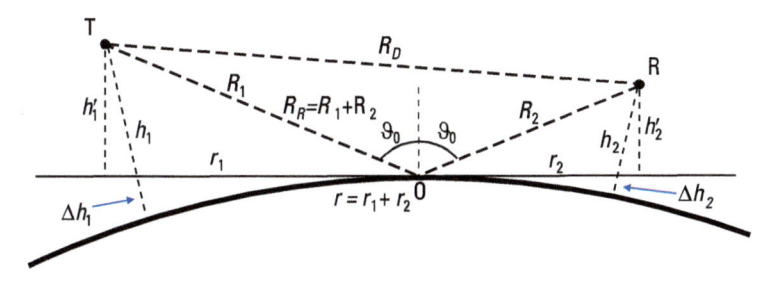

Figure 4.4 Radio link over a spherical Earth; two-ray model.

falls below the desired accuracy. In practice, very few iterations are sufficient to obtain a reasonable accuracy.

If $h_1 = h_2$, all of the procedure simplifies, since we have $r_1 = r_2 = r/2$ and hence $\Delta h_1 = \Delta h_2 = r^2/(8a)$. From this expression it can be verified that if the distance r does not exceed 2,250m, then Δh_1 and Δh_2 are smaller than 10 cm, so that in this case the Earth's curvature can be safely neglected.

4.3 Atmospheric Effect: Ray Curvature and Effective Earth Radius

The index of refraction of air n_a is very close to one, its exact value slightly depending on frequency and on the air temperature, pressure, and humidity. At radio and microwave frequencies, and in conditions of *standard atmosphere* (i.e., 15°C sea-level temperature, 1,013 hPa sea-level pressure, zero humidity, and prescribed vertical gradients of temperature and pressure), n_a is about 1.0003 at sea level, and its vertical gradient in the low atmosphere (up to a few km height) is

$$\frac{dn_a}{dz} \cong -4 \cdot 10^{-8}\,\mathrm{m}^{-1} \tag{4.20}$$

The air index of refraction has also a very small imaginary part, related to losses, which is considered in the next section. There are also very small spatial fluctuations of n_a caused by turbulence and water vapor content fluctuations, which usually have a negligible influence on propagation; these are considered in Section 4.3.2. Accordingly, the lower atmosphere can be modeled as a stratified medium (see Section 3.1.4), whose index of refraction decreases with the height z. The main effect of the atmosphere on propagation is then ray bending (*atmospheric refraction*). The radius of curvature R of the ray can be computed by using (3.27)

$$\frac{1}{R} = -\frac{d\ln n_a}{dz}\sin\vartheta = \frac{1}{n_a}\frac{dn_a}{dz}\sin\vartheta \cong -\frac{dn_a}{dz}\sin\vartheta \tag{4.21}$$

where ϑ is the angle formed by the ray with the vertical direction and where we have used $n_a \cong 1$. If we consider a radio link between two far-apart antennas close to the ground, $\vartheta \cong \pi/2$ and $\sin\vartheta \cong 1$, so that

$$\frac{1}{R} \cong -\frac{dn_a}{dz} \tag{4.22}$$

By using (4.20), we have that, for the standard atmosphere, the ray radius of curvature is $R \cong 0.25 \times 10^8 \text{m} = 25,000$ km. It is positive, and this means that the ray is bended downward, and it is quite larger than the Earth's radius, so that ray bending is appreciable only if the propagation path is at least some kilometers long.

Due to ray bending, the *radio horizon* (i.e., the locus of points at which direct rays from an antenna are tangential to the Earth's surface), is at a larger distance than the geometrical horizon. The effects of Earth and ray curvatures can be then simultaneously accounted for by still treating the rays as rectilinear, but replacing the Earth's radius a in (4.18) and (4.19) with the *effective Earth radius* a_e, defined as the radius of a hypothetical Earth (*effective Earth*) on which the distance to the geometrical horizon is the equal to the distance to the radio horizon on the actual Earth; see Figure 4.5. By considering an antenna at height h, whose distance to the radio horizon is r_{rh}, we can write (see triangles OCA and OC′A of Figure 4.5(a) and OC″T of Figure 4.5(b))

$$\begin{cases} r_{rh} \cong \sqrt{2a(h+\Delta)} & \Leftrightarrow & h \cong \dfrac{r_{rh}^2}{2a} - \Delta \\[2mm] r_{rh} \cong \sqrt{2R\Delta} & \Leftrightarrow & \Delta \cong \dfrac{r_{rh}^2}{2R} \\[2mm] r_{rh} \cong \sqrt{2a_e h} & \Leftrightarrow & h \cong \dfrac{r_{rh}^2}{2a_e} \end{cases} \qquad (4.23)$$

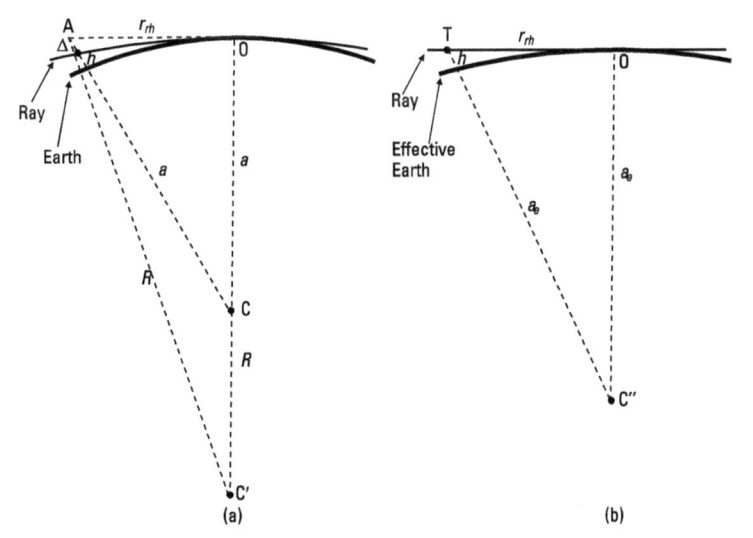

Figure 4.5 Radio horizon over a spherical Earth with atmospheric refraction: actual problem (a) and equivalent problem (b).

where Δ is defined in Figure 4.5. From (4.23), equating the first and last expressions of h, and substituting the value of Δ, we easily get

$$\frac{1}{a_e} = \frac{1}{a} - \frac{1}{R} \tag{4.24}$$

so that the curvature of the effective Earth is the difference between the curvatures of the actual Earth and of the ray. From (4.24) we can finally compute the effective Earth radius:

$$a_e = \frac{R}{R - a} a \tag{4.25}$$

For the standard atmosphere, $a_e \cong 4/3a \cong 1.33a$.

4.3.1 Example: Link Between Two Walkie-Talkies, Effects of Earth Curvature, and Atmospheric Refraction

Let us consider the same walkie-talkie system discussed in Section 4.1.1 and evaluate the received power at 5 km from the transmitter (which is the maximum range of operation declared by the manufacturer) by also accounting for Earth's curvature and atmospheric refraction. We have to use (4.14) in which $h_1 = h_2 \cong 1.6\text{m}$ are replaced by

$$h'_{1,2} \cong h_{1,2} - \Delta h_{1,2} \cong h_{1,2} - \frac{r^2}{8a_e} \cong 1.6 - \frac{25 \cdot 10^6}{8 \cdot 1.33 \cdot 6.37 \cdot 10^6}\text{m} \cong 1.24\text{m}$$

We then get

$$P_R \cong P_T G_T G_R \left(\frac{h'_1 h'_2}{r^2}\right)^2 \cong 0.45 \cdot 10^{-14}\,\text{W} \cong -113.5\ \text{dBm}$$

This value is in very good agreement with the receiver sensitivity (see Table 4.1), so that 5 km can be considered the walkie-talkies' maximum range of operation, in agreement with the value provided by the manufacturer.

4.3.2 Atmospheric Ducting and Tropospheric Scattering

Under nonstandard conditions, the index of refraction of the atmosphere may decrease with height with a faster or a slower rate than that predicted by (4.20). Accordingly, the ray curvature may be higher (*super-refraction*) or lower (*substandard refraction*) than the standard value. Usually, such variations are very small; however, in some parts of the world (e.g., oceans and very large plains), it sometimes happens that in the lowest atmospheric layer the refractive index decreases so rapidly that the ray curvature is greater than the Earth's. In this case, rays radiated upward are refracted back to the ground. These rays are then reflected by the ground and refracted back again repeatedly, in such a way that the field is guided in a thin layer of the atmosphere close to the Earth's surface [*surface duct*; see Figure 4.6(a)]. This phenomenon is known as *atmospheric ducting*; radio waves will propagate over long distances with an attenuation much lower than the free-space one, since the electromagnetic power flux is mostly confined within the duct. A similar phenomenon may occasionally occur at larger heights [*elevated duct*; see Figure 4.6(b)], if the refractive index is not monotonically decreasing with height, so that there is an air layer whose refractive index is higher than the one of the neighboring air layers. Atmospheric ducting may be exploited for long-range radio links in the VHF and UHF bands, but the reliability of these links is too low to be useful, due to their occasional nature. On the other hand, this phenomenon may cause occasional long-distance interference. However, this is not a significant problem for mobile radio systems in urban areas.

Finally, it is worth mentioning that spatial fluctuations of n_a caused by turbulences and by water vapor content fluctuations in the higher troposphere (about 5–10 km in height) may act as elevated scattering centers that reflect a (small) part of the upward-radiated electromagnetic power back to the ground. This phenomenon is known as *tropospheric scattering*, and it can be exploited to establish over-the-horizon links. However, these links have a high variability and a very low reliability, so that today they are only used by radio amateurs.

4.4 Atmospheric Attenuation: Clear Air, Fog, Rain

To this point, air has been considered as a lossless medium. However, as with all dielectric materials (see Section 2.2.2), air has a frequency-depending complex relative dielectric constant $\varepsilon_{ra} = \varepsilon'_{ra} - j\varepsilon''_{ra}$. Its real part ε'_{ra} is very close to one, whereas its imaginary part $-\varepsilon''_{ra}$, related to losses, is much smaller than

Figure 4.6 Atmospheric ducting: (a) surface duct and (b) elevated duct.

one in modulus. Accordingly, the propagation constant in the air is

$$k = k_0 n_a = k_0 \sqrt{\varepsilon'_{ra} - j\varepsilon''_{ra}} \cong k_0 \left(1 - j\varepsilon''_{ra}/2\right) = k_0 - j\alpha_a \qquad (4.26)$$

The imaginary part of the propagation constant describes an attenuation of the power density S as a wave propagates (see Section 2.3.1):

$$S(l) = \frac{1}{2\zeta}\left|E(l)\right|^2 = S(0)\exp\left(-2\alpha_a l\right) \qquad (4.27)$$

where l is the abscissa along the propagation path. The above equation is often expressed in decibels:

$$\left[S(l)\right]_{dB} = \left[S(0)\right]_{dB} - 20\alpha_a l \text{Log}(e) = \left[S(0)\right]_{dB} - \alpha_{dB/m} l \qquad (4.28)$$

where $\alpha_{dB/m} = 20\alpha_a \text{Log}(e) \cong 8.686\alpha_a$ is the specific attenuation in decibels per meter.

In view of (4.26), the specific attenuation is directly proportional to the imaginary part of the air relative dielectric constant, which is determined by the dielectric constants of atmosphere's main constituents. These are nitrogen (about 78%) and oxygen (about 21%), to which a variable amount of water vapor is added in the lower atmosphere. As noted in Section 2.2.2, peaks of the imaginary part of the dielectric constant are present at about 22 GHz for water, at 60 GHz for oxygen, and at much higher frequency for nitrogen. Accordingly, the specific attenuation of nitrogen is negligible at all the frequencies of interest for wireless communications, while the specific attenuations of oxygen and water vapor at sea level vary with frequency as illustrated in Figure 4.7. They are smaller than 0.01 dB/km at frequencies up to 10 GHz, and smaller than 1 dB/km from 10 to 100 GHz, except for the oxygen attenuation peak of about 10 dB/km placed at 60 GHz.

In conclusion, in clear air [i.e., in the absence of water droplets (fog or clouds, rain) or ice crystals (snow)], electromagnetic wave attenuation can be always ignored at frequencies up to 10 GHz; it can be ignored for short wireless links at frequencies up to 100 GHz, except that at frequencies close to 60 GHz; and it must be taken into account even for the design of very short links at frequencies close to 60 GHz.

4.4.1 Attenuation by Rain, Fog, and Snow

Electromagnetic waves propagating through rain are attenuated due to the power absorption in the lossy dielectric material represented by water and,

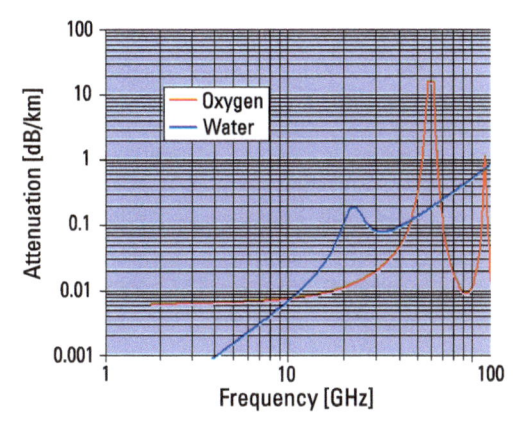

Figure 4.7 Specific attenuation by oxygen and water vapor at sea level, at 20°C temperature and for a water vapor content of 7.5 g/m^3 (reworked from [1, 2]).

in part, due to the scattering of the incident power out of the propagation direction by the rain droplets. Since the radius of raindrops is on the order of one millimeter, a rather simple theoretical evaluation of attenuation by rain at microwave frequencies can be obtained starting from the computation of the power dissipated and scattered by a spherical water droplet with radius much smaller than wavelength [1]. It turns out that, as expected, the attenuation increases with the droplets' number and sizes. However, this theoretical evaluation requires knowledge of the number of droplets per volume unit and of the droplet-size distribution, both functions of the rain rate R in millimeters of water per hour (mm/h). In practice, an empirical formula directly relating the specific attenuation to the rain rate can be more conveniently used:

$$\alpha_{dB/km} = aR^b \tag{4.29}$$

where the parameters a and b depend on frequency and (very slightly) on polarization. Their values for several frequencies are reported in Table 4.2 [8]. Plots of the specific attenuation as a function of rain rate are displayed in Figure 4.8 for some frequencies of interest for modern wireless communication systems. By considering that a rain rate of 20 mm/h already corresponds to a heavy rain, from Figure 4.8 it can be concluded that attenuation by rain can always be neglected for frequencies up to few gigahertz; that it can be neglected for short wireless links for frequencies up to about 10 GHz; and that it must be always considered for frequencies higher than 10 GHz. To include attenuation by rain in a wireless link design, knowledge of the rainfall statistics for

Table 4.2
Parameters a and b of (4.29) for Different Frequencies and for Horizontal (**H**) and Vertical (**V**) Polarizations (from [8])

Frequency (GHz)	a_H	b_H	a_V	b_V
1	0.0000259	0.9691	0.0000308	0.8592
1.5	0.0000443	1.0185	0.0000574	0.8957
2	0.0000847	1.0664	0.0000998	0.9490
2.5	0.0001321	1.1209	0.0001464	1.0085
3	0.0001390	1.2322	0.0001942	1.0688
3.5	0.0001155	1.4189	0.0002346	1.1387
4	0.0001071	1.6009	0.0002461	1.2476
4.5	0.0001340	1.6948	0.0002347	1.3987
5	0.0002162	1.6969	0.0002428	1.5317
5.5	0.0003909	1.6499	0.0003115	1.5882
6	0.0007056	1.5900	0.0004878	1.5728
7	0.001915	1.4810	0.001425	1.4745
8	0.004115	1.3905	0.003450	1.3797
9	0.007535	1.3155	0.006691	1.2895
10	0.01217	1.2571	0.01129	1.2156
11	0.01772	1.2140	0.01731	1.1617
12	0.02386	1.1825	0.02455	1.1216
13	0.03041	1.1586	0.03266	1.0901
14	0.03738	1.1396	0.04126	1.0646
15	0.04481	1.1233	0.05008	1.0440
16	0.05282	1.1086	0.05899	1.0273
17	0.06146	1.0949	0.06797	1.0137
18	0.07078	1.0818	0.07708	1.0025
19	0.08084	1.0691	0.08642	0.9930
20	0.09164	1.0568	0.09611	0.9847
21	0.1032	1.0447	0.1063	0.9771
22	0.1155	1.0329	0.1170	0.9700
23	0.1286	1.0214	0.1284	0.9630
24	0.1425	1.0101	0.1404	0.9561
25	0.1571	0.9991	0.1533	0.9491
26	0.1724	0.9884	0.1669	0.9421

Table 4.2 *(Continued)*

Frequency (GHz)	a_H	b_H	a_V	b_V
27	0.1884	0.9780	0.1813	0.9349
28	0.2051	0.9679	0.1964	0.9277
29	0.2224	0.9580	0.2124	0.9203
30	0.2403	0.9485	0.2291	0.9129
31	0.2588	0.9392	0.2465	0.9055
32	0.2778	0.9302	0.2646	0.8981
33	0.2972	0.9214	0.2833	0.8907
34	0.3171	0.9129	0.3026	0.8834
35	0.3374	0.9047	0.3224	0.8761
36	0.3580	0.8967	0.3427	0.8690
37	0.3789	0.8890	0.3633	0.8621
38	0.4001	0.8816	0.3844	0.8552
39	0.4215	0.8743	0.4058	0.8486
40	0.4431	0.8673	0.4274	0.8421
41	0.4647	0.8605	0.4492	0.8357
42	0.4865	0.8539	0.4712	0.8296
43	0.5084	0.8476	0.4932	0.8236
44	0.5302	0.8414	0.5153	0.8179
45	0.5521	0.8355	0.5375	0.8123
46	0.5738	0.8297	0.5596	0.8069
47	0.5956	0.8241	0.5817	0.8017
48	0.6172	0.8187	0.6037	0.7967
49	0.6386	0.8134	0.6255	0.7918
50	0.6600	0.8084	0.6472	0.7871
51	0.6811	0.8034	0.6687	0.7826
52	0.7020	0.7987	0.6901	0.7783
53	0.7228	0.7941	0.7112	0.7741
54	0.7433	0.7896	0.7321	0.7700
55	0.7635	0.7853	0.7527	0.7661
56	0.7835	0.7811	0.7730	0.7623
57	0.8032	0.7771	0.7931	0.7587
58	0.8226	0.7731	0.8129	0.7552

Table 4.2 *(Continued)*

Frequency (GHz)	a_H	b_H	a_V	b_V
59	0.8418	0.7693	0.8324	0.7518
60	0.8606	0.7656	0.8515	0.7486
61	0.8791	0.7621	0.8704	0.7454
62	0.8974	0.7586	0.8889	0.7424
63	0.9153	0.7552	0.9071	0.7395
64	0.9328	0.7520	0.9250	0.7366
65	0.9501	0.7488	0.9425	0.7339
66	0.9670	0.7458	0.9598	0.7313
67	0.9836	0.7428	0.9767	0.7287
68	0.9999	0.7400	0.9932	0.7262
69	1.0159	0.7372	1.0094	0.7238
70	1.0315	0.7345	1.0253	0.7215
71	1.0468	0.7318	1.0409	0.7193
72	1.0618	0.7293	1.0561	0.7171
73	1.0764	0.7268	1.0711	0.7150
74	1.0908	0.7244	1.0857	0.7130
75	1.1048	0.7221	1.1000	0.7110
76	1.1185	0.7199	1.1139	0.7091
77	1.1320	0.7177	1.1276	0.7073
78	1.1451	0.7156	1.1410	0.7055
79	1.1579	0.7135	1.1541	0.7038
80	1.1704	0.7115	1.1668	0.7021
81	1.1827	0.7096	1.1793	0.7004
82	1.1946	0.7077	1.1915	0.6988
83	1.2063	0.7058	1.2034	0.6973
84	1.2177	0.7040	1.2151	0.6958
85	1.2289	0.7023	1.2265	0.6943
86	1.2398	0.7006	1.2376	0.6929
87	1.2504	0.6990	1.2484	0.6915
88	1.2607	0.6974	1.2590	0.6902
89	1.2708	0.6959	1.2694	0.6889
90	1.2807	0.6944	1.2795	0.6876

Table 4.2 *(Continued)*

Frequency (GHz)	a_H	b_H	a_V	b_V
91	1.2903	0.6929	1.2893	0.6864
92	1.2997	0.6915	1.2989	0.6852
93	1.3089	0.6901	1.3083	0.6840
94	1.3179	0.6888	1.3175	0.6828
95	1.3266	0.6875	1.3265	0.6817
96	1.3351	0.6862	1.3352	0.6806
97	1.3434	0.6850	1.3437	0.6796
98	1.3515	0.6838	1.3520	0.6785
99	1.3594	0.6826	1.3601	0.6775
100	1.3671	0.6815	1.3680	0.6765
120	1.4866	0.6640	1.4911	0.6609
150	1.5823	0.6494	1.5896	0.6466
200	1.6378	0.6382	1.6443	0.6343
300	1.6286	0.6296	1.6286	0.6262
400	1.5860	0.6262	1.5820	0.6256
500	1.5418	0.6253	1.5366	0.6272
600	1.5013	0.6262	1.4967	0.6293

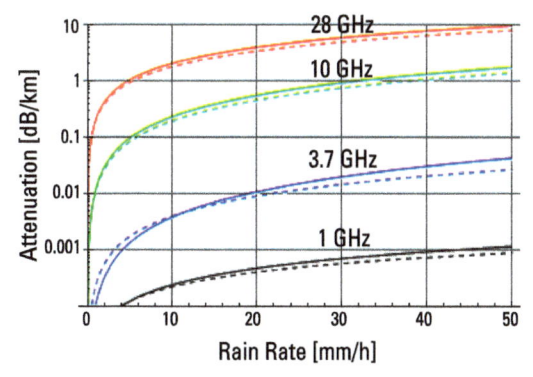

Figure 4.8 Specific attenuation by rain versus rain rate for selected values of frequency and for horizontal (solid lines) and vertical (dashed lines) polarizations.

the specific geographic area is needed. For instance, if it is required that the link remains active for at least 99.99% of time, in (4.29) R must be the rain rate that is exceeded for less than the 0.01% of time.

Attenuation by fog and clouds is governed by the same physical mechanisms as attenuation by rain. However, fog and clouds are made of very small, suspended water droplets with radii in the range 0.01–0.05 mm, much smaller than those of raindrops. It turns out that attenuation by fog is much smaller than attenuation by rain; even for very dense fog, the specific attenuation for microwaves and millimeter waves never exceeds 1 dB/km, and it is usually much smaller than that. Therefore, for wireless link designs including a sufficient power margin to overcome attenuation by rain, the attenuation by fog and clouds is not the limiting factor.

Finally, snow is made of ice crystals of various shapes, possibly mixed with liquid water (*wet snow*). Since losses in ice are much smaller than losses in liquid water, attenuation of microwaves in *dry snow* is at least one order of magnitude smaller than the one in rain at the same precipitation rate. However, attenuation in wet snow may be comparable to that of rain. Due to this strong variability, a simple relation between the snow precipitation rate and attenuation is not available. This is not a significant problem, since wireless link designs including a sufficient power margin to overcome attenuation by rain are usually able to overcome attenuation by snow, too.

4.5 Ionosphere

The ionosphere is a layer of the Earth's upper atmosphere extending from about 60 km to some thousand kilometers in altitude; see Figure 4.9. It consists of ionized gases and therefore can be modeled as a plasma (see Section 2.2.5). Since ionization is mainly caused by the Sun's radiation, the number of electrons per volume unit (or *electron concentration*) N_e varies according to solar illumination: It is higher during the day and lower during the night. In addition to the daily variations, ionization also varies yearly, with a maximum during the summer, and according to the 11-year solar cycle, with a maximum corresponding to the maximum of solar activity.

The ionosphere is subdivided into three sublayers, indicated with the letters D, E, and F. The D layer is the lowest, extending from about 60 to 90 km altitude. It has a low electron concentration but a rather high collision rate v, due to the relatively high air density at lower altitude. It completely disappears during the night. The E layer extends from about 90 to 150 km altitude and has a maximum electron concentration of about 10^{11} electrons per

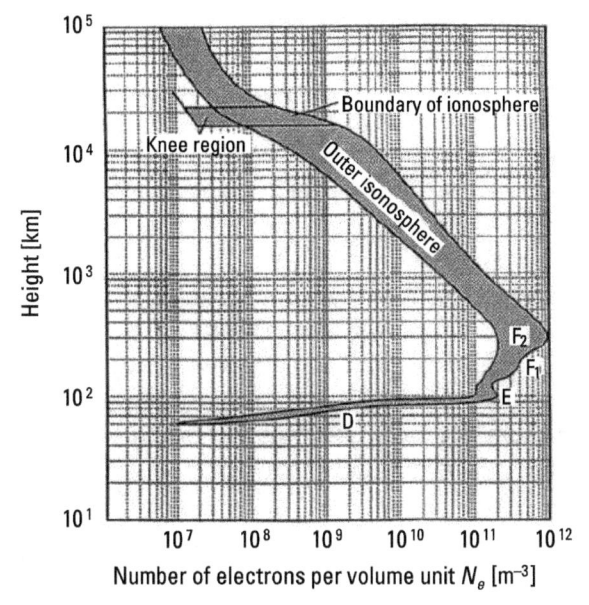

Figure 4.9 Daylight ionospheric electron concentration as a function of height.

m^3 and a low collision rate. It is strongly weakened during the night. Finally, the F layer extends from about 150 to 500 km, its maximum electron concentration is about 10^{12} electrons per m^3, and its collision rate is negligible. It is the only layer that persists also during the night with only a small reduction of ionization. Above the F layer, the electron concentration monotonically decreases with height.

4.5.1 Ionospheric Reflection and Sky Wave

Since the ionosphere can be modeled as a plasma, its relative dielectric constant is expressed by (2.42). Its imaginary part, related to losses, is only appreciable for frequencies on the order of the collision rate v or smaller. Accordingly, losses cause a significant attenuation of waves at frequencies in the LF band or smaller (i.e., less than 300 kHz), and, during the day, also in the MF band (300 kHz–3 MHz), due to the presence of the D layer, for which v is larger. At higher frequency, the imaginary part of the dielectric constant of the ionosphere can be neglected, and its propagation constant is then expressed as

$$k = k_0 n = k_0 \sqrt{1 - \frac{\omega_p^2}{\omega^2}} = k_0 \sqrt{1 - \frac{f_p^2}{f^2}} \tag{4.30}$$

where k_0 is the free-space propagation constant, $f_p = \omega_p/(2\pi)$ is the *plasma frequency*, and the plasma angular frequency ω_p, defined in Section 2.2.5, is expressed by (2.43). Considering the values of the ionosphere electron concentration N_e reported above, f_p is in the range from about 4 to 9 MHz.

If frequency is lower than plasma frequency (i.e., $f < f_p$), then the propagation constant is purely imaginary: Electromagnetic waves cannot propagate into the ionosphere; they exponentially attenuate. Therefore, the power density carried by a wave radiated toward the sky by an Earth-based transmitting antenna is totally reflected back toward the ground by the ionosphere, regardless of the incidence angle (*ionospheric reflection*).

If frequency is higher than plasma frequency (i.e., $f > f_p$), then the propagation constant is real and electromagnetic waves may propagate through the ionosphere. However, the ionosphere index of refraction n is lower than 1, so that if a wave radiated toward the sky impinges onto the ionosphere with an incidence angle larger that the critical angle (see Section 2.3.5),

$$\vartheta_c = \arcsin n = \arcsin\sqrt{1 - \frac{f_p^2}{f^2}} \tag{4.31}$$

then its power density is again totally reflected back toward the ground. This also implies that for each fixed incidence angle ϑ, a *maximum usable frequency* exists, above which ionospheric reflection does not occur. It can be computed by inverting (4.31):

$$f_{max} = \frac{f_p}{\sqrt{1 - \sin^2\vartheta}} = \frac{f_p}{\cos\vartheta} \tag{4.32}$$

As the incidence angle increases, the maximum usable frequency increases. However, due to the Earth's curvature, the incidence angle on the ionosphere cannot arbitrarily approach $\pi/2$; see Figure 4.10. Therefore, ionospheric reflection never happens for frequencies higher than about 30 MHz.

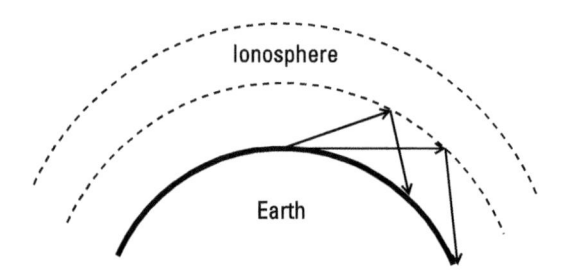

Figure 4.10 Ionospheric reflection.

Note that, in view of the analysis of stratified media reported in Section 3.1.4, these evaluations of critical angle and maximum usable frequency are still valid even considering that N_e (and hence f_p) gradually increases to its maximum value, and there is no abrupt discontinuity.

Ionospheric reflection was widely used from 1920 to about 1960 for over-the-horizon HF (3–30-MHz) radio links to provide intercontinental, transoceanic telephone and telegraph services, and for long-distance radio broadcasting (*sky-wave propagation*). Due to its rather low reliability, and to the necessity of using higher frequencies to support higher bandwidths and transmission rates, this technology was gradually abandoned in favor of satellite communications, although it is still used by some radio broadcasting stations, by radio amateurs, and for some long-distance transmission services requiring a very narrow bandwidth.

For frequencies much higher than the plasma frequency (i.e., $f \gg f_p$), we have $n \cong 1$, and the propagation constant of the ionosphere is very close to that of free space. Therefore, in the design of microwave satellite links, the presence of the ionosphere can be ignored (apart from the Faraday rotation effect, analyzed in Section 4.5.2, which may play a role at frequencies up to a few gigahertz). However, in global navigation satellite systems (GNSSs), such as the global positioning system (GPS), operating at L band (1–2 GHz), the propagation velocity must be very precisely known, in order to estimate the satellite-to-receiver range from the propagation time with high accuracy. The group velocity in the ionosphere is then of interest in this case, and it can be computed by using (4.30) in (2.72):

$$v_g = \left(\frac{dk}{d\omega}\right)^{-1} = \left(\frac{1}{c}\frac{d}{d\omega}\sqrt{\omega^2 - \omega_p^2}\right)^{-1} = \left(\frac{1}{c}\frac{\omega}{\sqrt{\omega^2 - \omega_p^2}}\right)^{-1}$$

$$= c\sqrt{1 - \frac{f_p^2}{f^2}} \cong c\left(1 - \frac{f_p^2}{2f^2}\right) \tag{4.33}$$

Due to the variability of N_e, f_p is not *a priori* precisely known, but it can be retrieved simultaneously with the range by measuring the propagation delays at two different frequencies.

4.5.2 Effect of the Earth's Magnetic Field on Ionospheric Propagation

Due to the presence of the Earth's magnetostatic field $\mathbf{H}_0$, the ionosphere should be more accurately modeled as a magnetized plasma; see Section 2.2.5.

Therefore, it is an anisotropic medium whose relative dielectric constant is a matrix: By using a Cartesian reference system whose z-axis coincides with the direction of $\mathbf{H}_0$ (i.e., $\mathbf{H}_0 = H_0 \hat{\mathbf{z}}$), the relative dielectric constant matrix $\boldsymbol{\varepsilon}_r$ is given by (2.44–45), with $\omega_c \cong 2\pi \cdot 1.4$ MHz being the cyclotron angular frequency. This does not substantially change the results on ionospheric reflection reported in Section 4.5.1. However, propagation through the ionosphere must be analyzed by using the results presented in Section 2.3.7 for magnetized plasma; the electric field of a homogeneous plane wave propagating through the ionosphere must satisfy (2.114), which here we rewrite as

$$\mathbf{ME} = 0 \quad \text{with} \quad \mathbf{M} = k^2\left(\mathbf{I} - \hat{\mathbf{k}}\hat{\mathbf{k}} \cdot\right) - k_0^2 \boldsymbol{\varepsilon}_r \tag{4.34}$$

and the propagation constant k of the wave must obey (2.116); in other words,

$$\text{Det}\left[\mathbf{M}\right] = 0 \tag{4.35}$$

so that (4.34) has nonnull solutions for the vector $\mathbf{E}$. The wave magnetic field is finally obtained from the electric one by using (2.117), which we rewrite here as

$$\mathbf{H} = \frac{k}{\omega\mu_0}\hat{\mathbf{k}} \times \mathbf{E} \tag{4.36}$$

4.5.2.1 Direction of Propagation Perpendicular to the Magnetostatic Field

Let us first consider a wave propagating along a direction perpendicular to the magnetostatic field: $\hat{\mathbf{k}} = \hat{\mathbf{x}}$. Then, we have

$$\mathbf{M} = k^2\begin{pmatrix} 0 & 0 & 0 \\ 0 & 1 & 0 \\ 0 & 0 & 1 \end{pmatrix} - k_0^2\begin{pmatrix} \varepsilon_1 & -j\varepsilon_2 & 0 \\ j\varepsilon_2 & \varepsilon_1 & 0 \\ 0 & 0 & \varepsilon_3 \end{pmatrix} = \begin{pmatrix} -k_0^2\varepsilon_1 & jk_0^2\varepsilon_2 & 0 \\ -jk_0^2\varepsilon_2 & k^2 - k_0^2\varepsilon_1 & 0 \\ 0 & 0 & k^2 - k_0^2\varepsilon_3 \end{pmatrix} \tag{4.37}$$

Accordingly, (4.34) can be written as

$$\mathbf{M}'\mathbf{E}' = 0 \tag{4.38a}$$

$$\left(k^2 - k_0^2\varepsilon_3\right)E_z = 0 \tag{4.38b}$$

with

$$M' = \begin{pmatrix} -k_0^2\varepsilon_1 & jk_0^2\varepsilon_2 \\ -jk_0^2\varepsilon_2 & k^2 - k_0^2\varepsilon_1 \end{pmatrix} \quad \text{and} \quad \mathbf{E}' = E_x\hat{\mathbf{x}} + E_y\hat{\mathbf{y}}$$

In addition, we have

$$\text{Det}[M] = \left(k^2 - k_0^2\varepsilon_3\right)\left[k_0^2\varepsilon_1\left(k^2 - k_0^2\varepsilon_1\right) + k_0^4\varepsilon_2^2\right] = \left(k^2 - k_0^2\varepsilon_3\right)\text{Det}[M']$$

$$(4.39)$$

so that (4.35) is satisfied either if $k^2 - k_0^2\varepsilon_3 = 0$ or if $\text{Det}[M'] = 0$.

In the former case we obtain the solution

$$k = k_{\text{ord}} = k_0\sqrt{\varepsilon_3} = k_0\sqrt{1 - \frac{\omega_p^2}{\omega^2}} \tag{4.40}$$

where ε_3 has been expressed via (2.45). This propagation constant is coincident with the propagation constant of the nonmagnetized plasma. In this case, $\text{Det}[M'] \neq 0$, and (4.38a) is only satisfied by $E_x = E_y = 0$; conversely, (4.38b) is satisfied for any E_z. By also using (4.36) we then get the following electromagnetic field:

$$\mathbf{E} = E\hat{\mathbf{z}}$$

$$\mathbf{H} = \frac{k_0\sqrt{\varepsilon_3}}{\omega\mu_0}\hat{\mathbf{x}} \times \mathbf{E} = -\frac{\sqrt{\varepsilon_3}}{\zeta_0}E\hat{\mathbf{y}} \tag{4.41}$$

so that the wave is TEM. This wave is called the *ordinary wave*.

The second solution for k, obtained by imposing $\text{Det}[M'] = 0$, is

$$k = k_{\text{extr}} = k_0\sqrt{\varepsilon_1 - \frac{\varepsilon_2^2}{\varepsilon_1}} \tag{4.42}$$

In this case, $k^2 - k_0^2\varepsilon_3 \neq 0$, and (4.38b) is only satisfied by $E_z = 0$; conversely, (4.38a) is satisfied if $E_x = j(\varepsilon_2/\varepsilon_1)E_y$. Accordingly, by also using (4.36), we get the following electromagnetic field:

$$\mathbf{E} = E_y\left(\hat{\mathbf{y}} + j\frac{\varepsilon_2}{\varepsilon_1}\hat{\mathbf{x}}\right)$$

$$\mathbf{H} = \frac{k_{\text{extr}}}{\omega\mu_0}\hat{\mathbf{x}} \times \mathbf{E} = \frac{1}{\zeta_0}\sqrt{\varepsilon_1 - \frac{\varepsilon_2^2}{\varepsilon_1}}E_y\hat{\mathbf{z}} \tag{4.43}$$

so that the wave is TM. This wave is called the *extraordinary wave*.

Ordinary and extraordinary waves travel with different propagation constants and hence with different velocities. However, for ω much larger than ω_c and ω_p, $\varepsilon_2 \ll \varepsilon_1$ so that $E_x \cong 0$ and also the extraordinary wave is TEM; in addition, we can write [see (2.45)]

$$k_{\text{ord}} = k_0 \sqrt{1 - \frac{\omega_p^2}{\omega^2}} \cong k_0 \left(1 - \frac{\omega_p^2}{2\omega^2} - \frac{\omega_p^4}{8\omega^4} \right)$$

$$k_{\text{extr}} = k_0 \sqrt{\varepsilon_1 - \frac{\varepsilon_2^2}{\varepsilon_1}} \cong k_0 \left(1 - \frac{\omega_p^2}{2\omega^2} - \frac{\omega_p^2 \omega_c^2}{2\omega^4} - \frac{\omega_p^4}{8\omega^4} \right)$$

$$(4.44)$$

so that the phase difference between the two waves after propagating through a ionospheric layer of thickness l is

$$\Delta\varphi = l\left(k_{\text{ord}} - k_{\text{extr}} \right) \cong k_0 l \frac{\omega_p^2 \omega_c^2}{2\omega^4} = \frac{l}{c} \frac{\omega_p^2 \omega_c^2}{2\omega^3} = \frac{\pi l}{c} \frac{f_p^2 f_c^2}{f^3} \qquad (4.45)$$

where $f_c = \omega_c/(2\pi)$ is the *cyclotron frequency*. Even considering an ionospheric layer with a thickness l on the order of 1,000 km, this phase difference is totally negligible for frequencies higher than about 100 MHz. At these frequencies, both propagation constants can be safely approximated with the propagation constant of the nonmagnetized plasma (4.30).

4.5.2.2 Direction of Propagation Parallel to the Magnetostatic Field

Let us now consider a wave propagating along a direction parallel to the magnetostatic field: $\hat{\mathbf{k}} = \hat{\mathbf{z}}$. Then, we have

$$M = k^2 \begin{pmatrix} 1 & 0 & 0 \\ 0 & 1 & 0 \\ 0 & 0 & 0 \end{pmatrix} - k_0^2 \begin{pmatrix} \varepsilon_1 & -j\varepsilon_2 & 0 \\ j\varepsilon_2 & \varepsilon_1 & 0 \\ 0 & 0 & \varepsilon_3 \end{pmatrix} = \begin{pmatrix} k^2 - k_0^2\varepsilon_1 & jk_0^2\varepsilon_2 & 0 \\ -jk_0^2\varepsilon_2 & k^2 - k_0^2\varepsilon_1 & 0 \\ 0 & 0 & -k_0^2\varepsilon_3 \end{pmatrix}$$

$$(4.46)$$

In this case, (4.34) can be written as

$$M'' \mathbf{E}' = 0 \qquad (4.47a)$$

$$-k_0^2 \varepsilon_3 E_z = 0 \qquad (4.47b)$$

with

$$M'' = \begin{pmatrix} k^2 - k_0^2\varepsilon_1 & jk_0^2\varepsilon_2 \\ -jk_0^2\varepsilon_2 & k^2 - k_0^2\varepsilon_1 \end{pmatrix} \quad \text{and} \quad \mathbf{E}' = E_x\hat{\mathbf{x}} + E_y\hat{\mathbf{y}}$$

In addition, we have

$$\text{Det}[M] = -k_0^2\varepsilon_3\left[k^4 - 2k^2k_0^2\varepsilon_1 + k_0^4\left(\varepsilon_1^2 - \varepsilon_2^2\right)\right] = -k_0^2\varepsilon_3\text{Det}[M''] \quad (4.48)$$

so that (4.35) is satisfied if $\text{Det}[M''] = 0$. This equation has two possible solutions for k^2, and hence there are two possible propagation constants:

$$k_+ = k_0\sqrt{\varepsilon_1 + \varepsilon_2} = k_0\sqrt{1 - \frac{\omega_p^2}{\omega(\omega + \omega_c)}}$$

$$k_- = k_0\sqrt{\varepsilon_1 - \varepsilon_2} = k_0\sqrt{1 - \frac{\omega_p^2}{\omega(\omega - \omega_c)}}$$

$$(4.49)$$

In both cases, (4.47b) is only satisfied by $E_z = 0$, so that both waves are TEM, whereas (4.47a) is satisfied if $E_y^\pm = \pm jE_x^\pm$. Accordingly, by also using (4.36), we get the following electromagnetic field:

$$\mathbf{E}^\pm = E^\pm\left(\hat{\mathbf{x}} \pm j\hat{\mathbf{y}}\right)$$

$$\mathbf{H}^\pm = \frac{k_\pm}{\omega\mu_0}\hat{\mathbf{z}} \times \mathbf{E}^\pm = \frac{\sqrt{\varepsilon_1 \pm \varepsilon_2}}{\zeta_0}E^\pm\left(\hat{\mathbf{y}} \mp j\hat{\mathbf{x}}\right)$$

$$(4.50)$$

so that the two waves have left-circular and right-circular polarization, respectively.

Let us now consider a linearly polarized wave, whose electric field is directed along the x-axis, that propagates along the direction $\hat{\mathbf{z}}$ of the Earth's magnetic field and enters the ionosphere at $z = 0$:

$$\mathbf{E}(0) = 2E_0\hat{\mathbf{x}} = E_0\left(\hat{\mathbf{x}} + j\hat{\mathbf{y}}\right) + E_0\left(\hat{\mathbf{x}} - j\hat{\mathbf{y}}\right) = \mathbf{E}^+(0) + \mathbf{E}^-(0) \quad (4.51)$$

This wave propagates through the ionosphere layer, of thickness l, as two circularly polarized waves with different propagation constants. Accordingly, at the exit plane $z = l$ the electric field is:

$$\mathbf{E}(l) = \mathbf{E}^+(l) + \mathbf{E}^-(l) = E_0\left(\hat{\mathbf{x}} + j\hat{\mathbf{y}}\right)\exp\left(-jk_+l\right) + E_0\left(\hat{\mathbf{x}} - j\hat{\mathbf{y}}\right)\exp\left(-jk_-l\right)$$

$$= E_0\left[\exp\left(-jk_+l\right) + \exp\left(-jk_-l\right)\right]\hat{\mathbf{x}} + jE_0\left[\exp\left(-jk_+l\right) - \exp\left(-jk_-l\right)\right]\hat{\mathbf{y}}$$

$$= E_0\exp\left(-j\frac{k_++k_-}{2}l\right)\left\{\begin{array}{l}\left[\exp\left(-j\frac{k_+-k_-}{2}l\right) + \exp\left(j\frac{k_+-k_-}{2}l\right)\right]\hat{\mathbf{x}} \\ + j\left[\exp\left(-j\frac{k_+-k_-}{2}l\right) - \exp\left(j\frac{k_+-k_-}{2}l\right)\right]\hat{\mathbf{y}}\end{array}\right\}$$

$$= 2E_0\exp\left(-j\frac{k_++k_-}{2}l\right)\left[\cos\left(\frac{k_+-k_-}{2}l\right)\hat{\mathbf{x}} + \sin\left(\frac{k_+-k_-}{2}l\right)\hat{\mathbf{y}}\right]$$

$$(4.52)$$

Therefore, the wave is still linearly polarized, but the direction of the electric field is rotated by an angle

$$\gamma = \frac{k_+ - k_-}{2}l \qquad (4.53)$$

See Figure 4.11 (*Faraday rotation*). For ω much larger than ω_c and ω_p, we can write

$$k_+ = k_0\sqrt{1 - \frac{\omega_p^2}{\omega(\omega + \omega_c)}} \cong k_0\left(1 - \frac{\omega_p^2}{2\omega^2} + \frac{\omega_p^2\omega_c}{2\omega^3} - \frac{\omega_p^4}{8\omega^4} + \frac{\omega_p^2\omega_c^2}{2\omega^4}\right)$$

$$\qquad (4.54)$$

$$k_- = k_0\sqrt{1 - \frac{\omega_p^2}{\omega(\omega - \omega_c)}} \cong k_0\left(1 - \frac{\omega_p^2}{2\omega^2} - \frac{\omega_p^2\omega_c}{2\omega^3} - \frac{\omega_p^4}{8\omega^4} - \frac{\omega_p^2\omega_c^2}{2\omega^4}\right)$$

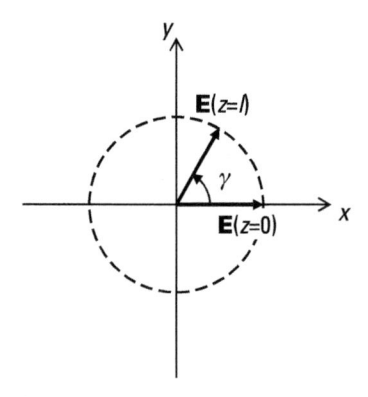

Figure 4.11   Faraday rotation.

so that

$$\gamma = \frac{k_+ - k_-}{2}l \cong \frac{k_0 l}{2}\frac{\omega_p^2 \omega_c}{\omega^3} = \frac{l}{2c}\frac{\omega_p^2 \omega_c}{\omega^2} = \frac{\pi l}{c}\frac{f_p^2 f_c}{f^2} \qquad (4.55)$$

Considering that the ionospheric layer thickness is on the order of 1,000 km, this angle may be of several degrees even at frequencies as high as a few gigahertz. Its exact value is not easily predictable, due to the variability of f_p, and it may cause polarization mismatch in satellite links using antennas with linear polarization. This is one of the reasons why circular polarization is often preferred for satellite links.

Conversely, the phase delay is

$$\varphi = \frac{k_+ + k_-}{2}l \cong k_0 l\left(1 - \frac{\omega_p^2}{2\omega^2} - \frac{\omega_p^4}{8\omega^4}\right) \cong k_0 l\sqrt{1 - \frac{\omega_p^2}{\omega^2}} \qquad (4.56)$$

so that for frequency higher than about 100 MHz, it can be safely approximated with the one of nonmagnetized ionosphere.

4.5.2.3 Arbitrary Direction of Propagation and High Frequency

Finally, at a frequency higher than about 100 MHz, if the wave direction of propagation forms an angle ψ with the Earth's magnetostatic field direction, it can be shown that the Faraday rotation angle is

$$\gamma \cong \frac{\pi l}{c}\frac{f_p^2 f_c}{f^2}\cos\psi \qquad (4.57)$$

and that the propagation constant can be safely approximated with that of the nonmagnetized ionosphere.

4.5.3 Ionosphere and Electromagnetic Wave Propagation: Summary

In summary, ionospheric reflection can be used for long-distance over-the-horizon links in the HF band. Conversely, the presence of the ionosphere does not affect satellite links at microwaves, except for the Faraday rotation at few gigahertz, and for an additional propagation delay that is negligible for all purposes, except for high-precision GNSS geolocalization.

References

[1]   Collin, E. R., *Antennas and Radiowave Propagation*, New York: McGraw-Hill, 1985.

[2]   Parsons, J. D., *The Mobile Radio Propagation Channel*, Chichester: John Wiley & Sons Inc, 2000.

[3]   Bertoni, H. L., *Radio Propagation for Modern Wireless Systems*, Hoboken, NJ: Prentice Hall, 2000.

[4]   Dolukhanov, M., *Propagation of Radio Waves*, London: Central Books Ltd., 1971.

[5]   Sommerfeld, A., "The Propagation of Waves in Wireless Telegraphy," *Ann. Phys.*, Vol. 28, 1909, p. 665.

[6]   Norton, K. A., "The Propagation of Radio Waves over the Surface of the Earth and in the Upper Atmosphere (Part 1)," *Proc. Inst. Radio. Eng.*, Vol. 24, 1936, pp. 1367–1387.

[7]   Norton, K. A., "The Propagation of Radio Waves over the Surface of the Earth and in the Upper Atmosphere (Part 2)," *Proc. Inst. Radio. Eng.*, Vol. 25, 1937, pp. 1203–1236.

[8]   *Specific attenuation model for rain for use in prediction methods,* document ITU-R P.838-3, International Telecommunication Union, Radiowave Propagation P Series, Mar. 2005.

5

Propagation in Complex Environments

This chapter describes the general properties that characterize electromagnetic propagation in complex environments, such as urban areas, which are of primary interest here, as well as irregular terrains and vegetation. A complex environment can be modeled as a homogeneous background (air or vacuum) in which there are several objects (*obstacles*), made of different dielectric and/or conducting materials. In this case, as noted in Chapter 3, closed-form analytical solutions of Maxwell's equations are not available, and numerical methods are not practical for those environments whose dimensions are much larger than the operating wavelength. Therefore, asymptotic methods introduced in Chapter 3 and applied in Chapter 4 can be more conveniently used. A practical alternative is to resort to the empirical models presented in Section 6.2, for the case of urban areas. Accordingly, detailed description of both theoretical and empirical models for propagation in urban areas is deferred to Chapter 6. This chapter presents some general properties of the electromagnetic field behavior in complex environments, where the obstacles are large compared to the wavelength; such properties can be derived by modeling the field at the receiver as the superposition of waves that propagate from the transmitter to the receiver both directly and via interactions with the obstacles located in the scene. These properties significantly change depending on whether the LOS is obstructed or not (see Section 5.1), and they are mainly generated by the existence of several paths along which the waves propagate from the transmitter to the receiver (*multipath*; see Section 5.2). In fact, multipath causes fast spatial (and temporal) fluctuations of the signal level (*fading*; see Sections 5.2.1 and 5.3) and the appearance at the receiver of multiple replicas of the transmitted signal, each one arriving with a different delay and a different attenuation (*delay spread*; see Sections 5.2.2 and 5.4). This chapter's physical description of the electromagnetic waves and environment interaction clarifies the features of the final wave propagation.

5.1   LOS and Non-LOS (NLOS) Propagation

When no obstacle is present along the LOS from the transmitter to the receiver [see Figure 5.1(a)], the dominant contribution to the received signal is provided by the direct wave if the antennas are very directive, or by the ground wave if the antennas are nondirective; see Chapter 4. In the former case, this dominant contribution can be evaluated by using the free-space formula, (2.150), while in the latter case the two-ray model, (4.4), should be used. In addition to the direct or ground wave, minor contributions are provided by waves reaching the receiver after interactions with the surrounding obstacles: specular reflections on their smooth exterior surfaces, to be evaluated by using GO (see Section 3.1.3); diffractions by the edges of the obstacles, that can be examined by using GTD/UTD (see Section 3.5); and scatterings by obstacle irregularities of the order of wavelength, which are, however, usually negligible compared to specular reflections and edge diffractions. As fully illustrated in Section 5.2, all these minor contributions coherently interfere with the dominant one, producing small fluctuations of the field level around the free-space, or ground wave, value. Occasionally, specular reflections from obstacles located not very far from the LOS may produce contributions whose amplitudes are comparable to the one of the dominant LOS contribution; in this case, field level fluctuations may be large.

When the LOS is obstructed by obstacles [see Figure 5.1(b)], the electromagnetic wave may reach the receiver via alternative lines: diffraction on, and transmission through, the obstructing obstacles, and/or reflections on and diffractions from the surrounding obstacles. In this case, for the same transmitter-to-receiver distance, the received signal level is usually much smaller than the one achievable via the LOS. In addition, usually amplitudes of all the interfering contributions are on the same order of magnitude, so that, as fully illustrated in Section 5.2, fast spatial and temporal fluctuations of the received signal level are present. Therefore, in summary, NLOS propagation is usually characterized by low signal levels and fast fluctuations.

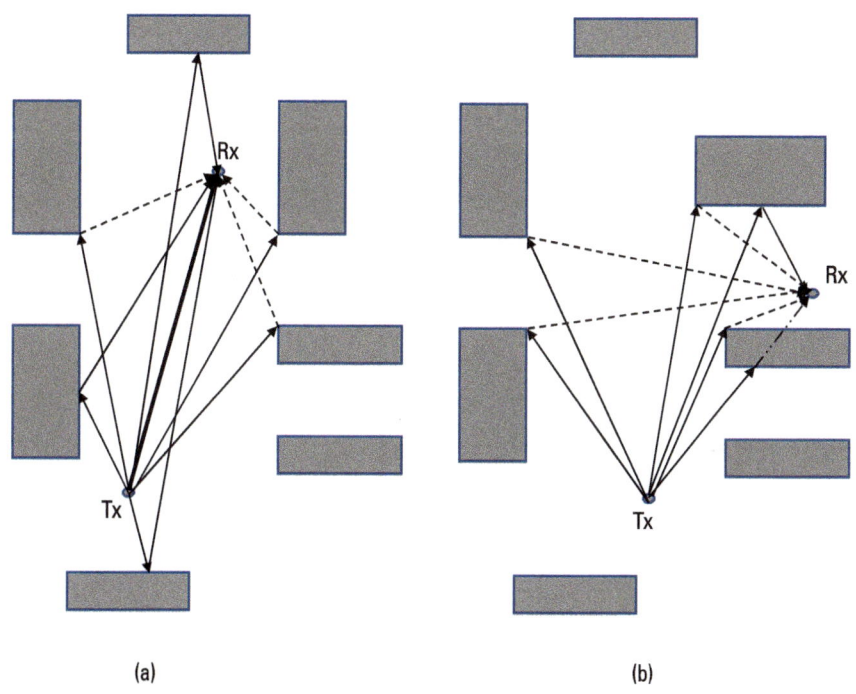

(a)   (b)

Figure 5.1   (a) LOS and (b) NLOS propagation. Thick solid line: direct ray; thin solid lines: incident and reflected rays; dashed lines: diffracted rays; and dot-dashed line: transmitted ray.

5.1.1   Reflection on and Transmission Through a Homogeneous Wall

As already noted, reflection on, and transmission through, obstacles can be dealt with by using ray optics (i.e., GO). Building walls are typical obstacles to be considered in the frame of wave propagation in urban areas. In this case, however, ray optics may not be a convenient procedure to evaluate the effects of a wall on propagation, due to the presence of multiple reflections within the wall. As a matter of fact, a ray emerging from one side of the wall is actually the superposition of all the rays that have traveled back and forth a different number of times within the wall before emerging (see Figure 5.2). A compact formulation, including all multiple reflections at the same time, can be obtained if the source is far from the wall, so that the wave impinging on the wall can be locally assumed to be a plane wave. In this case, in fact, the results reported in Section 2.3.6 can be used, so that reflection on and transmission through the wall can be accounted for with no need to explicitly consider the multiple reflections within the wall. For each ray incident onto

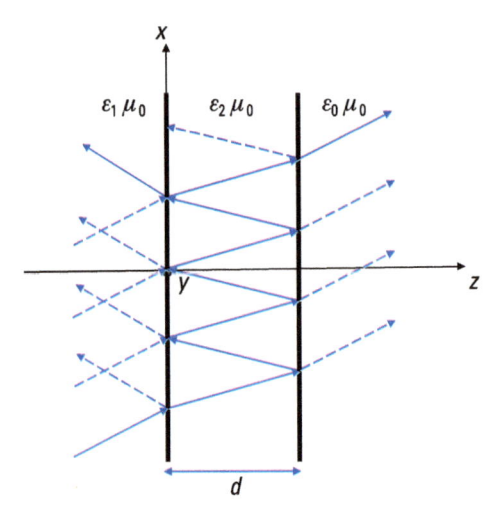

Figure 5.2   Reflection on and transmission through a homogeneous layer; ray optics and multiple bounces within the layer.

the wall, we can simply consider a reflected and a transmitted (beyond the wall) ray, whose electric fields are related to the one of the incident ray via the layer reflection coefficient Γ_A of (2.109) and the layer transmission coefficient T_{AB} of (2.104), respectively. In fact, Γ_A and T_{AB} implicitly account for the multiple reflections within the wall.

5.2 Multipath

Let us assume that the electric field $\mathbf{E}(\mathbf{r})$ at point $\mathbf{r}$ is the superposition of N components $\mathbf{E}_n(\mathbf{r})$ associated to N different rays coming from the transmitting antenna along different paths to $\mathbf{r}$. The open-circuit voltage $V_0(\mathbf{r})$ at a receiving antenna placed at $\mathbf{r}$ is:

$$V_0(\mathbf{r}) = \sum_{n=1}^{N} \mathbf{E}_n(\mathbf{r}) \cdot \mathbf{h}_{Rn} = \sum_{n=1}^{N} \frac{jk\zeta V_T L_n \mathbf{h}_{Tn} \cdot \mathbf{h}_{Rn}}{Z_T \, 4\pi r_n} \exp\!\left(-jkr_n\right)$$

$$= V_T \sum_{n=1}^{N} A_n(\mathbf{r}) \exp\!\left[-jkr_n(\mathbf{r})\right] \tag{5.1}$$

where:

     V_T is the transmitting antenna input voltage;

     ζ is the intrinsic impedance of vacuum;

$k = 2\pi/\lambda = \omega/c$ is the free-space propagation constant;

λ is the wavelength;

c is the velocity of light in vacuum;

Z_T is the transmitting antenna input impedance;

$\mathbf{h}_{Tn}$ and $\mathbf{h}_{Rn}$ are the transmitting and receiving antenna effective lengths at the directions of radiation and incidence of the nth path;

r_n is the length of the nth path from the transmitting antenna to the point $\mathbf{r}$;

L_n is a complex attenuation matrix accounting for the reflection, transmission, diffraction, and scattering events undergone by the nth ray, which cause attenuation and modify wave polarization.

The complex attenuation factors $A_n(\mathbf{r})$ are implicitly defined by (5.1):

$$A_n = \frac{jk\zeta\, L_n \mathbf{h}_{Tn} \cdot \mathbf{h}_{Rn}}{Z_T\, 4\pi r_n} \tag{5.2}$$

5.2.1 Narrowband Characterization of the Multipath Channel

Let us first assume that the transmitted signal can be approximated by a sinusoidal time-varying function. Precise conditions under which this is possible are specified in Section 5.4. Equation (5.1) shows that in this case the received signal is the sinusoidal time-varying function resulting from the sum of N sinusoidal functions, each one attenuated by the factor $|A_n(\mathbf{r})|$ and phase-shifted by the phase $\mathrm{Arg}\{A_n(\mathbf{r})\} + kr_n(\mathbf{r})$ with respect to the transmitted one. It can be noted that both amplitude and phase of the received signal are dependent on the receiver position $\mathbf{r}$. In particular, with regard to spatial variations of the amplitude, we can consider two different spatial scales; over distances on the order of several wavelengths, but much smaller than the transmitter-to-receiver distance, $A_n(\mathbf{r})$ can be considered constant, and amplitude fluctuations are due to the fact that the contributions to the received signal sum up with different phase relations at different points, so that they may interfere constructively at one point and disruptively at another point at short distance. This phenomenon of fast spatial variability of the signal level, typical of multipath propagation, is called *fast fading*, or *short-term fading*. Conversely, over distances much larger than wavelength and nonnegligible with respect to the transmitter-to-receiver distance, once short-term fading is averaged out, we are left with slow fluctuations related to variations of $A_n(\mathbf{r})$. This phenomenon, not strictly related to multipath, is called *slow fading*, or *long-term fading*, or *shadow fading*, since the largest fluctuations of this type are caused by passages from LOS to NLOS propagation.

Let us now focus on fast fading. Let's consider a volume Ω including the origin O of the reference system, placed far from the transmitter and from the obstacles, whose linear size is several wavelengths and much smaller than the transmitter-to-receiver distance. We can then let

$$A_n(\mathbf{r}) \cong A_n(O), \quad r_n(\mathbf{r}) \cong r_n(O) + \hat{\mathbf{s}}_n \cdot \mathbf{r} \tag{5.3}$$

where $\hat{\mathbf{s}}_n = \sin\vartheta_n \cos\varphi_n \hat{\mathbf{x}} + \sin\vartheta_n \cos\varphi_n \hat{\mathbf{y}} + \cos\vartheta_n \hat{\mathbf{z}}$ is the unit vector tangent to the nth ray in O, so that the field of each ray is locally approximated by a plane wave. By using (5.3), the value $V_0(\mathbf{r})$ of the signal (5.1) at the receiver is formally expressed as

$$V_0(\mathbf{r}) = V_T \sum_{n=1}^{N} A_{0n} \exp\left(-jk\hat{\mathbf{s}}_n \cdot \mathbf{r}\right) \tag{5.4}$$

where $A_{0n} = A_n(O)\exp[-jkr_n(O)]$.

The intensity of the received signal [i.e., the square modulus of (5.4)] is then given by

$$
\begin{aligned}
|V_0(\mathbf{r})|^2 &= |V_T|^2 \sum_{n=1}^{N}\sum_{m=1}^{N} A_{0n} A_{0m}^* \exp\left[-jk\left(\hat{\mathbf{s}}_n - \hat{\mathbf{s}}_m\right)\cdot\mathbf{r}\right] \\
&= |V_T|^2 \sum_{n=1}^{N}\sum_{m=1}^{N} |A_{0n}||A_{0m}| \exp\left[-jk\left(\hat{\mathbf{s}}_n - \hat{\mathbf{s}}_m\right)\cdot\mathbf{r} + j\left(\varphi_n - \varphi_m\right)\right] \\
&= |V_T|^2 \left\{ \sum_{n=1}^{N} |A_{0n}|^2 + \sum_{n=1}^{N-1}\sum_{m=n+1}^{N} 2|A_{0n}||A_{0m}| \cos\left[k\left(\hat{\mathbf{s}}_n - \hat{\mathbf{s}}_m\right)\cdot\mathbf{r} + \left(\varphi_m - \varphi_n\right)\right] \right\}
\end{aligned}
$$

$$\tag{5.5}$$

where $\varphi_n = \mathrm{Arg}\{A_{0n}\}$ and $\varphi_m = \mathrm{Arg}\{A_{0m}\}$.

The first (single) summation at the right-hand side of (5.5)—last line—is a constant term equal to the sum of the intensities of the individual contributions, whereas the second (double) summation is the sum of spatially oscillating terms representing interference. In particular, each term of the double summation is constant over the planes perpendicular to the direction of $\hat{\mathbf{s}}_n - \hat{\mathbf{s}}_m$ and varies in a sinusoidal way along the direction of $\hat{\mathbf{s}}_n - \hat{\mathbf{s}}_m$, oscillating between $+2|A_{0n}||A_{0m}|$ and $-2|A_{0n}||A_{0m}|$ with a spatial frequency $k|\hat{\mathbf{s}}_n - \hat{\mathbf{s}}_m|/(2\pi)$; that is, a spatial period

$$\Lambda_{nm} = \frac{\lambda}{\left|\hat{\mathbf{s}}_n - \hat{\mathbf{s}}_m\right|} = \frac{\lambda}{2\sin\left(\alpha_{nm}/2\right)} \tag{5.6}$$

where α_{nm} is the angle between $\hat{\mathbf{s}}_n$ and $\hat{\mathbf{s}}_m$ or

$$\alpha_{nm} = \arccos\left(\hat{\mathbf{s}}_n \cdot \hat{\mathbf{s}}_m\right) \tag{5.7}$$

and we have used the relation (see Figure 5.3)

$$\left|\hat{\mathbf{s}}_n - \hat{\mathbf{s}}_m\right| = 2\sin\left(\alpha_{nm}/2\right) \tag{5.8}$$

Therefore, the signal intensity has fast (i.e., at wavelength scale) spatial fluctuations around a constant average value equal to the sum of the intensities of the individual contributions. The maximum amplitude of these fluctuations is on the order of $\max_{n,m} \{2|A_{0n}||A_{0m}|\}$. Accordingly, if there is one dominant contribution, whose intensity is much larger than those of the other contributions (as is usually the case for LOS propagation), the amplitude of these fluctuations is much smaller than the average intensity (which is practically coincident with the one of the dominant contribution); see Figure 5.4(a). Conversely, if the intensities of all contributions are on the same order of magnitude (as it is usually the case for NLOS propagation), then the amplitude of the above-mentioned fluctuations is on the same order of the average intensity; see Figure 5.4(b). In this case, it is useful to employ a *spatial diversity* technique at the receiver [i.e., to use two or more receiving antennas, spaced by at least one wavelength (so that the received signal values can be considered each independent of the other)]. In this case, in fact, the probability that the signal level for at least one of the receiving antennas is high enough to ensure a sufficient SNR is much greater than the same probability for a single receiving antenna.

A particular case occurs if two rays have very close directions; this happens for the direct ray and a ray that is specularly reflected from an obstacle placed not very far from the LOS, as, for instance, in the case of ground wave propagation. In this case, since $\alpha_{nm} \cong 0$, (5.6) shows that the interference of

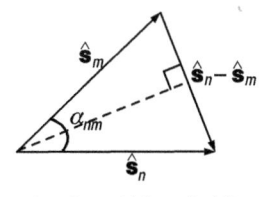

Figure 5.3 Geometry for the evaluation of $|\hat{\mathbf{s}}_n - \hat{\mathbf{s}}_m|$ (i.e., the length of $\hat{\mathbf{s}}_n - \hat{\mathbf{s}}_m$).

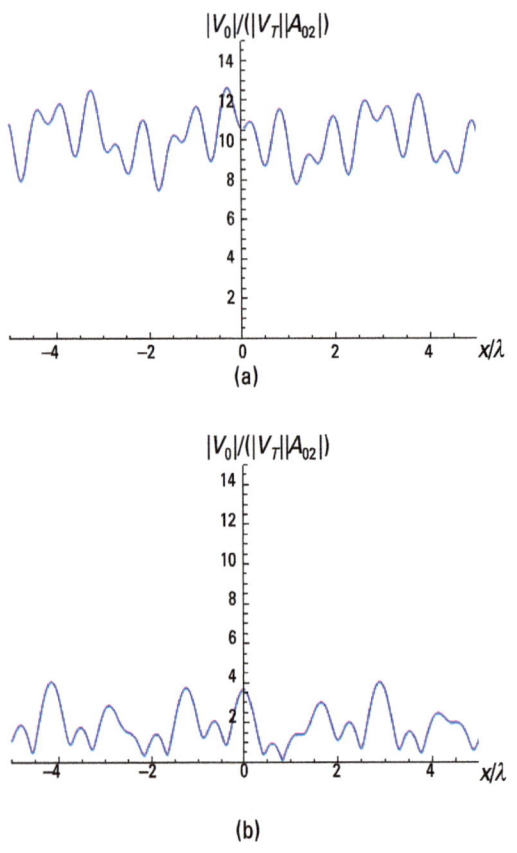

Figure 5.4   Normalized modulus of the open-circuit voltage at the receiving antenna as a function of the receiving antenna position for a multipath channel: (a) LOS case with $|A_{01}| = 10|A_{02}|$ and $|A_{0n}| = |A_{02}|$ for $n > 2$ and (b) NLOS case with $|A_{0n}| = |A_{02}|$ for all n.

the two rays has oscillations over a scale much larger than wavelength, comparable to the transmitter-to-receiver distance. Therefore, these oscillations can be classified as slow fading.

If the receiver and/or the transmitter—or also the obstacles—move, then spatial fluctuations become time fluctuations: The received signal is no longer a pure sinusoid, because its amplitude and phase are time-varying. Since spatial variations are at the wavelength scale, time variations are at a scale on the order of the wavelength divided by the receiver (or transmitter, or obstacles) speed. This phenomenon can be also interpreted in terms of Doppler shifts of the multipath components; by assuming that the receiver moves with velocity

v along the direction $\hat{\mathbf{v}}$, and that at time $t = 0$ it traverses the point O, then $\mathbf{r} = vt\hat{\mathbf{v}}$, and (5.4) becomes

$$V_0(t) = V_T \sum_{n=1}^{N} A_{0n} \exp\left(-jkvt\,\hat{\mathbf{v}} \cdot \hat{\mathbf{s}}_n\right) = V_T \sum_{n=1}^{N} A_{0n} \exp\left[-j2\pi\frac{v}{\lambda}\cos(\alpha_n)t\right]$$

$$= V_T \sum_{n=1}^{N} A_{0n} \exp\left[-j\omega\frac{v}{c}\cos(\alpha_n)t\right] \tag{5.9}$$

where α_n is the angle between the receiver motion direction and the nth ray. Therefore, each contribution has a frequency shift (*Doppler shift*) equal to $\cos(\alpha_n)v/\lambda = \cos(\alpha_n)fv/c$. Note that the maximum Doppler shift is v/c times the frequency of the transmitted signal; even if the receiver is on a vehicle moving at a 30 m/s speed (i.e., about 110 km/h), the maximum Doppler shift is 10^{-7} times the transmitted frequency (at 1 GHz, it is 100 Hz), so that it does not have significant effect on modern digital communications, although it might have audible effects on voice analog communications. However, handheld devices carried by pedestrians (velocity not exciding about 3 m/s) always experience negligible Doppler shift. Conversely, it becomes important when very high mobility of the terminals is experienced, such as in high-speed trains.

We finally note that it is often more realistic to model the received field as a continuous distribution of rays, rather than a finite collection of discrete rays. In this case, the summation in (5.4) is replaced by integration, so that (5.4) is replaced by

$$V_0(x,y,z) = V_T \int_0^{2\pi}\int_0^{\pi} A_0(\vartheta,\varphi)\exp\left[-jk\left(\sin\vartheta\cos\varphi x + \sin\vartheta\sin\varphi y + \cos\vartheta z\right)\right]d\vartheta\,d\varphi$$

$$\tag{5.10}$$

where $A_0(\vartheta,\varphi)$ is called the *angle spread function*. However, the results obtained above for the discrete-ray case remain substantially unchanged.

A more complete analysis of fading, with its statistical characterization, is presented in Section 5.3.

5.2.2 Wideband Characterization of the Multipath Channel

If the transmitted signal is an arbitrarily time-varying function that cannot be approximated by a sinusoidal function, the expression (5.1) of the received signal at $\mathbf{r}$ still applies, provided that the quantities appearing in it are considered as ω-dependent Fourier transforms of their time-domain counterparts (see Section 2.1.2):

$$V_0(\mathbf{r},\omega) = V_T(\omega)\sum_{n=1}^{N} A_n(\mathbf{r},\omega)\exp\left[-j\frac{\omega}{c}r_n(\mathbf{r})\right]$$

$$= V_T(\omega)\sum_{n-1}^{N} A_n(\mathbf{r},\omega)\exp\left[-j\omega t_n(\mathbf{r})\right] = V_T(\omega)G(\mathbf{r},\omega) \tag{5.11}$$

where $t_n(\mathbf{r}) = r_n(\mathbf{r})/c$ is the *time delay* of the nth contribution and

$$G(\mathbf{r},\omega) = \sum_{n=1}^{N} A_n(\mathbf{r},\omega)\exp\left[-j\omega t_n(\mathbf{r})\right] \tag{5.12}$$

is the *transfer function* of the multipath channel. By performing the inverse Fourier transform (IFT) of (5.11) and using the convolution theorem [see (2.12)], we get

$$v_0(\mathbf{r},t) = \int_{-\infty}^{+\infty} g(\mathbf{r},t')v_T(t-t')dt' \tag{5.13}$$

where $g(\mathbf{r}, t')$ is the IFT of $G(\mathbf{r}, \omega)$ and is called the *impulse response* of the multipath channel. Equation (5.13) shows that, in general, in a multipath channel, the received signal is not a perfect copy of the transmitted one (i.e., the signal is *distorted* by the channel). This is because the different frequency components of the signal are modified in different ways by the channel; see (5.11) and (5.12).

We explicitly note that transfer function and impulse response of the multipath channel are spatially fast-varying, for the same reasons illustrated in Section 5.2.1. However, for the time being, we assume that the receiver position is fixed, so that we can focus on the frequency dependence of the received signal and ignore its dependence on $\mathbf{r}$.

Let us now assume that the bandwidth $\Delta\omega$ of the transmitted signal is much smaller than its central frequency ω_0 (narrowband signal; see Section 2.3.4). In this case, the frequency dependence of the transmitting and receiving antennas, and of the complex dielectric constants of the obstacles, can be ignored, so that in (5.11) and (5.12) the dependence of A_n on ω can be neglected. Therefore, for $\omega > 0$ we can write

$$V_0(\omega) = V_T(\omega)\sum_{n=1}^{N} A_{0n}\exp\left[-j(\omega-\omega_0)t_n\right] \tag{5.14}$$

where

$$A_{0n} = A_n(\omega_0)\exp\left[-j\omega_0 t_n\right] \tag{5.15}$$

In addition, in narrowband signals (see Section 2.3.4), the information is conveyed by the modulating signal (i.e., by the complex envelope $\tilde{v}_T(t)$ of the transmitted signal $v_T(t)$):

$$v_T(t) = \left|\tilde{v}_T(t)\right|\cos\left[\omega_0 t + \arg\{\tilde{v}_T(t)\}\right]$$

$$\text{with } \tilde{v}_T(t) = \frac{1}{\pi}\int_{-\Delta\omega/2}^{\Delta\omega/2} V_T(\omega + \omega_0)\exp(j\omega t)\,d\omega \tag{5.16}$$

See (2.74). The received signal is then

$$v_0(t) = \left|\tilde{v}_0(t)\right|\cos\left[\omega_0 t + \arg\{\tilde{v}_0(t)\}\right]$$

with

$$\begin{aligned}
\tilde{v}_0(t) &= \frac{1}{\pi}\int_{-\Delta\omega/2}^{\Delta\omega/2} V_0(\omega + \omega_0)\exp(j\omega t)\,d\omega \\
&= \frac{1}{\pi}\int_{-\Delta\omega/2}^{\Delta\omega/2} V_T(\omega + \omega_0)\sum_{n=1}^{N} A_{0n}\exp\left[j\omega(t - t_n)\right]\,d\omega \\
&= \sum_{n=1}^{N} A_{0n}\frac{1}{\pi}\int_{-\Delta\omega/2}^{\Delta\omega/2} V_T(\omega + \omega_0)\exp\left[j\omega(t - t_n)\right]\,d\omega = \sum_{n=1}^{N} A_{0n}\tilde{v}_t(t - t_n)
\end{aligned}$$

$$\tag{5.17}$$

Therefore, the received modulating signal is the superposition of attenuated and delayed replicas of the transmitted one (*delay spread*). The *average delay* $\bar{t}$ is defined as the weighted average of time delays t_n, in which the weight of each delay is the intensity of the corresponding contribution:

$$\bar{t} = \frac{\sum_{n=1}^{N}\left|A_{0n}\right|^2 t_n}{\sum_{n=1}^{N}\left|A_{0n}\right|^2} \tag{5.18}$$

Similarly, the *root mean square (rms) delay spread* σ_t is defined as:

$$\sigma_t = \sqrt{\frac{\sum_{n=1}^{N}\left|A_{0n}\right|^2(t_n - \bar{t})^2}{\sum_{n=1}^{N}\left|A_{0n}\right|^2}} \tag{5.19}$$

The rms delay spread is an important parameter describing the multipath channel, since it is a measure of the time interval over which replicas are distributed. Only if it is much smaller than the time scale of variation of the transmitted signal, the received signal is not distorted. This concept is detailed in Section 5.4.

We finally note that it is often useful to consider a continuous distribution of delays, rather than a finite collection of discrete delays: in this case, the summations in (5.17) to (5.19) are replaced by integrals, so that (5.17) to (5.19) are substituted with

$$v_0(t) = |\tilde{v}_0(t)| \cos\left[\omega_0 t + \arg\left\{\tilde{v}_0(t)\right\}\right]$$

with

$$\tilde{v}_0(t) = \int_{-\infty}^{+\infty} A_0(t')\tilde{v}_T(t-t')dt' \tag{5.20}$$

$$\bar{t} = \frac{\int_{-\infty}^{+\infty}|A_0(t')|^2 t'\,dt'}{\int_{-\infty}^{+\infty}|A_0(t')|^2\,dt'} \tag{5.21}$$

$$\sigma_t = \sqrt{\frac{\int_{-\infty}^{+\infty}|A_0(t')|^2 (t'-\bar{t})^2\,dt'}{\int_{-\infty}^{+\infty}|A_0(t')|^2\,dt'}} \tag{5.22}$$

and $A_0(t')$ is called *delay spread function*.

5.3 Fading

In wireless communications, fading is the deviation of the attenuation affecting a signal over certain propagation media. A *fading channel* is a communication channel comprising fading. As noted in Section 5.2.1, fading may either be due to multipath propagation (fast fading), or due to shadowing from obstacles affecting the wave propagation (shadow fading).

Positions, sizes, shapes, and electromagnetic parameters of the obstacles present in a complex environment are often not known deterministically. Therefore, the phenomenon of fading requires a statistical description [1–6]. The fluctuations of the signal level are statistically characterized by modeling the open-circuit voltage V_0 at the receiver as a random variable, described by the pdfs (see Section 3.6) of its real and imaginary parts or, equivalently, of its modulus and phase.

5.3.1 NLOS: Rayleigh Fading

In view of (5.1), V_0 is the coherent summation of several contributions. In a NLOS multipath scenario, it is reasonable to assume that all these contributions $V_1...V_N$ are independent identically distributed zero-mean complex random variables. According to a classical result of statistics, namely the *central limit theorem*, this implies that the real and imaginary parts of V_0 are independent zero-mean Gaussian random variables whose standard deviation is indicated with σ. Under this hypothesis, the phase of V_0 is uniformly distributed in the interval $0, 2\pi$, and its modulus is distributed according to the *Rayleigh* pdf:

$$p_{|V_0|}(x) = \frac{x}{\sigma^2} \exp\left(-\frac{x^2}{2\sigma^2}\right) \tag{5.23}$$

Therefore, the probability that the signal amplitude $|V_0|$ exceeds a given threshold V_{th} is

$$\text{prob}\left(|V_0| > V_{th}\right) = \int_{V_{th}}^{\infty} p_{|V_0|}(x)\,dx = \int_{V_{th}}^{\infty} \frac{x}{\sigma^2} \exp\left(-\frac{x^2}{2\sigma^2}\right) dx = \exp\left(-\frac{V_{th}^2}{2\sigma^2}\right) \tag{5.24}$$

The mean value of the signal amplitude is

$$\langle |V_0| \rangle = \int_0^{\infty} x\,p_{|V_0|}(x)\,dx = \int_0^{\infty} x\,\frac{x}{\sigma^2} \exp\left(-\frac{x^2}{2\sigma^2}\right) dx = \sqrt{\frac{\pi}{2}}\sigma \cong 1.2533\sigma \tag{5.25}$$

its mean square value is

$$\langle |V_0|^2 \rangle = \int_0^{\infty} x^2\,p_{|V_0|}(x)\,dx = \int_0^{\infty} x^2 \frac{x}{\sigma^2} \exp\left(-\frac{x^2}{2\sigma^2}\right) dx = 2\sigma^2 \tag{5.26}$$

and its variance is

$$\sigma_{|V_0|} = \langle |V_0|^2 \rangle - \langle |V_0| \rangle^2 = \left(2 - \frac{\pi}{2}\right)\sigma^2 \cong 0.4292\sigma^2 \tag{5.27}$$

In deriving (5.25) and (5.26), integration by parts and (C.1) have been used.

We recall that (see Section 2.5.2), the received power P_R is related to $|V_0|$ by $P_R = |V_0|^2/(8R_{in})$, where R_{in} is the input resistance of the receiving antenna. In addition (see Section 2.6), the received power is often expressed in decibels: $[P_R]_{dB} = 10 \, Log(P_R)$. From the pdf (5.23) it is possible to obtain some of the statistical parameters of P_R and $[P_R]_{dB}$ in the case of Rayleigh fading [1], namely, their mean values

$$\langle P_R \rangle = \frac{\sigma^2}{4R_{in}}, \quad \langle [P_R]_{dB} \rangle = 10Log\left(\frac{\sigma^2}{4R_{in}}\right) - \frac{10\gamma}{\ln 10} \cong \left[\langle P_R \rangle\right]_{dB} - 2.51 \quad (5.28)$$

where $\gamma \cong 0.5772$ is the Euler constant; and the variance σ_{dB}^2 of $[P_R]_{dB}$

$$\sigma_{dB}^2 = var\left(\left[P_R\right]_{dB}\right) = \left(\frac{10\pi}{\sqrt{6}\ln 10}\right)^2 \quad (5.29)$$

so that the standard deviation of Rayleigh fading in decibels is

$$\sigma_{dB} = \frac{10\pi}{\sqrt{6}\ln 10} \cong 5.57 \text{ dB} \quad (5.30)$$

5.3.2 LOS: Rician Fading

If, in a multipath scenario, the LOS component is present, then one of the components of V_0 in (5.1) is deterministically known, and it is often dominant with respect to the other multipath components. Let us assume that the LOS contribution is V_1 and that, with no loss of generality, it is real. With regard to the remaining components $V_2 \ldots V_N$, they can be still modeled as independent identically distributed zero-mean random variables, so that real and imaginary parts of their sum ΔV_0 are independent zero-mean Gaussian random variables with standard deviation σ. Under these hypotheses, the signal amplitude $|V_0|$ $= |V_1 + \Delta V_0|$ is distributed according to the Rician pdf:

$$p_{|V_0|}(x) = \frac{x}{\sigma^2}\exp\left(-\frac{x^2 + V_1^2}{2\sigma^2}\right)I_0\left(\frac{xV_1}{\sigma^2}\right) \quad (5.31)$$

where I_0 is the zero-order modified Bessel function of the first kind. If $V_1 = 0$, the Rician distribution (5.31) reduces to the Rayleigh one (5.23), whereas if the Rician parameter $K = V_1^2/(2\sigma^2)$ is large (i.e., the deterministic component is dominant), then the Rician pdf is well approximated by a Gaussian pdf with a mean equal to V_1 and standard deviation equal to σ; see Figure 5.5. In this last case, the received power P_R and its value in decibels $[P_R]_{dB}$ are also approximately Gaussian. In fact,

$$
\begin{aligned}
P_R &= \frac{|V_0|^2}{8R_{in}} = \frac{|V_1 + \Delta V_0|^2}{8R_{in}} = \frac{\left(V_1 + \mathrm{Re}\{\Delta V_0\}\right)^2 + \mathrm{Im}\{\Delta V_0\}^2}{8R_{in}} \\
&\cong \frac{V_1^2 + 2V_1\,\mathrm{Re}\{\Delta V_0\}}{8R_{in}} = P_{R1} + 2P_{R1}\frac{\mathrm{Re}\{\Delta V_0\}}{V_1}
\end{aligned}
\tag{5.32}
$$

so that P_R is (approximately) a Gaussian random variable with mean value equal to the power of the deterministic component P_{R1} and standard deviation equal to $2P_{R1}\sigma/V_1$. In addition,

$$
\begin{aligned}
\left[P_R\right]_{dB} &\cong 10\mathrm{Log}\left(P_{R1} + 2P_{R1}\frac{\mathrm{Re}\{\Delta V_0\}}{V_1}\right) = 10\mathrm{Log}\left[P_{R1}\left(1 + 2\frac{\mathrm{Re}\{\Delta V_0\}}{V_1}\right)\right] \\
&= \left[P_{R1}\right]_{dB} + 10\mathrm{Log}\left[\left(1 + 2\frac{\mathrm{Re}\{\Delta V_0\}}{V_1}\right)\right] \\
&= \left[P_{R1}\right]_{dB} + \frac{10}{\ln 10}\ln\left[\left(1 + 2\frac{\mathrm{Re}\{\Delta V_0\}}{V_1}\right)\right] \cong \left[P_{R1}\right]_{dB} + \frac{20}{\ln 10}\frac{\mathrm{Re}\{\Delta V_0\}}{V_1}
\end{aligned}
\tag{5.33}
$$

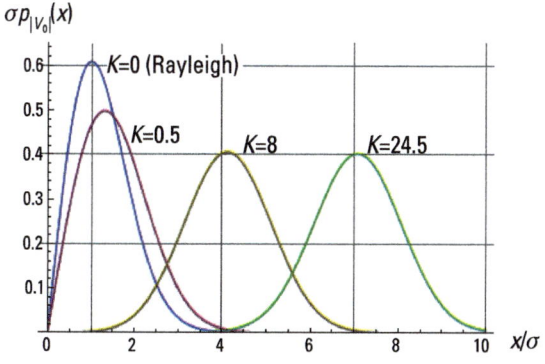

Figure 5.5 Rician pdf for V_1/σ equal to 0, 1, 4, 7 (i.e., K equal to 0, 0.5, 8, 24.5).

where the approximation $\ln(1 + x) \cong x$ for $x \ll 1$ has been used; therefore, $[P_R]_{\text{dB}}$ is (approximately) a Gaussian random variable with mean value equal to the value in decibels of the power of the deterministic component, and standard deviation equal to

$$\sigma_{\text{dB}} = \frac{20}{\ln 10} \frac{\sigma}{V_1} = \frac{10\sqrt{2}}{\ln 10} \frac{1}{\sqrt{K}} \cong \frac{6.142}{\sqrt{K}} \qquad (5.34)$$

5.3.3 Slow Fading: Lognormal Distribution

Statistics of fast fading are observed over areas whose linear size is small with respect to the transmitter-to-receiver distance. As already noted, fluctuations over larger spatial scales that remain after averaging out the fast fading are termed slow fading. This is usually statistically described by modeling the corresponding signal level as a lognormal random variable (i.e., as a random variable whose logarithm is a Gaussian random variable). Accordingly, its value in decibels is a Gaussian random variable.

5.3.4 Example: Outage Probability in Rayleigh Fading

The *outage probability* p_{out} is the probability that the received power P_R is lower than the receiver sensitivity (i.e., lower than the minimum power $P_{R\min}$ such that a sufficient SNR is obtained at the receiver) (see Section 4.1.1). Let us consider a link affected by Rayleigh fading. We then have, using (5.24) and (5.26)

$$p_{\text{out}} = \text{prob}\left(P_R < P_{R\min}\right) = 1 - \text{prob}\left(P_R \geq P_{R\min}\right) = 1 - \text{prob}\left(|V_0| \geq V_{tb}\right)$$

$$= 1 - \exp\left(-\frac{V_{tb}^2}{2\sigma^2}\right) = 1 - \exp\left(-\frac{P_{R\min}}{\langle P_R \rangle}\right) \qquad (5.35)$$

Let us now compute the mean received power $< P_R >$ needed to obtain a given small p_{out}. Inverting (5.35), we get

$$\exp\left(-\frac{P_{R\min}}{\langle P_R \rangle}\right) = 1 - p_{\text{out}} \rightarrow -\frac{P_{R\min}}{\langle P_R \rangle} = \ln\left(1 - p_{\text{out}}\right) \cong -p_{\text{out}} \rightarrow \langle P_R \rangle = \frac{P_{R\min}}{p_{\text{out}}}$$

Accordingly, if we require that p_{out} is 10^{-n}, the mean received power must be 10^n times the receiver sensitivity $P_{R\min}$ (i.e., $10n$ dB larger than $[P_{R\min}]_{\text{dB}}$).

5.4 Delay Spread

As noted in (5.17), the received modulating signal is the superposition of attenuated and delayed replicas of the transmitted one, and the time interval over which the time delays of replicas are distributed is well represented by the rms delay spread σ_τ; see (5.19) or (5.22). If over a time interval on the order of the rms delay spread the transmitted (modulating) signal can be considered constant, then the overall received signal is an attenuated and delayed, but substantially undistorted, version of the transmitted one; otherwise, the signal is significantly distorted by the multipath channel (see Figure 5.6). For

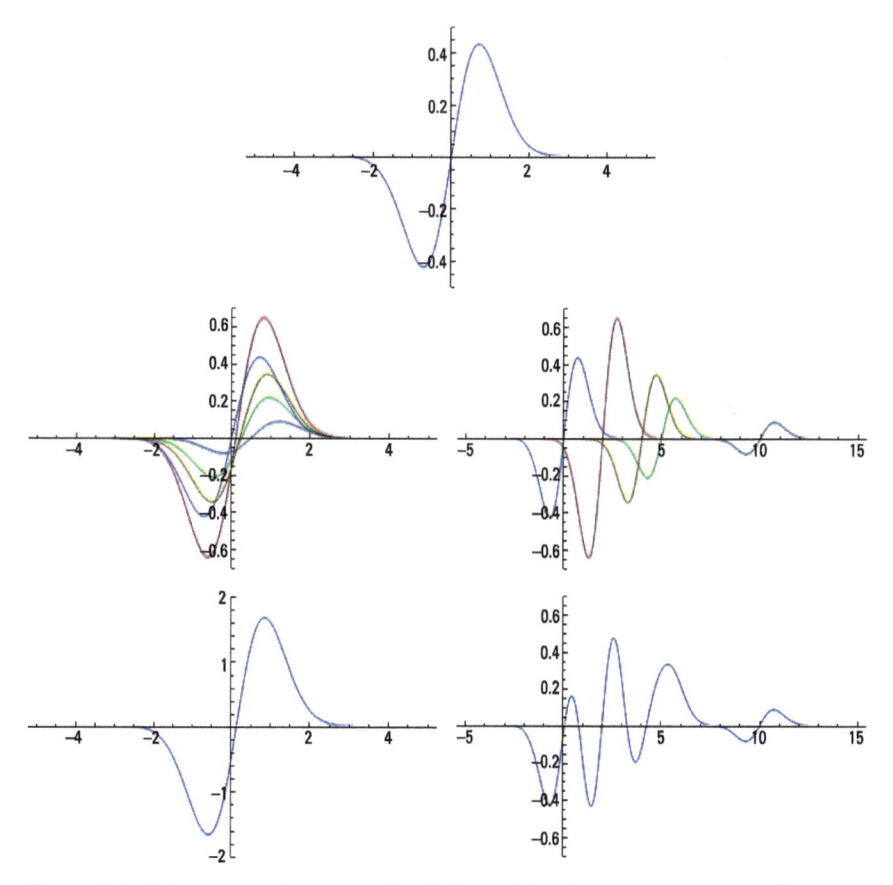

Figure 5.6 Delay spread. Top: transmitted signal; left column: multipath contributions (top) and corresponding received signal (bottom) for an rms delay spread much smaller than the inverse of the signal bandwidth; and right column: multipath contributions (top) and corresponding received signal (bottom) for an rms delay spread of the order of the inverse of the signal bandwidth.

analog audio signals, the effect is the presence of multiple echoes; for analog video signals, the presence of double or multiple images; and for digital signals, the effect is *intersymbolic interference*, which can completely disrupt the received signal. (See Figure 5.7.) To provide a quantitative condition to be satisfied to avoid signal distortion, let us evaluate the received signal at time $t + \sigma_t$ by using (5.17):

$$\tilde{v}_0\left(t + \sigma_t\right) = \frac{1}{\pi} \int_{-\Delta\omega/2}^{\Delta\omega/2} V_0\left(\omega + \omega_0\right)\exp\left(j\omega t\right)\exp\left(j\omega\sigma_t\right)d\omega \qquad (5.36)$$

It will be $\tilde{v}_0\left(t + \sigma_t\right) \cong \tilde{v}_0\left(t\right)$ if $|\omega|\sigma_t << 2\pi$ (i.e., $\Delta\omega\sigma_t << 2\pi$, or $\Delta f\sigma_t << 1$, so that $\Delta f << 1/\sigma_t$). In other words, distortion is avoided if the signal bandwidth Δf is much smaller than the inverse of the rms delay spread.

The condition above can be expressed in terms of another parameter describing the multipath channel and strictly related to σ_t, namely the channel *coherence bandwidth* B_c (i.e., the size of the frequency interval over which the modulus of the channel transfer function can be considered constant). In order to evaluate the coherence bandwidth, let us rewrite the channel transfer function (5.12) by neglecting the dependence of A_n on ω:

$$G(\omega) = \sum_{n=1}^{N} A_n \exp\left(-j\omega t_n\right) = \exp\left(-j\omega\bar{t}\right)\sum_{n=1}^{N} A_n \exp\left[-j\omega\left(t_n - \bar{t}\right)\right] \qquad (5.37)$$

Let us now evaluate the modulus of the transfer function at a different angular frequency $\omega + \eta$:

$$\left|G(\omega + \eta)\right| = \left|\sum_{n=1}^{N} A_n \exp\left[-j\omega\left(t_n - \bar{t}\right)\right]\exp\left[-j\eta\left(t_n - \bar{t}\right)\right]\right| \qquad (5.38)$$

It will be $\left|G(\omega + \eta)\right| \cong \left|G(\omega)\right|$ if $\eta\left(t_n - \bar{t}\right) << 2\pi$ (i.e., if $\eta\sigma_t << 2\pi$, or $\eta << 2\pi/\sigma_t$). Accordingly, the coherence bandwidth B_c is on the order of the inverse of the rms delay spread, and the condition to avoid distortion can be expressed as $\Delta f << B_c$. Therefore, the signal is not distorted by the multipath channel if its bandwidth is much smaller than the channel coherence bandwidth. In fact, in this case, all the frequency components of the signal are affected in the same way by the channel; see Figure 5.8.

In conclusion, if $\Delta f << B_c$, the delay spread effect can be ignored, and the signal can be approximated by a sinusoidal function. In this case, the only

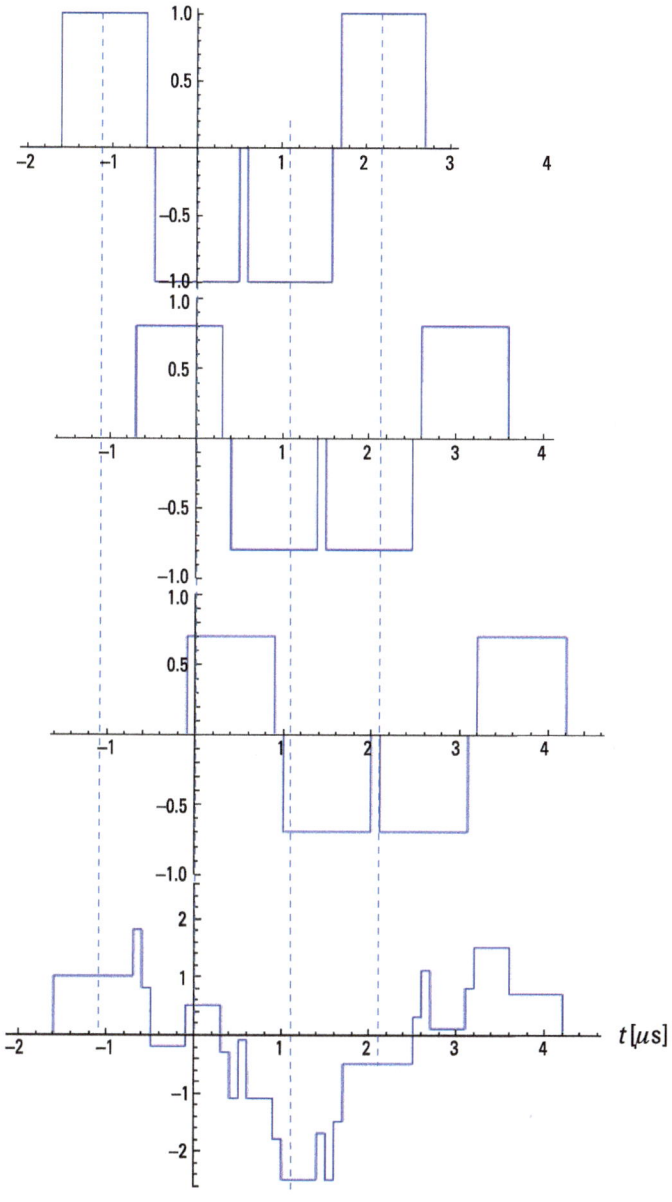

Figure 5.7 Intersymbolic interference caused by delay spread. (From top to bottom: original signal, two delayed and attenuated replicas, and overall received signal.) The transmitted four-bit sequence is 1001 (positive pulse = 1, negative pulse = 0), whereas the received one, detected at times −1.1, 0, 1.1, 2.2 μs, is 1100.

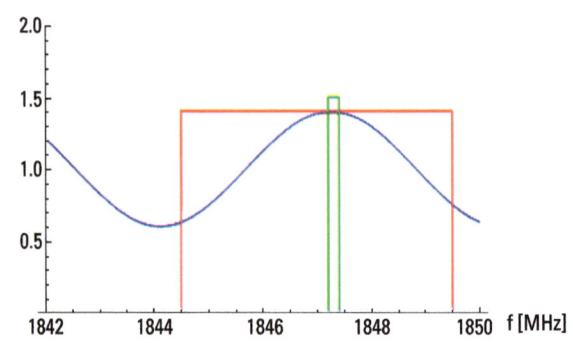

Figure 5.8 Modulus of the multipath channel transfer function (blue), signal bandwidth of a 2G mobile phone system (green), signal bandwidth of a 3G or 4G mobile phone system (red). For the 2G system, the signal bandwidth is much smaller than the channel coherence bandwidth, so that all the frequency components of the signal are affected in the same way by the channel. This is not the case for the more recent systems, for which the frequency components of the signal are affected in different ways by the channel.

effect of multipath is fading. Otherwise, the delay spread must be accounted for in the link design.

5.4.1 Example: Delay Spread in Urban Areas and Mobile Telephone Systems

The linear sizes of cells of mobile telephone systems in urban areas may vary from about one hundred meters to a few kilometers, so that it is reasonable to assume that the differences between different transmitter-to-receiver paths may assume values between a few tens of meters (for example, 30m) to a few hundred meters (e.g., 300m). Considering that the velocity of light is about 3×10^8 m/s, the rms delay spread may assume values from about 100 ns to about $1\,\mu$s, corresponding to coherence bandwidths from about 1 to about 10 MHz.

Signals of mobile telephone systems up to the second generation (GSM— see Chapter 1) had a bandwidth not exceeding 200 kHz, so that they were not affected by the delay spread. However, for third-generation systems and subsequent systems, the bandwidth is on the order of some megahertz, and therefore, such systems must use coding and modulation techniques that are able to reduce the effects of delay spread. For instance, a single high-speed stream of bits, on the order of some megabits per second, can be decomposed into several lower-speed streams, on the order of few hundreds of kilobits per second, each modulating a different subcarrier. In this way, each low-speed stream has a bandwidth much smaller than the coherence bandwidth. This is the main concept behind the use of orthogonal frequency division multiplexing (OFDM) in 4G systems.

References

[1]   Parsons, J. D., *The Mobile Radio Propagation Channel*, Chichester: John Wiley & Sons, Inc., 2000.

[2]   Bertoni, H. L., *Radio Propagation for Modern Wireless Systems*, Hoboken, NJ: Prentice Hall, 2000.

[3]   Rice, S. O., "Statistical Properties of a Sine Wave Plus Random Noise," *The Bell System Technical Journal*, Vol. 27, No. 1, Jan. 1948, pp. 109–157.

[4]   Davenport, W. B., and W. L. Root, *An Introduction to the Theory of Random Signals and Noise*, New York: McGraw-Hill, 1958.

[5]   Longley, A. G., *Radio Propagation in Urban Areas*, OT Report 78-144, April 1978.

[6]   Papoulis, A., *Probability, Random Variables, and Stochastic Processes*, Third Edition, New York: McGraw-Hill, 1991.

6

Propagation in Urban Areas

This chapter describes the methods that can be used to predict the electromagnetic field levels generated by a transmitting antenna in an urban area. In this framework, an important classification of propagation scenarios is between *outdoor* and *indoor propagation*; in the former case, the transmitter and receiver are both outside the buildings, while in the latter they are both inside a building. In addition, in many cases of interest, the transmitter and receiver may be apart, with one inside and the other outside a building; this situation is sometimes termed *outdoor-indoor propagation*. Section 6.1 discusses these different scenarios.

There are two main classes of methods for predicting the electromagnetic field levels generated by a transmitting antenna in an urban area. In some cases, a detailed description of the specific considered urban environment is not available (or it is considered too complex to be efficiently used), and only some average parameters are possibly known (e.g., average building height, and/or average street width). In these situations, as illustrated in Chapter 5, the electromagnetic field level can be modeled as a random variable with a proper pdf, characterized by appropriate parameters: usually, mean value and standard deviation. The evaluation of these parameters as a function of the above-cited environmental parameters, as well as those of the transmitter (position, frequency, power, gain), is obtained by resorting to empirical expressions, established by researchers via extensive field measurement campaigns.

Section 6.2 details the methods based on this approach, which are termed *empirical* or *statistical methods*.

If a detailed description of the considered urban site is available and the obstacle sizes are much larger than the operating wavelength, the asymptotic methods introduced in Chapter 3 (namely, GO and GTD/UTD) can be used. Accordingly, propagation in urban areas is described in terms of rectilinear rays and their reflections, transmissions, and diffractions. This approach leads to the conception of *ray-tracing algorithms*, which are implemented in appropriate software codes. These algorithms, described in Section 6.4, have the advantage, compared to empirical methods, of not requiring measurement campaigns and of achieving a higher accuracy. However, their computational load may be large, and as they require large databases describing the terrain, buildings, and other possible obstacles located in the considered area, their reliability is strictly related to the database accuracy.

Intermediate situations exist between empirical/statistical methods and deterministic ray-tracing algorithms. For instance, only the buildings' main body may be deterministically described, whereas minor architectonical elements (e.g., balconies and windows) and other smaller obstacles may be ignored; in this case, the total electromagnetic field can be considered as the superposition of a deterministic component, computed via a ray-tracing algorithm, and a (small) random component, due to the unmodeled minor obstacles. Further simplifications in the description of the environment may lead, in some cases, to closed-form expressions of the electromagnetic field. For instance, electromagnetic propagation along streets with high-rise buildings at their sides (sometimes called *urban canyons*, or *street canyons*) can be evaluated by using concepts and techniques borrowed from waveguide theory. An example of such an approach is described in Section 6.3.

6.1 Urban Area Propagation Scenarios: Outdoor and Indoor

Propagation scenarios in urban areas can be classified based on the positions of the transmitter and receiver. A first obvious classification is between outdoor and indoor propagation. In the former case, both the transmitter and the receiver are outside the buildings, whereas in the latter they are both inside.

In addition to the direct wave (present if transmitter and receiver are in LOS), the main outdoor propagation paths are via reflections and diffractions either along the streets [see Figure 6.1(a)], or over the buildings [see Figure 6.1(b)]. Transmission through an entire building is also possible in principle,

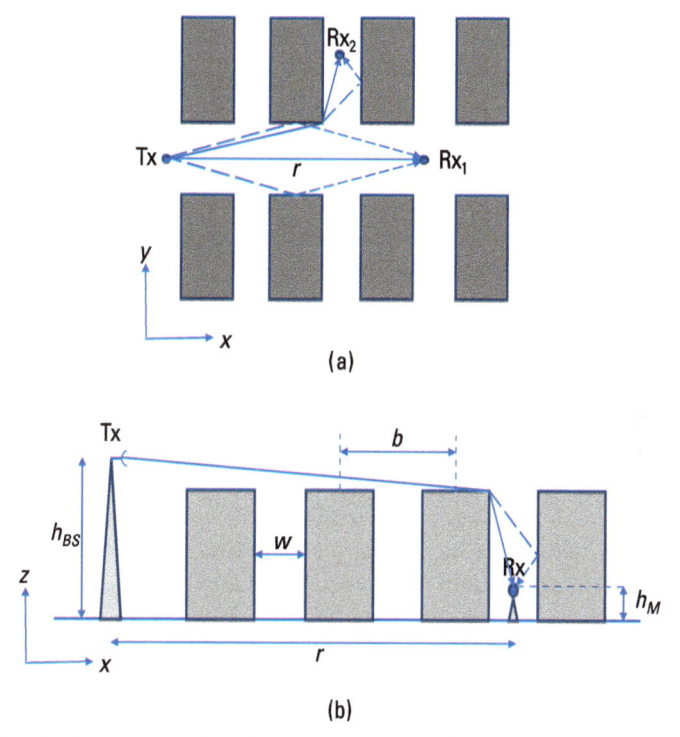

Figure 6.1 Outdoor propagation: (a) street-canyon and around-the-corner propagation and (b) over-the-roof propagation.

but the intensity of the corresponding wave is usually negligible with respect to the previous ones.

Within the outdoor case, a further classification is based on the distance of the antennas from the ground. A common situation in mobile phone systems is that one antenna (the *base-station* antenna) is at rooftop level, and the other (the *mobile* antenna) is at street level. In a *downlink*, the former is the transmitting and the latter the receiving antenna, and the opposite happens for an *uplink*. In this case (either *uplink* or *downlink*), usually the main propagation path is within the vertical plane containing the two antennas, via single or multiple diffraction on the roofs' edges (*over-the-roof propagation*) [see Figure 6.1(b)]. Conversely, if transmitting and receiving antennas are both at street level, or anywhere well below the roof level, the main propagation paths in the LOS case are via the direct and ground-reflected waves (see Section 4.1) and via reflections on the buildings' vertical walls (*street-canyon propagation*) [see Figure 6.1(a), receiver Rx_1]. In the NLOS case, the main propagation mechanisms are reflections and diffractions on the buildings' vertical walls and edges, respectively (*around-the-corner propagation*) [see Figure 6.1(a), receiver Rx_2].

Regarding the size of the considered outdoor urban area, with reference to mobile-phone cellular networks, it is customary to distinguish among *macrocells*, whose radius is larger than about 1 km, and *microcells*, whose radius is smaller than about 1 km. Cells whose radius is smaller than about 100m are sometimes called *picocells*.

In the indoor case, in addition to the direct wave, the main propagation paths are via reflections on and transmission through the internal walls of the building. Diffractions by the edges of doors and windows, as well as by pieces of furniture and by people moving around the building, are also present, but their effects are usually smaller (except that for millimeter waves, when blockage by pieces of furniture and people is significant—see Chapter 8) and also difficult to predict, due to their time-varying nature. A noteworthy feature of indoor propagation is the waveguiding behavior of corridors, thanks to which propagation losses even smaller than those of free-space can be obtained along a corridor.

Finally, if the transmitter and receiver are apart, with one outside and the other inside a building (outdoor-indoor propagation), transmission through the external wall of the building must also be considered (see Section 5.1.1 for a discussion of *through-the-wall propagation*), in addition to the outdoor and indoor propagation paths mentioned above.

6.2 Empirical Propagation Models

As already noted, when a detailed description of the specific considered urban environment is not available, the electromagnetic field level can be modeled as a random variable with a proper pdf, characterized by appropriate parameters (usually, mean value and standard deviation). Evaluation of these parameters can be performed by resorting to empirical propagation models, which are usually obtained as follows. A parametric analytical expression of the mean value L of the propagation loss in decibels is heuristically assumed, often in analogy with the propagation loss expression of a simple canonical case [e.g., free-space loss, as in (2.169); the two-ray model, as discussed in Section 4.1; or edge diffraction, described in Sections 3.4 and 3.5]. This parametric expression is a function of few variables describing the system (e.g., transmitter-to-receiver distance r and frequency f) and possibly the scenario [e.g., buildings' average height (h_{roof})], and contains some constant coefficients (A, B, etc.) whose values need to be adjusted to fit measurements. Therefore, a large measurement campaign is performed, in which the propagation loss is measured for a huge number N of values of the variables (e.g., r, f, and h_{roof}). The values of the coefficients (A, B, etc.), that are selected are those minimizing the following cost function:

$$F(A,B,\ldots) = \sum_{n=1}^{N} \left[\hat{L}_n\left(r_n, f_n, h_{\text{roof}n}, \ldots\right) - L\left(r_n, f_n, h_{\text{roof}n}, \ldots ; A, B, \ldots\right) \right]^2 \quad (6.1)$$

where $\hat{L}_n$ is the nth propagation loss measurement. Once the values of the coefficients are identified, the mean value of the received power in decibels can be expressed in analogy with the Friis formula (2.168) as

$$\left[P_R \right]_{\text{dB}} = \left[P_T \right]_{\text{dB}} + \left[G_T \right]_{\text{dB}} + \left[G_R \right]_{\text{dB}} - L \quad (6.2)$$

where P_T is the transmitted power, and G_T and G_R are the gains of transmitting and receiving antennas, respectively. The variance of the received power in decibels can be estimated as the minimum value of the cost function F divided by N.

The advantage of an empirical model is that, once the analytical expression is available, its use requires a negligible computational load. However, its results are reliable only in urban areas with characteristics similar to the ones of the area where the measurement campaign has been conducted.

Sections 6.2.1 and 6.2.2, respectively, describe the main available empirical propagation models for outdoor and indoor environments. Other empirical models, specifically developed to deal with millimeter-wave propagation in 5G systems, are briefly illustrated in Chapter 8.

6.2.1 Outdoor

The simplest empirical models for outdoor propagation are the so-called ABC, or alpha-beta-gamma (ABG), models, which are developed starting from the expression of the free-space loss. The mean value (or sometimes the *median value*—i.e., the value that is exceeded in the 50% of time) of propagation loss in decibels is given by

$$L = A + B\mathrm{Log}(r) + C\mathrm{Log}(f) \quad (6.3)$$

These models are defined by the values of the coefficients A, B, and C and by the loss standard deviation in decibels σ. For free-space loss [see (2.169)],

$B = C = 20$, and $A = 32.4$ if f is measured in megahertz and r in kilometers (or, also, if f is measured in gigahertz and r in meters).

ABC models can be employed in a variety of urban scenarios, when no quantitative information about the system and environment are available, except for transmitter-to-receiver distance and frequency. For instance, the propagation research group of the COST231 project [1] suggests, for LOS propagation in urban areas, an ABC model based on measurements conducted in the city of Stockholm, Sweden, and characterized by $A = 42.6$, $B = 26$, $C = 20$, with f measured in megahertz and satisfying 800 MHz $< f <$ 2000 MHz, and r measured in kilometers and satisfying 20m $< r <$ 5 km. The value for σ is not provided in [1]. Similarly, the International Telecommunication Union (ITU) recommends the use of ABC models in several urban scenarios, with parameters listed in Table 6.1 [2].

Note that in all cases the C coefficient, related to the dependence of loss on frequency, is very close to the free-space one, while the B coefficient, related to the dependence of loss on distance, is slightly higher than the free-space one in the LOS case, and about 40 dB (similar to the ground wave value at large distances; see Section 4.1) in the NLOS case.

Table 6.1
Propagation Loss Coefficients for ITU ABC Models

Frequency Range (Gigahertz)	Distance Range (Meters)	Type of Environment	Propagation Type	A	B	C	σ
0.8–82	5–660	Urban high-rise, urban low-rise/ suburban	Street canyon, LOS	29.2	21.2	21.1	5.06
0.8–82	30–715	Urban high-rise	Street canyon, NLOS	10.2	40.0	23.6	7.60
0.8–73	30–170	Residential	Street canyon, NLOS	18.8	30.1	20.7	3.07
2.2–73	55–1,200	Urban high-rise, urban low-rise/ suburban	Above rooftop, LOS	28.6	22.9	19.6	3.48
2.2–66.5	260–1,200	Urban high-rise	Above rooftop, NLOS	−6.27	43.9	23.0	6.89

An enhanced version of ABC model, which also accounts for the heights h_{BS} and h_M of base station and mobile antennas, is the Okumura-Hata model [3]. It is based on a large measurement campaign carried out by Okumura in the city of Tokyo and its surroundings [4], and it was first provided in graphical form. Later, Hata [5] devised empirical mathematical expressions to describe the graphical information provided by Okumura. According to this model, the median value of the propagation loss in decibels is expressed as follows:

$$L = 69.55 - 13.82\text{Log}(h_{BS}) - a(h_M)$$
$$+ \left[44.9 - 6.55\text{Log}(h_{BS}) \right]\text{Log}(r) + 26.16\text{Log}(f) \tag{6.4}$$

where:

$150 \leq f \leq 1{,}500$ (f in megahertz);
$1 \leq r \leq 20$ (r in kilometers);
$30 \leq h_{BS} \leq 200$ (h_{BS} in meters);
$1 \leq h_M \leq 10$ (h_M in meters).

and

$$a(h_M) = \begin{cases} [1.1\text{Log}(f) - 0.7]h_M - [1.56\text{Log}(f) - 0.8] & \text{for small or medium-sized cities} \\ 8.29[\text{Log}(1.54h_M)]^2 - 1.1 & \text{for } f \leq 200 \text{ MHz and large cities} \\ 3.2[\text{Log}(11.75h_M)]^2 - 4.97 & \text{for } f \geq 200 \text{ MHz and large cities} \end{cases}$$

COST231 group has then extended the Okumura-Hata model to the range of frequency from 1,500 to 2,000 MHz by modifying (6.4) as follows [1]:

$$L = 46.3 - 13.82\text{Log}(h_{BS}) - a(h_M)$$
$$+ \left[44.9 - 6.55\text{Log}(h_{BS}) \right]\text{Log}(r) + 33.9\text{Log}(f) + C \tag{6.5}$$

where $C = 0$ for small and medium-sized cities and $C = 3$ for large cities.

The Okumura-Hata model only holds if the base station is above the rooftop level, and it is often used for fast field-coverage predictions in urban macrocells, while it cannot be used in microcells, when the base station is below the rooftop level.

A more accurate prediction of over-the-roof NLOS propagation, which also accounts for average building height h_{roof}, average street width w, and average building separation b [see Figure 6.1(b)], can be obtained by using the COST231-Walfish-Ikegami model. This model is the work of researchers at the

COST231 project [1]; it started from a theoretical computation of diffraction by multiple absorbing screens devised by Walfish and Bertoni [6] and from a model by Ikegami substantially based on the edge-diffracted field evaluation (see Sections 3.4 and 3.5). According to the COST231-Walfish-Ikegami model, the propagation loss in decibels is given by

$$L = \begin{cases} L_{FS} + L_{rts} + L_{msd} & \text{for } L_{rts} + L_{msd} \geq 0 \\ L_{FS} & \text{for } L_{rts} + L_{msd} < 0 \end{cases} \tag{6.6}$$

where L_{FS} is the free-space loss (2.169), and L_{rts} and L_{msd} are the roof-to-street and multiscreen-diffraction losses. In particular,

$$L_{rts} = -16.9 - 10\text{Log}(w) + 10\text{Log}(f) + 20\text{Log}(h_{roof} - h_M) + L_{ori} \tag{6.7}$$

where

$$L_{ori} = \begin{cases} -10 + 0.354\varphi & \text{for } 0° \leq \varphi < 35° \\ 2.5 + 0.075(\varphi - 35) & \text{for } 35° \leq \varphi < 55° \\ 4.0 - 0.114(\varphi - 55) & \text{for } 55° \leq \varphi < 90° \end{cases}$$

with φ being the angle, in degrees, between the vertical plane containing the two antennas and the direction of the street where the mobile antenna is located; and

$$L_{msd} = L_{bsh} + k_a + k_r\text{Log}(r) + k_f\text{Log}(f) - 9\text{Log}(b) \tag{6.8}$$

where

$$L_{bsh} = \begin{cases} -18\text{Log}(1 + h_{BS} - h_{roof}) & \text{for } h_{BS} \geq h_{roof} \\ 0 & \text{for } h_{BS} < h_{roof} \end{cases}$$

$$k_a = \begin{cases} 54 & \text{for } h_{BS} \geq h_{roof} \\ 54 - 0.8(h_{BS} - h_{roof}) & \text{for } h_{BS} < h_{roof} \text{ and } r \geq 0.5 \text{ km} \\ 54 - 1.6(h_{BS} - h_{roof})r & \text{for } h_{BS} < h_{roof} \text{ and } r < 0.5 \text{ km} \end{cases}$$

$$k_r = \begin{cases} 18 & \text{for } h_{BS} \geq h_{roof} \\ 18 - 15\dfrac{h_{BS} - h_{roof}}{h_{roof}} & \text{for } h_{BS} < h_{roof} \end{cases}$$

$$k_f = 4 + \begin{cases} 0.7\left(\dfrac{f}{925} - 1\right) & \text{for medium-sized cities} \\[2ex] 1.5\left(\dfrac{f}{925} - 1\right) & \text{for metropolitan centers} \end{cases}$$

In (6.7) and (6.8), f is measured in megahertz, r in kilometers, and h_M, h_{BS}, h_{roof}, w, and b are all measured in meters. Some measurements [1] have shown that the standard deviation of the model errors is 6–8 dB, but a large (10–17-dB) mean error is obtained, which can be eliminated via calibration with a few measurement points.

For urban microcells, when both antennas are below the rooftop level, empirical models based on the two-ray model are usually employed in the LOS case. For distances shorter than the breakpoint one [see (4.15)], oscillations around the free-space received power level are expected, whereas at larger distances, the received power decreases as the fourth power of distance. Accordingly, ITU recommends using the following expression for the median value of the propagation loss [2]:

$$L = L_{bp} + 6 + \begin{cases} 20\mathrm{Log}\left(\dfrac{r}{r_{bp}}\right) & \text{for } r \leq r_{bp} \\[2ex] 40\mathrm{Log}\left(\dfrac{r}{r_{bp}}\right) & \text{for } r > r_{bp} \end{cases} \tag{6.9}$$

where the breakpoint distance r_{bp} is given by (4.15), which we rewrite here as

$$r_{bp} = \frac{4h_{BS}h_M}{\lambda} \tag{6.10}$$

and

$$L_{bp} = 20\mathrm{Log}\left(\frac{8\pi h_{BS}h_M}{\lambda^2}\right) \tag{6.11}$$

Instead of a standard deviation of loss σ, in this case, ITU provides upper and lower bounds for the loss, which, for $r > r_{bp}$, are 14 dB larger and 6 dB lower than the median value (6.9), respectively. In Figure 6.2, the median value of the received power computed via (6.9) to (6.11), as well as its upper and lower bounds, are compared with the groundwave two-ray model value obtained via (4.6).

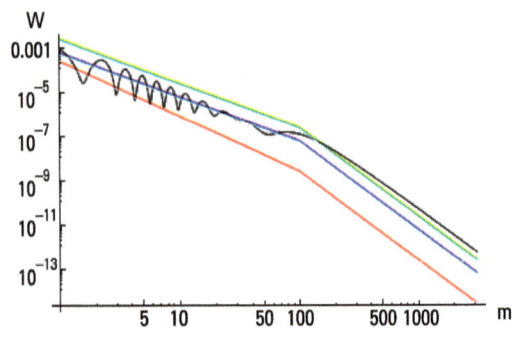

Figure 6.2 Received power at 1 GHz for 1-W transmitted power and unitary-gain antennas, $h_{BS}=5m$, $h_M=1.5m$: ITU street-canyon model average (blue), maximum (green), and minimum (red) values. The two-ray model value, for $\varepsilon_r=20$ and vertical polarization, is plotted in black for comparison purposes.

Equations (6.9) to (6.11) can be used for any frequency from 300 MHz to 15 GHz. However, for frequencies higher than 3 GHz, h_{BS} and h_M in (6.10) and (6.11) should be replaced by $h_{BS}-h_S$ and h_M-h_S, where h_S is the effective road height, which accounts for the effect on the wave propagation of vehicles on the road and pedestrians near the roadway. Hence h_S depends on the traffic on the road and is empirically determined via measurements [2].

Finally, for NLOS around-the-corner propagation (see Figure 6.3), ITU recommends the following expression for the propagation loss in decibels in the frequency range from 800 to 2,000 MHz:

$$L = -10\text{Log}\left(10^{-L_r/10} + 10^{-L_d/10}\right) \tag{6.12}$$

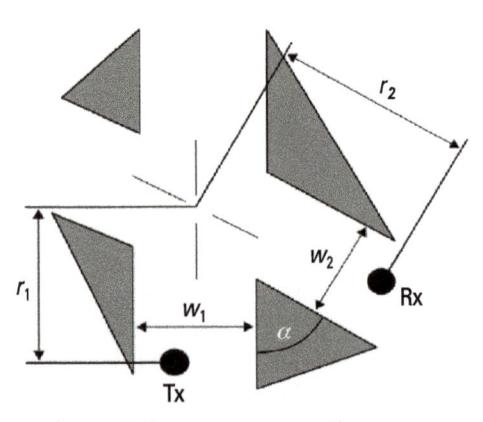

Figure 6.3   Geometry of around-the-corner propagation.

where L_r is the corner reflection loss, given by

$$L_r = 20\text{Log}(r_1 + r_2) + \frac{3.86 r_1 r_2}{\alpha^{3.5} w_1 w_2} + 20\text{Log}\left(\frac{4\pi}{\lambda}\right) \tag{6.13}$$

with α measured in radians, $0.6 < \alpha < \pi$; and L_d is the corner diffraction loss, given by

$$L_d = 10\text{Log}\left[r_1 r_2 (r_1 + r_2)\right] + 2D_a - 0.1\left(90 - \alpha\frac{180}{\pi}\right) + 20\text{Log}\left(\frac{4\pi}{\lambda}\right) \tag{6.14}$$

$$\text{with } D_a = \frac{40}{2\pi}\left[\arctan\left(\frac{r_2}{w_2}\right) + \arctan\left(\frac{r_1}{w_1}\right) - \frac{\pi}{2}\right]$$

6.2.2 Indoor

A very simple empirical model for indoor propagation is the *linear attenuation model*, according to which the indoor environment is schematized by an equivalent lossy medium, so that the field undergoes an exponential attenuation. Therefore, the power propagation loss in decibels is given by

$$L = L_{FS} + \alpha r \tag{6.15}$$

where L_{FS} is the free-space loss (2.169), and α is the attenuation coefficient, measured in decibels per meter and determined empirically via measurement campaigns. Employed α values, obtained via measurements at 1,800 MHz [1], range from 0.22 to 0.62 if transmitting and receiving antennas are on the same floor, and are close to 3 if they are on different floors. Comparison of model predictions and measurements made at 856, 1,800, and 1,900 MHz leads to mean errors up to ±7 dB and to error standard deviations ranging from 4.5 to 7.5 dB.

Another popular and very simple model, based on the free-space loss expression, is the *one-slope model*, according to which the propagation loss in decibels is

$$L = A + B\text{Log}(r) \tag{6.16}$$

where r is measured in meters, A is the propagation loss at 1m from the transmitting antenna, and the coefficient B is usually expressed as $B = 10n$, where n is the power decay index (in the free space, $n = 2$). Employed values of A range from about 33 to about 40 dB [1], whereas the employed value of n strongly depends on the indoor environment. If transmitting and receiving antennas are in different rooms on the same floor, it varies from about 2 to 4 for environments made of very large to very small rooms; if the antennas are at different points along the same corridor, n is about 1.4, so that in this case, due to the waveguiding property of the corridor, the wave attenuation is smaller than in the free space. Comparison of one-slope model predictions and measurements made at 856, 1,800, and 1,900 MHz leads to mean errors ranging from −1.6 to 20.7 dB and to error standard deviations ranging from 5.7 to 10 dB.

If the map of the building is available, along with some information on the walls' materials and thickness, it is possible to use the *multiwall model*, according to which the power loss in decibels is:

$$L = L_{FS} + L_0 + \sum_{i=1}^{I} k_{wi} L_{wi} + k_f^{\left(\left(k_f+2\right)/\left(k_f+1\right)\right)-b} L_f \qquad (6.17)$$

where:

L_{FS} is the free-space loss;

L_0 is an additional constant loss, often assumed negligible;

k_{wi} is the number of penetrated walls of type i;

k_f is the number of penetrated floors;

L_{wi} is the loss of wall of type i;

L_f is the loss between adjacent floors;

b is an empirical parameter;

I is the number of wall types.

Reference [1] suggests that we consider only two types of walls: light wall ($w1$) [i.e., plasterboard, particle board, or thin (less than 10-cm-thick) light concrete walls] and heavy wall ($w2$) [i.e., load-bearing walls or other thick (more than 10-cm-thick) concrete or brick walls]. The third term in (6.17) expresses the total wall loss as the sum of the losses of walls between the transmitter and receiver. It must be noted that the loss factors in (6.17) are not physical wall losses, but model coefficients that are optimized via measured path loss data. Accordingly, they implicitly include the effect of furniture as well as the effect of the waveguiding property of corridors. The following values of the model coefficients are suggested in [1]: $L_{w1} = 3.4$ dB, $L_{w2} = 6.9$ dB, $L_f = 18.3$ dB, and $b = 0.46$. Comparison of multiwall model predictions and measurements made at 856, 1,800, and 1,900 MHz leads to mean errors ranging from

−10.3 to 2.2 dB and to error standard deviations ranging from 2 to 9.5 dB. Figure 6.4 shows examples of received power level maps obtained with one-slope and multiwall models.

6.2.3 Example: Downlink in a 4G LTE Mobile Phone System

Let us assume that the base-station antenna serving a 4G LTE urban mac-rocell in a medium-sized city is placed at 40m above the ground plane, over a tower, and that its gain G_T is 16 dB; in addition, the carrier frequency f is 1,800 MHz, and the transmitted power P_T is 43 dBm (i.e., 20W). We want to compute the median value of the power received by a user's mobile device placed at a distance of 1 km. We can reasonably assume that the receiving antenna is at 1.6m above the ground and that its gain is close to unity. Since we only know the heights of the antennas, and no other scene parameter, and considering the employed frequency, we can use the COST231-Hata model (6.5), with f = 1,800 MHz, r = 1 km, h_{BS} = 40m, and h_M = 1.6m. We get L = 134 dB, which, by using (6.2), leads to a received power P_R = −75 dBm. Note that the Friis formula would lead to a received power of about −38 dBm, and that the typical receiver sensitivity of mobile devices is −95 to −100 dBm.

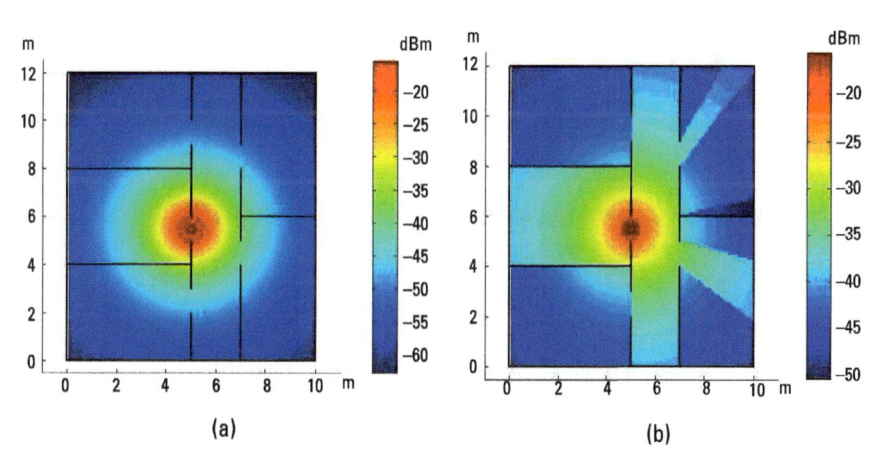

(a)   (b)

Figure 6.4 Received power maps at 2.4 GHz, 0.1-W transmitted power, unitary-gain antennas, obtained via (a) the one-slope model and (b) the multiwall model. The received power in decibels per meter is represented in false colors (see the color bars), and the maps are superimposed to the floor plan of the considered apartment. This makes it possible to appreciate that the multiwall model explicitly accounts for the specific propagation environment, at variance with the one-slope model.

6.2.4 Coverage Area and Location Probability

The *coverage area* of a broadcasting antenna or of a base station in a mobile telephony system is the set of locations in which a sufficient signal strength is received (i.e., for which the received power P_R is greater than the receiver sensitivity $P_{R\min}$). In this case the location is said to be *covered* by the antenna/ base station. If the received power is modeled as a random variable, we can only compute the probability that a given location $\mathbf{r}$ is covered. Let $x = [P_R]_{\mathrm{dB}}$, $x_0 = [P_{R\min}]_{\mathrm{dB}}$; in addition, $\bar{x} = < [P_R]_{\mathrm{dB}} >$ is the received power mean value computed via (6.2) with one of the empirical models of Section 6.2.1, and σ is the standard deviation, also provided by empirical models (see, for example, Table 6.1). The probability that the location is covered (*location probability*) is then

$$P(\mathbf{r}) = \mathrm{prob}(x \geq x_0) = \int_{x_0}^{\infty} p_x(x)\,dx \tag{6.18}$$

where $p_x(x)$ is the pdf of the received power expressed in decibels. The latter is usually assumed Gaussian, which is appropriate in case of slow fading (see Section 5.3.3), or in the case of multipath with dominant LOS contribution (see Section 5.3.2):

$$p_x(x) = \frac{1}{\sqrt{2\pi}\sigma}\exp\left[-\frac{(x-\bar{x})^2}{2\sigma^2}\right] \tag{6.19}$$

For a given area A (e.g., a cell of a mobile-phone system), the fraction F_c of covered locations can be computed as [3]:

$$F_c = \frac{1}{A}\iint_A P(\mathbf{r})\,dA \tag{6.20}$$

If $\bar{x}$ and σ, and hence $P(\mathbf{r})$, only depend on the distance r from the base station, and assuming a circular cell of radius R, (6.20) can be rewritten as

$$F_c = \frac{1}{\pi R^2}\int_0^{2\pi}\int_0^{R} P(r)\,r\,d\vartheta\,dr = \frac{2}{R^2}\int_0^{R} P(r)\,r\,dr \tag{6.21}$$

In general, the computation of this integral, whose value depends on the employed empirical model, must be performed numerically.

A deterministic evaluation of the coverage area can be performed by using ray-tracing methods, described in Section 6.4.

6.3 Urban Canyon as a Roofless Waveguide

An examination of the empirical methods commonly used to model street-canyon propagation, such as those illustrated in Section 6.2.1, seems to show that they do not fully account for the intrinsic guiding nature of the built-up scenario, determined by the channeling nature of streets and their intersections (see Figure 6.5). In fact, they provide path-loss formulations that do not even depend on the street width w, which is the most important geometric parameter describing the urban canyon scenario. To explicitly account for the waveguiding property of urban canyons, [7–9] suggest the use of concepts and techniques borrowed from the theory of waveguides. The basic guiding structure, modeling a straight street limited by buildings, is a *roofless rectangular waveguide* of constant width (see Figure 6.5). By making the simplifying assumptions of constant height of the lateral buildings and homogeneous internal medium, the field in the guiding structure can be expressed as a superposition of propagating *modes* (*modal-expansion approach*), each one characterized by an integer index $n > 0$ and by a different longitudinal (i.e., along the direction of the street) complex propagation constant $k_{zn} = \beta_n - j\alpha_n$, whose imaginary part α_n implies that the modes are attenuated as they propagate along the street (with α_n as the attenuation constant).

The number of propagating modes is large, since it is on the order of the ratio between street width and wavelength; however, the attenuation constant

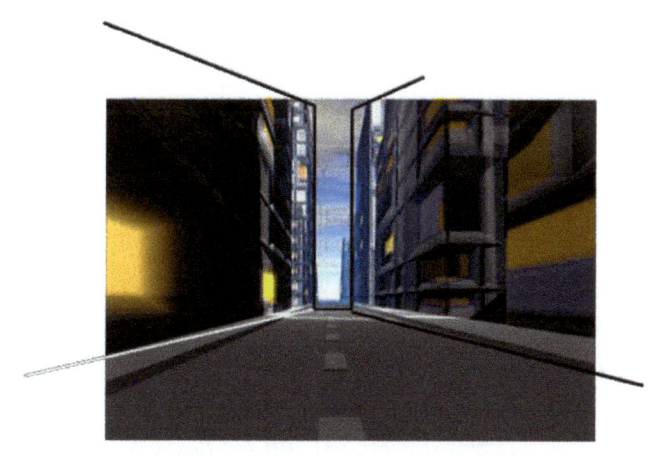

Figure 6.5 Street canyon and its roofless rectangular waveguide schematization.

of the modes increases with n, so that the number of nonnegligible modes decreases as the distance from transmitter to receiver increases. For finite height of the lateral walls of the roofless waveguide, determining the modes' propagation constants requires the numerical solution of an integral equation [7]. However, if transmitting and receiving antennas are well below the rooftop level, it is reasonable to assume that the building walls are infinitely high [see Figure 6.6(a)]; in this case, real and imaginary parts of modes' propagation constants can be analytically computed as [8]

$$\beta_n \cong \sqrt{k^2 - \left(\frac{n\pi}{w}\right)^2}, \quad \alpha_n^{v,h} \cong \frac{\varepsilon_{v,h} n^2 \lambda^2}{2w^3 \sqrt{\varepsilon_{wr} - 1}} \tag{6.22}$$

where k is the propagation constant in vacuum, ε_{wr} is the real part of the relative dielectric constant of the wall, v and h stand for vertical and horizontal polarization, and $\varepsilon_v = 1$, $\varepsilon_h = \varepsilon_{wr}$. An analytical expression of the field, and hence of the received power, is then obtained. The received power turns out to depend also on the position of the receiver within the street canyon cross section: its average over the street width leads to [8]

$$P_R = P_{FS} L_{2R} G_{WG} L_{Wv,h} \tag{6.23}$$

where P_{FS} is the free-space received power, given by the Friis formula,

$$L_{2R} = \begin{cases} 1 & \text{for } r < r_{2R} \\ \left(\dfrac{r_{2R}}{r}\right)^2 & \text{for } r \geq r_{2R} \end{cases} \quad \text{with } r_{2R} = \frac{4\pi h_{BS} h_M}{\lambda} = \pi r_{bp} \tag{6.24}$$

is the two-ray loss,

$$G_{WG} = \frac{2r}{w} \tag{6.25}$$

is the waveguiding gain, and

$$L_{Wv,h} \cong \begin{cases} 1 & \text{for } r < r_{W1v,h} \\ \dfrac{1}{2}\sqrt{\dfrac{\pi\omega\sqrt{\varepsilon_{wr}-1}}{4r\varepsilon_{v,h}}} & \text{for } r_{W1v,h} \leq r < r_{W2v,h} \\ \dfrac{\lambda}{2w}\exp\left(-2\alpha_1^{v,h} r\right) & \text{for } r \geq r_{W2v,h} \end{cases} \tag{6.26}$$

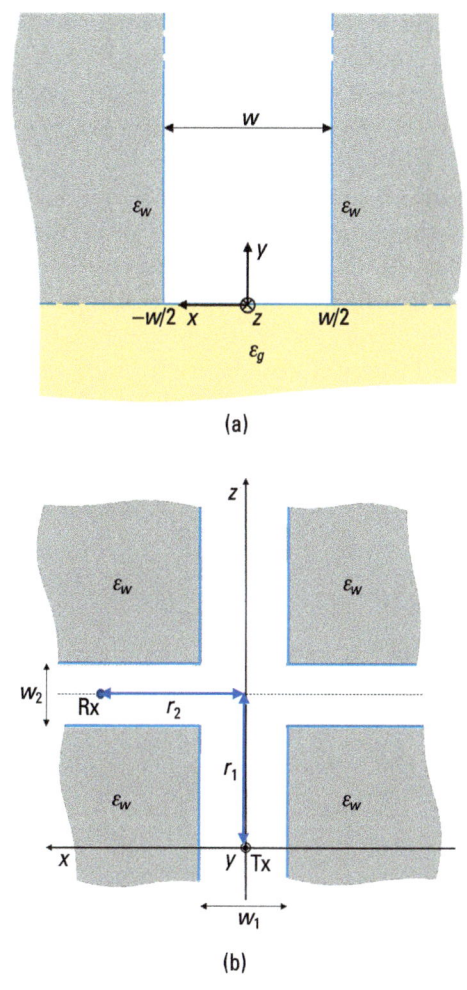

Figure 6.6 Geometry of (a) the canonical street canyon consisting of smooth walls of infinite height with complex relative permittivity ε_w, and a smooth street with complex relative permittivity ε_g and width w, and (b) the around-the-corner propagation.

with $r_{W1v,h} = \pi w \sqrt{\varepsilon_{wr} - 1}/\left(16\varepsilon_{v,h}\right)$, $r_{W2v,h} = w^3 \sqrt{\varepsilon_{wr} - 1}/\left(2\lambda^2\varepsilon_{v,h}\right)$, is the loss due to the power transmitted through (and partly dissipated in) the lateral walls.[1] Most of distances of interest are in the range $r_{W1v,h} \lesssim r < r_{W2v,h}$. In this

---

1. Note that all relations in this Section 6.3 are not expressed in decibels, but in linear units, so that losses are here factors not larger than unity.

range, according to this model, the received power decreases as $r^{-1.5}$ and $r^{-3.5}$ for distances shorter and longer than r_{2R}, respectively. In addition, an overall decrease of the received power with the square root of the street width is obtained. Finally, the received power is larger at vertical than at horizontal polarization.

The modal expansion approach also makes it possible to consider the problem of around-the-corner propagation [Figure 6.6(b)], by using a mode-matching method, similar to the one used to deal with waveguide intersections [8, 9]. Also in this case, the field in the two crossing streets is expressed as a superposition of modes, and a rather simple expression of the average (over the street width) received power can be obtained, which, in the case of vertically polarized antennas, is [9]:

$$P_R = P_T G_T G_R \left[\frac{\lambda}{4\pi(r_1 + r_2)} \right]^2 L_{2R} D_{\text{cross}} L_{Wv} \tag{6.27}$$

where

$$L_{2R} = \begin{cases} 1 & \text{for } r_1 + r_2 < r_{2R} \\ \left(\dfrac{r_{2R}}{r_1 + r_2} \right)^2 & \text{for } r_1 + r_2 \geq r_{2R} \end{cases} \quad \text{with } r_{2R} = \frac{4\pi h_{BS} h_M}{\lambda} = \pi r_{bp} \tag{6.28}$$

is the two-ray loss,

$$D_{\text{cross}} = \frac{4\lambda(r_1 + r_2)}{3\pi^2 w_1 w_2} \tag{6.29}$$

is the street-crossing diffraction factor, and

$$L_{Wv} \cong \min \left\{ 1, \frac{3\pi(\varepsilon_{wr} - 1)}{32} \frac{w_1 w_2}{4 r_1 r_2} \right\} \tag{6.30}$$

is the lateral wall loss.

6.4 Ray-Tracing Methods

If a digital description of the propagation environment is available, ray-tracing methods can be employed. According to such methods, based on the asymptotic techniques of Chapter 3, propagation in urban areas is described in terms of rectilinear rays and their reflections, transmissions, and diffractions. Reflections and transmissions are treated by using the GO formulation (see Section 3.1), whereas diffractions are described by using GTD or UTD (see Section 3.5). Therefore, the mathematical formulation is computationally very efficient, but a significant computational load is required by the geometrical problem of determining the rays radiated by the source and reaching the points where the field must be computed. In addition, the accuracy of the field prediction of ray-tracing techniques is in general much higher with respect to that of the empirical methods, but ray tracing requires large databases describing terrain, buildings, and other possible obstacles located in the considered area. In addition, the field prediction accuracy is strongly dependent on the accuracy of such obstacle descriptions. Transmitting antenna position, pointing, gain, polarization, radiation pattern, and input power must be also available.

Ray-tracing techniques can be subdivided into two main classes: *ray-launching*, or *direct ray-tracing*, algorithms, and *inverse ray-tracing*, or *image-based*, algorithms. In the former case, rays are launched from the source in all directions, with a given (small) angular sampling step, and each ray is followed in its possible reflections, transmissions, and diffractions. Every time a ray intersects the *capture sphere* (or, more in general, the *capture volume*) of a point where the field must be computed, the field associated to that ray is summed to the already accumulated field in the considered point. Each ray is followed until either it goes outside the area of interest, or it fulfills a *stop criterion*; this may mean that it has reached a given maximum number of reflection or diffraction events, or that the intensity of the field associated to the ray has fallen below a specified threshold.

Conversely, instead of using the "brute force" approach of launching rays in all directions, inverse ray-tracing algorithms only search for the rays actually connecting the transmitter to the points where the field must be computed. This is obtained by considering all obstructions as potential reflectors and calculating their effect by using the method of images (see Section 3.1.3 and Figure 6.7). A maximum number of reflection events must be fixed in this case.

If the field must be calculated only in a few points and/or the environment is simple (e.g., a small indoor environment), the inverse algorithms are more efficient than the direct ones. However, if we are interested in computing the field in an entire area (i.e., if we want to determine the coverage area of a given base-station antenna), so that the field must be computed on a regular dense grid of points, direct ray-launching algorithms are more convenient.

It is important to note that, since ray-tracing algorithms necessarily compute ray path lengths and directions, they make it possible to evaluate

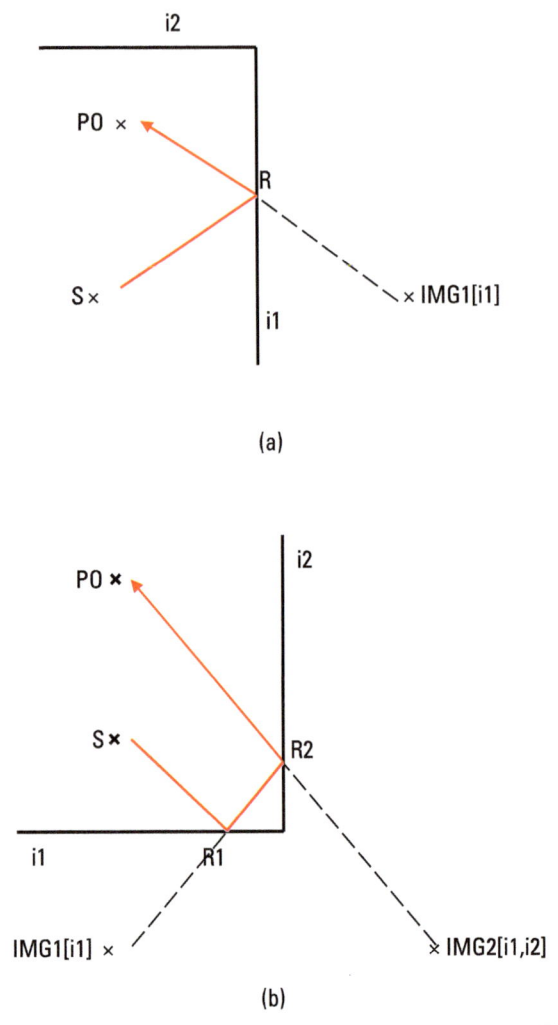

Figure 6.7 Image method: (a) single reflection, where S is the source, PO the point of observation; IMG1[i1] the image of the source with respect to wall i1, and R is the reflection point (i.e., the intersection of the line segment IMG1-PO with the wall i1). (b) Double reflection, where IMG2[i1,i2] is the image of IMG1[i1] with respect to wall i2, and R1 and R2 are the reflection points on walls i1 and i2.

not only the electromagnetic field level, but also wideband parameters, such as delay spread, (see Sections 5.2.2 and 5.4) and angular spread (see Section 5.2.1). In addition, since these algorithms imply repeatedly evaluating intersections between line segments, they benefit from the adoption of techniques of *computational geometry* [10], widely employed in computer graphics. Also, each ray or group of rays can be independently elaborated by a different computer processor, and therefore ray-tracing algorithms can be easily implemented on parallel computing architectures: multicore machines, graphical processing units (GPUs), and computer grids. All this has led, in the last two decades, to the development of efficient electromagnetic solvers based on ray tracing, a technology that is now commercially available.

Finally, a significant computational load reduction can be achieved when, as it is often the case in built-up environments, all reflecting surfaces are either horizontal or vertical. In fact, all rays belonging to a vertical plane, that are reflected on a horizontal or vertical plane surface, generate reflected rays all belonging to a single vertical plane, too; see Figure 6.8. This suggests reducing the full tridimensional problem to two bidimensional problems via the *vertical plane launching* approach [11, 12], summarized as follows. A two-dimensional ray-launching method is first used in the horizontal plane, so that each ray represents the trace of a vertical plane on the horizontal one, and buildings are represented by polygons (i.e., by their floor plans); reflections on the vertical walls of the buildings and diffractions on their vertical edges are all identified in this two-dimensional ray-launching stage. Then, propagation in the vertical planes is evaluated by using an efficient inverse ray-tracing algorithm, also accounting for the building heights (i.e., for the "city skyline" in each considered vertical plane).

An example of a ray-tracing algorithm, based on the vertical plane launching approach, is described in detail in Chapter 7. An example of a field coverage map obtained via a ray-tracing algorithm is shown in Figure 6.9. A review of available ray-tracing algorithms can be found in [13].

In conclusion, ray-tracing methods are very useful tools to predict electromagnetic field levels (and delay and angular spread) in deterministically

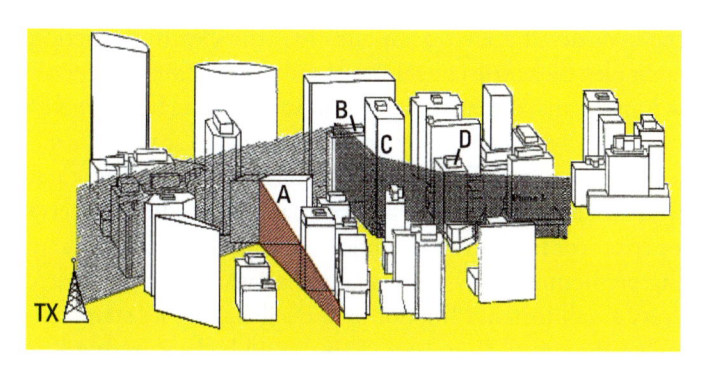

Figure 6.8 Vertical plane launching (A and B: reflections on vertical walls; and C and D: diffractions on vertical edges).

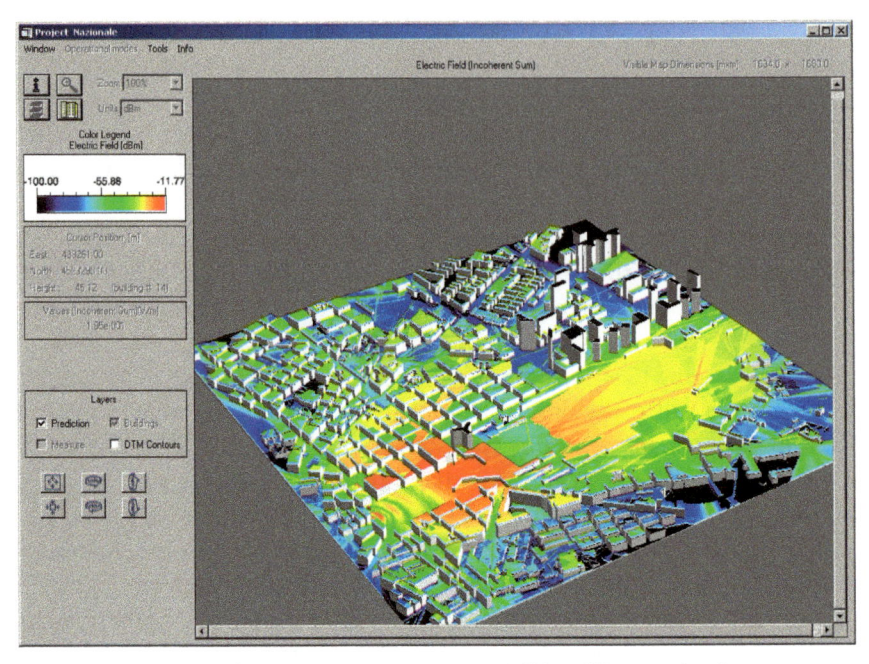

Figure 6.9 Example of a received power map at 1.8 GHz, 5-W transmitted power, 17.3-dB antenna gain, obtained via a ray-tracing tool. The received power in dBm is represented in false colors (see the color bar), and the map is superimposed onto a 3-D representation of the considered urban scenario.

known urban scenarios. Therefore, they are sometimes called *deterministic methods*, as opposed to the statistical or empirical methods of Section 6.2. However, the available description of the environment necessarily neglects some geometrical details of obstacles, not to mention people and vehicles moving within the considered scenario. In addition, the electromagnetic properties (i.e., the complex dielectric constants) of soil and building materials are often not known in detail, so that some default values of such parameters are usually selected, based on the typology of the considered scenario (e.g., historical area, residential area, or office area). Finally, the use of asymptotic techniques itself implies a certain degree of approximation. All these uncertainties are sources of errors for field-level prediction. Comparisons of measurements and ray-tracing predictions show that usually the mean value of the error on field levels is very small (often a fraction of a decibel), but its standard deviation is not smaller than 4–6 dB at best. This should be kept in mind when using the results of such methods. Accordingly, the actual total electromagnetic field can be considered as the superposition of a deterministic component, computed via the ray-tracing algorithm, and a small random component, due to the unmodeled minor obstacles and other error sources.

References

[1]   *Digital Mobile Radio Towards Future Generation Systems—COST 231* Final Report—Chapter 4 "Propagation Prediction Models," European Union, 1999.

[2]   *Propagation data and prediction methods for the planning of short-range outdoor radio-communication systems and radio local area networks in the frequency range 300 MHz to 100 GHz,* document ITU-R P.1411-11, International Telecommunication Union, Radiowave Propagation P Series, Sep. 2021.

[3]   Parsons, J. D., *The Mobile Radio Propagation Channel,* Chichester: John Wiley & Sons, Inc., 2000.

[4]   Okumura, Y., et al., "Field Strength and Its Variability in VHF and UHF Land-Mobile Radio Service," *Review of the Electrical Communication Laboratory,* Vol. 16, No. 9–10, 1968, pp. 825–73.

[5]   Hata, M., "Empirical Formula for Propagation Loss in Land Mobile Radio Services," *IEEE Transactions on Vehicular Technology,* Vol. 29, No. 3, 1980, pp. 317–25.

[6]   Bertoni, H. L., *Radio Propagation for Modern Wireless Systems,* Hoboken, NJ: Prentice Hall, 2000.

[7]   Franceschetti, G., and A. Iodice, "Toward the Development of an Equivalent Microwave Circuit for Built-up Areas," *Proceedings of the 2016 IEEE Antennas and Propagation Society International Symposium (APS-URSI),* Fajardo, Puerto Rico, 2016, pp. 1523–1524.

[8]   Di Simone, A., and A. Iodice, "Modal Expansion Approach for Electromagnetic Propagation in Street Canyons," *IEEE Transactions on Antennas and Propagation,* Vol. 67, No. 4, 2019, pp. 2103–2117.

[9]   Di Simone, A., and A. Iodice, "Reciprocal Formulation of the Modal Expansion Approach for Around-the-Corner Propagation in Street Canyons," *Radio Science,* Vol. 55, 2020, pp. 1–19.

[10]   Preparata, F. P., and M. I. Shamos, *Computational Geometry—An Introduction,* Springer-Verlag, 1985.

[11]   Rossi, J.-P., et al., "A Ray Launching Method for Radio-Mobile Propagation in Urban Area," *Proc. Antennas Propagat. Soc. Symp.,* 1991, pp. 1540–1543.

[12]   Liang. G., and H. L. Bertoni, "A New Approach to 3-D Ray Tracing for Propagation in Cities," *IEEE Trans. Antennas Propagat.,* Vol. 46, No. 6, 1998, pp. 853–863.

[13]   Yun, Z., and M. F. Iskander, "Ray Tracing for Radio Propagation Modeling: Principles and Applications," *IEEE Access,* Vol. 3, 2015, pp. 1089–1100.

7

Ray-Tracing Tool Example[1]

Chapter 6 presented methods that can be used to predict the electromagnetic field levels generated by a transmitting antenna in an urban area. Among these methods, ray-tracing techniques, described in Section 6.4, are assuming greater and greater importance, as the need for planning new-generation wireless networks progressively increases [1, 2], and the computer hardware evolution continues to reduce computational times.

This chapter describes a possible implementation of the ray-tracing technique, by presenting a deterministic solver for the evaluation of electromagnetic coverage in an urban environment [3]. This solver implements a ray-launching technique, namely vertical-plane launching, following in the footsteps of pioneering works by Rossi et al. [4] and Liang and Bertoni [5]; see Section 6.4. It is based on appropriate physical models whose implementation provides time and precision performance that is adequate for cell planning.

This deterministic solver takes into account any relevant electromagnetic contribution. Direct contribution—along with its implemented reflections and diffractions—is accounted for, without the simplifying assumption that the number of diffractions and reflections is limited. Each ray is followed

---

1. This chapter was written by the authors of this book and Giuseppe Ruello, who is with the University of Naples "Federico II."

until either it goes outside the area of interest, or the intensity of its field falls below a specified threshold, which can be set according to user needs. Scattering from rough soil and outdoor-indoor transmission can be also accounted for. The computational efficiency is due to the fact that closed-form GO and UTD solutions (see Chapter 3) are employed to represent the electromagnetic phenomena. This ray-launching technique also makes it possible to evaluate delay and angular spread; see Chapter 5.

The remainder of this chapter is organized as follows. Section 7.1 presents the implemented vertical-plane launching algorithm. Section 7.2 illustrates the input and output of the solver and discusses the computational time and the main limitations of the approach. Section 7.3 covers a measurement campaign performed in a site composed of areas with different urban propagation characteristics. Finally, Section 7.4 compares measured data with solver predictions.

7.1 Vertical-Plane Launching Implementation

There are different algorithmic implementations of the basic rationale for the vertical-plane launching outlined in Section 6.4 and pictorially illustrated in Figure 6.8. The solver described in this chapter is aimed at the evaluation of the field radiated by a transmitting antenna over a regular grid of points placed at a fixed height with respect to ground (or with respect to a building roof, if the considered grid point lays in correspondence of a building). This height, and the grid spacing, are selected by the user. Typical values are 1.5m for the height and 5m for the grid spacing. In addition, the solver can evaluate the delays and the directions of arrival of the different contributions to the total field. Hence, it also provides wideband parameters (e.g., angle spreading function and rms delay spread).

The urban scenario is described in terms of buildings and topography. The buildings are approximated by right prisms with smooth vertical walls. The base polygons of the prisms can be of any—however irregular or complex—shape. A digital terrain model (DTM) is required to account for the topography. Figure 7.1 shows an example of the digital description of an urban scenario. Information on the transmitting antenna must also be provided, as detailed in Section 7.2.

The implementation of a ray-launching algorithm requires three main steps:

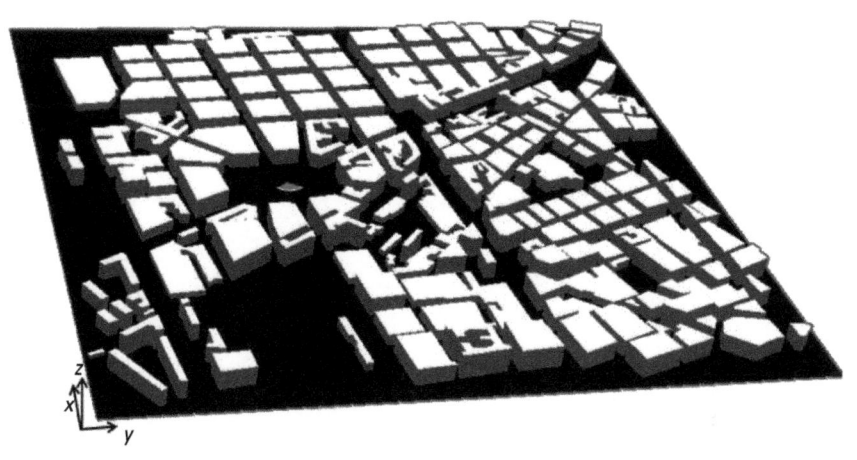

Figure 7.1 Typical 3-D representation of the urban scene provided as input to the presented prediction tool.

- Ray definition;
- Space scanning;
- Electromagnetic modeling.

The corresponding solutions implemented in the software are detailed next.

7.1.1 Ray Definition

In the employed ray-launching algorithm, each single ray actually represents a ray beam with a finite angular aperture. In particular, since a vertical-plane launching algorithm is used, the ray beams are fully described by their horizontal and vertical angular sizes. In the horizontal plane (HP), the area illuminated by the antenna (the *primary source*) is divided into angular sectors (referred to as *anxels*, the abbreviation of *angular elements*), whose bisectrices are the rays (each one representing the trace of a vertical plane on the horizontal one; see Section 6.4) and whose width $d\theta$ is set in agreement to the maximum separation between two rays dr, selected by the user; see Figure 7.2. In the vertical plane (VP), the antenna radiation diagram defines the ray beam angular aperture α, see Figure 7.2. It is assumed that the radiation diagram is negligible outside this aperture; however, if a nondirective antenna is considered, the beam may be represented by the entire vertical half-plane (so that the vertical aperture angle is 180°).

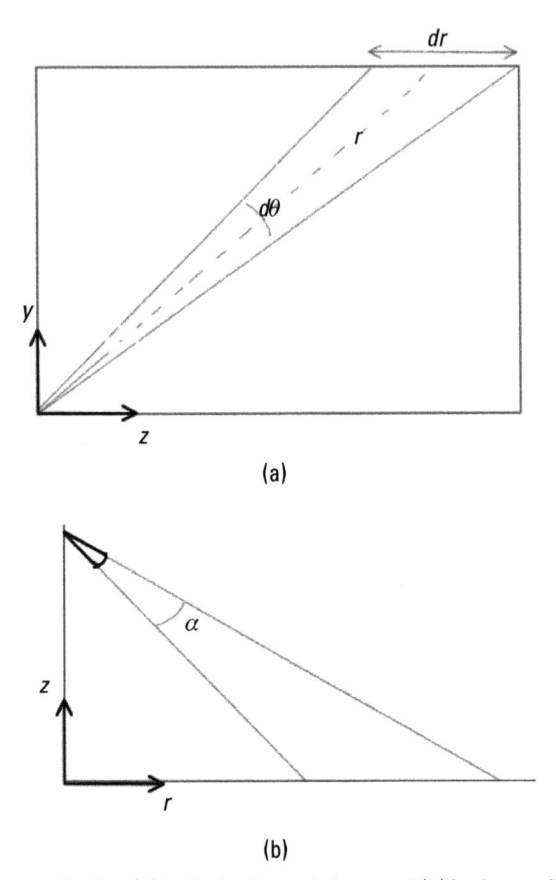

(a)

(b)

Figure 7.2 Anxel definition (a) in the horizontal plane and (b) in the vertical plane.

7.1.2   Space Scanning

The 3-D space is scanned with a two-step procedure. First, in the HP, all the output grid points and wall vertexes included in each anxel, and all the walls intersecting it, are identified and stored in a list ordered according to their distance from the source. (See Figure 7.3(a), where the grey area represents an anxel originating at the antenna A.) Therefore, at the end of this horizontal scanning, for each anxel (i.e., for each launched vertical plane), we have an ordered list of grid points, building walls, and building vertexes that "belong" to that anxel.

Then, the scene profile in each vertical plane is analyzed. This means that for each anxel all the elements of the list, from the nearest to the source to the farthest, are considered. If the considered element is a grid point, and if it is illuminated by the antenna pattern, the field at that point is updated by summing a field value computed by using the two-ray model (see Section 4.1), so that both the direct and the ground-reflected contributions are considered. If the considered element is a wall, and if it is illuminated by the antenna pattern, it reflects the incident field, and this is accounted for by defining and storing an image source (see Figure 3.5) and computing and storing a reflection coefficient. Optionally, based on the user's choice, outdoor-indoor transmission through the wall is computed by using the approach described in Sections 2.3.6 and 5.1.1, so that an estimate of the indoor field in proximity to the perimetral wall is obtained, one for each building floor. In addition, a wall may shadow part of the subsequent grid points, and this is accounted for by modifying the vertical angular aperture of the beam. As an example, in Figure 7.3(b),

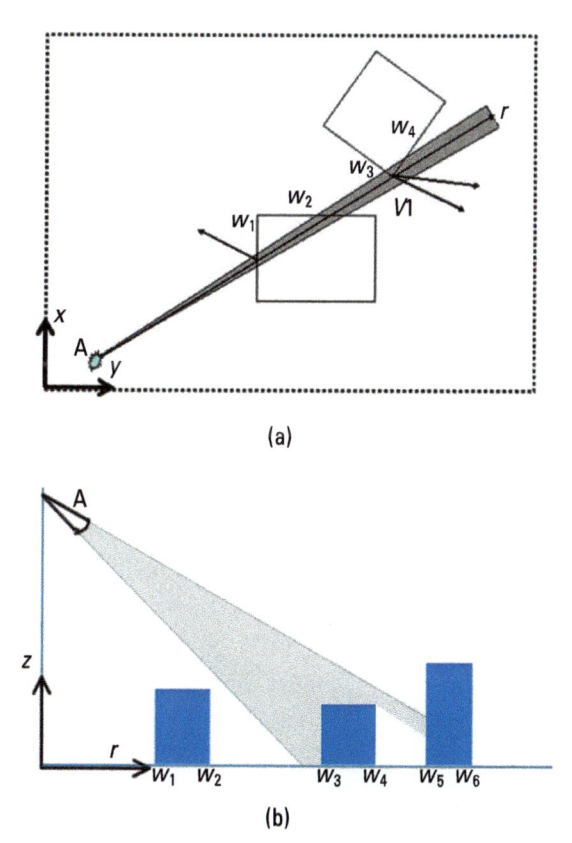

(a)

(b)

Figure 7.3 (a) Horizontal and (b) vertical scanning. Symbols $w_1,..., w_6$ represent buildings' walls intersected by the considered anxel; and the symbol V1 represents the vertex included in the considered anxel.

the walls w_1 and w_2 are not hit by the incident ray; w_3 reflects the incident field; w_4 changes the vertical angular aperture of the ray beam; and the wall w_5 completely occludes the field to possible farther buildings and grid points.

The interaction between a ray and a wall can also produce diffraction from the edges of the wall. According to the GTD/UTD, when a ray impinges on the edge of a vertical wall, a new set of rays departs from the edge (see Section 3.4). In the horizontal plane, each diffracted ray can be seen as generated by an image source s_i, which is identified and stored, together with the corresponding UTD diffraction coefficient (see Section 3.5). Figure 7.4 shows the HP representation of the diffracting sources $(s_1,...s_n)$ generated by the field radiated by the primary source A, placed at distance r from the edge. Diffractions from vertical edges are therefore rigorously computed by the ray-launching algorithm. Conversely, an approximated evaluation is performed for diffractions from horizontal roof edges: the interested reader is referred to [5] for details on this subject.

When all the anxels of the primary source have been analyzed, we end up with a set of stored secondary sources of reflections and diffractions from vertical walls and edges, and corresponding reflection and diffraction coefficients. The algorithm is then iterated by considering all secondary sources and scanning them according to the preceding rationale. Sources of higher and higher order are then iteratively considered; the procedure stops considering new sources when the field level associated to the currently considered ray decreases below a threshold value. No limits are set on the number of possible reflections and diffractions. The threshold value can be appropriately set, in accordance with the goal of the prediction; if the user needs to evaluate the coverage area of the transmitting antenna, the threshold may be lower or equal to the receiver sensitivity. Conversely, if it is necessary to verify that the field

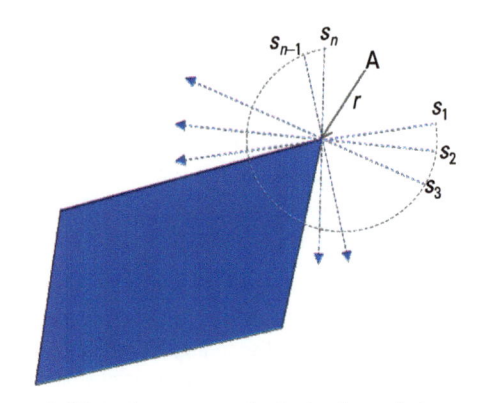

Figure 7.4 Positions of diffraction sources in the horizontal plane.

intensity is compliant with the exposure limits imposed by regulations (see Chapter 9), then only the highest field levels are of interest, and the threshold may be much higher, thereby strongly reducing the computational time.

7.1.3 Electromagnetic Modeling

Reflection, scattering, and diffraction phenomena can be dealt with by using asymptotic techniques, as illustrated in Chapter 3. However, their implementation in a numerical tool is not always straightforward. The solver illustrated in this chapter employs only closed-form solutions, with a significant computational time reduction, detailed as follows.

Reflection on building walls is accounted for by means of the Fresnel coefficients [see Section 3.1.3 and (3.20)] in accordance with the hypothesis of flat building walls.

With regard to reflection on the ground surface, usually the street floor can be considered flat, and the reflection is still described by the Fresnel coefficients. Whenever natural surfaces are of concern, surface roughness must be accounted for, as illustrated in Section 4.1.2. In this case, the roughness height standard deviation σ_z must be provided as an input parameter.

Electromagnetic diffraction plays a key role in urban propagation, mainly in cities where the density of buildings is extremely high. In this tool, an appropriate formulation of the UTD is used, as detailed in Chapter 3 and Appendix D. In particular, (3.82) to (3.86) are employed to compute UTD diffraction coefficients, but their use in a numerical tool calls for further development. In fact, the transition function $F(x)$ appearing in these expressions implies the numerical evaluation of an integral; see (3.48). Indeed, the number of diffraction sources is high, and repeated numerical evaluation of integrals would be too time-consuming. Accordingly, closed-form approximations of the transition function are used. In fact, it can be shown that a series expansion for small values of the dimensionless x argument makes it possible to express the transition function as [6]:

$$F(x) \approx \left[\sqrt{\pi x} - 2x \exp\left(j\frac{\pi}{4} \right) - \frac{2}{3}x^2 \exp\left(-j\frac{\pi}{4} \right) \right] \tag{7.1}$$

whereas a series expansion for the large x argument makes it possible to express the transition function as [6]:

$$F(x) \approx 1 + j\frac{1}{2x} - \frac{3}{4}\frac{1}{x^2} - j\frac{15}{8}\frac{1}{x^3} + \frac{75}{16}\frac{1}{x^4} \tag{7.2}$$

In addition, for intermediate values of x, [6]

$$F(x) \approx \sqrt{\frac{\pi x}{2}} \left[\frac{1+0.926x_p}{1+0.896x_p+1.552x_p^2} + j\frac{1}{1+2.07x_p+1.746x_p^2+3.335x_p^3} \right]$$

(7.3)

where $x_p = \sqrt{2x/\pi}$.

Figure 7.5 shows a comparison between the exact numerical evaluation of (3.86) and its approximate expression via (7.1) to (7.3). The curves are almost indiscernible, leading to the conclusion that the use of the approximated function does not cause significant loss of precision, but it does allow a significant processing time reduction.

7.1.4 Coherent Versus Incoherent Summation

Let us now analyze the best way to implement the superposition of the different multipath contributions at an output grid point (i.e., of the fields associated to the ray beams reaching the grid point). In principle, the field phases of each contribution can be deterministically evaluated from the computed path lengths; therefore, coherent summation of these fields, accounting for their vector and phase relations (i.e., complex vector summation) can be accomplished, thereby fully evaluating the (constructive or disruptive) interference of different contributions in the specific considered grid point. However, we recall that, as illustrated in Sections 5.2.1 and 5.3, in the presence of multipath, the field intensity has significant fluctuations at the wavelength scale (fast fading), so that in points close to the considered grid point the field level may assume very different values. If the aim of the solver is to obtain a continuous field coverage map of the transmitting antenna, then it is desirable that the field value computed at a given point of the grid be representative of the field behavior in a surrounding area (a "resolution cell") of linear size equal to the grid spacing. The latter is usually on the order of a few meters, and it is much larger than the electromagnetic wavelength (on the order of no more than tens of centimeters at the frequencies employed in modern mobile communications). Accordingly, within a resolution cell, the field intensity has many spatial fluctuations around its average value; and the latter is equal to the sum of the intensities of the individual contributions (i.e., their incoherent summation), as shown in Section 5.2.1 and (5.5). This average is certainly a better representation of the overall field level in the resolution cell. Therefore, the presented solver performs the incoherent summation of the different contributions, hence averaging the fast fading over the resolution cell.

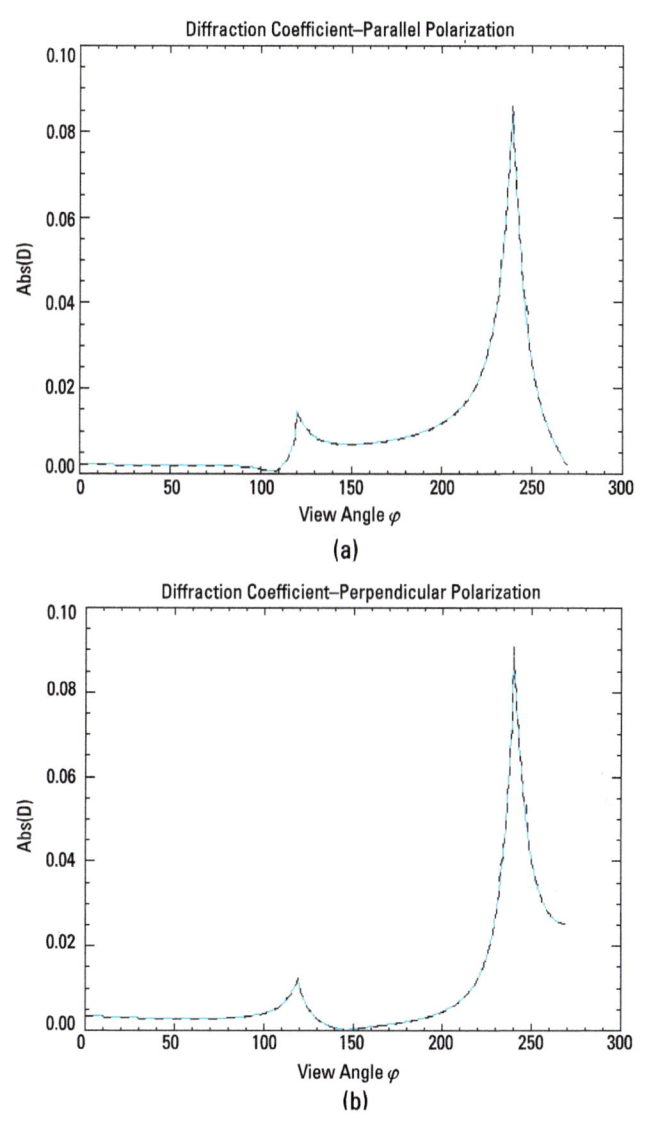

Figure 7.5 Comparison between exact (green solid line) and approximated (black dashed line) diffraction coefficient at 2 GHz in (a) parallel (or hard) and (b) perpendicular (or soft) polarizations.

7.2   Input, Output, and Processing Time

7.2.1   Input

The electromagnetic solver input is a digital description of the scene and of the transmitting antenna. The scene description is provided by a vector file in Planet or Keyhole Markup Language (kml) formats[2] describing the buildings and a raster file describing the terrain topography (DTM). The buildings are geometrically described only by their heights and by polygons representing their floor plan. The positions of the polygons' vertexes are expressed by using metric Universal Transverse Mercator (UTM) geographic coordinates. A building's walls and terrain relative permittivity and conductivity can also be provided in separate files, if available. More often, default values are selected, based on the considered scenario (e.g., historical area, residential area, or office area). Terrain roughness is only accounted for in rural areas, and it is described by its height standard deviation.

The input scenario is also graphically displayed for the user's convenience. For example, in the case study in Figure 7.6, the input data is relevant to a scenario concerning a portion of the urban area of Naples, Italy. The scene is composed of areas with different building densities, heights, and shapes.

As for the antenna, its horizontal position (x, y), altitude over the ground h, pointing direction (i.e., the direction of maximum radiation), radiated power, polarization, and 3-D radiation diagram must be provided. They are stored in a properly formatted file. Antenna information is also graphically displayed for the user's convenience; see Figure 7.7.

Horizontal and vertical cuts of the radiation diagram are provided, as shown in Figure 7.7, and factorization of the dependence on azimuth and elevation angles is assumed. Apart from this, the solver can accept any radiation diagram.

7.2.2   Output

The final field intensity map is stored in an output file, and a user-friendly, quantitative field graphical representation in false colors is also displayed, superimposed onto the considered urban scene. An example is shown in Figure 7.8, which features the coverage map corresponding to the scenario presented in Section 7.2.1. In Figure 7.8, the effect of ray propagation is clearly visible.

---

2. This is a text file containing the coordinates of the vertexes and the height of each building in the scene.

Figure 7.6   3-D representation of the input scenario. The red spot identifies the transmitting antenna position.

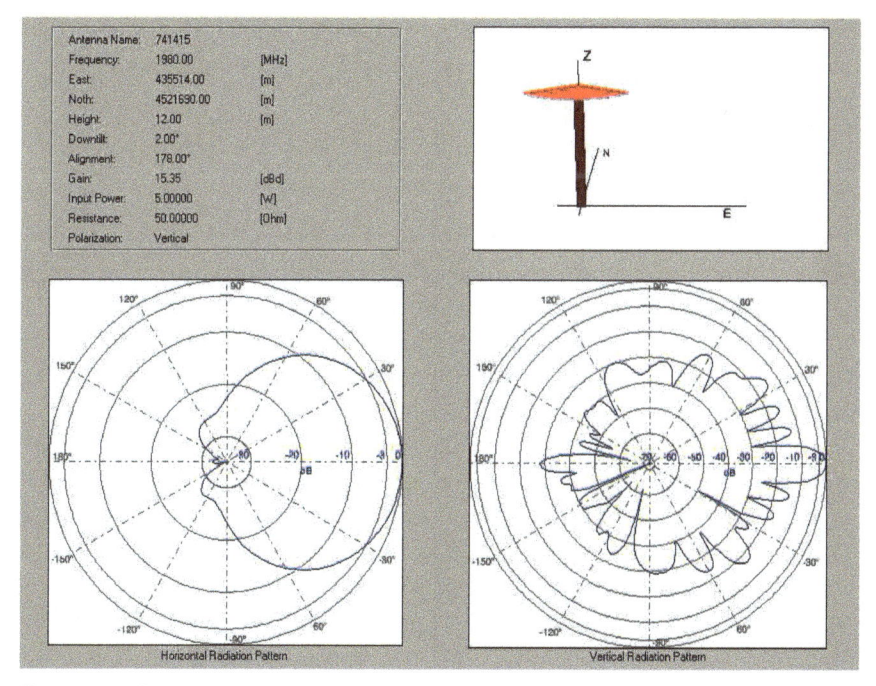

Figure 7.7   Graphical representation of the antenna data, with horizontal and vertical cuts of the antenna pattern.

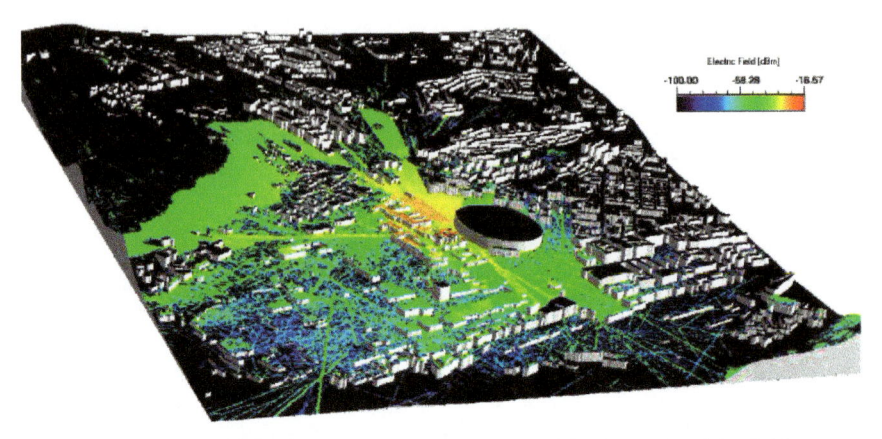

Figure 7.8   Solver output; color scale representation of the intensity of the field predicted by the software.

The field intensity distribution appears to be consistent with the distances from the antenna and the building positions.

Optionally, an rms delay spread map can be produced, and in this case, the user, by clicking on a point of the output map, can analyze the local delay and angular spread, visually displayed by using a "radar-like" representation; see Figure 7.9.

7.2.3   Processing Time

The computational time needed to obtain a simulation depends on building density, scene size, field threshold, anxel width, output grid spacing, and, of course, on the employed hardware. In order to provide significant information on the computational time, some experiments were performed by considering a $1,130 \times 1,050$ m^2 scene as a benchmark. Different building densities were simulated, by considering 10–165 buildings placed in the given area; see Figure 7.10. The selected transmitting antenna is that shown in Figure 7.7 with 5-W input power, located 30m above the ground level. The field level threshold is set considering a receiver sensitivity of −90 dBm. The output grid spacing is set to 5m, and the anxel width is set by choosing $dr = 10$m; (see Figure 7.2(a)). Output regarding delay and angular spread was not computed.

Experiments have been conducted by using a medium- to high-level laptop computer commercially available at time of writing. Processing time varies from about one to about 15 minutes, according to the number of buildings. It scales approximately linearly with this number, but significant deviations

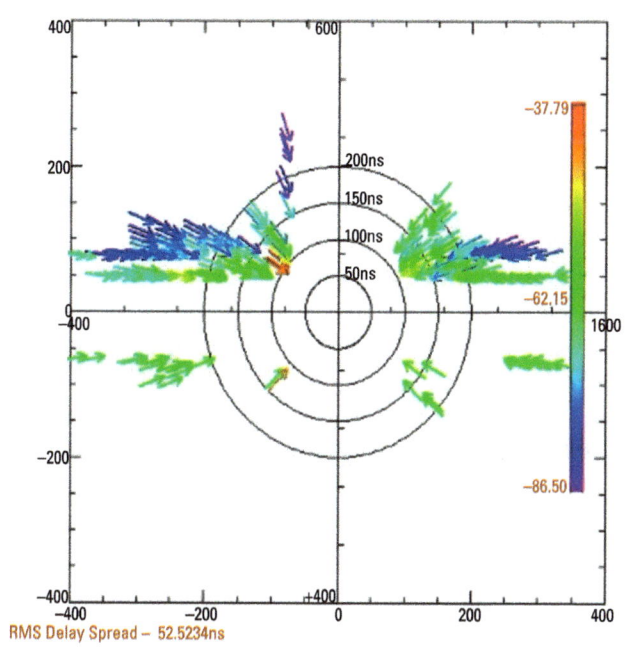

RMS Delay Spread − 52.5234ns

Figure 7.9 Solver output: A "radar-like" representation of the contributions of all the rays to the field in a specific location; the color and the distance of the arrow from the center indicate the field intensity and the delay, respectively.

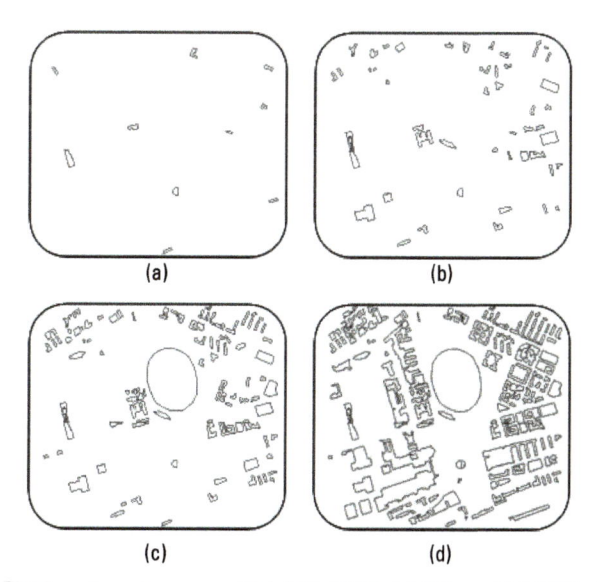

(a)   (b)

(c)   (d)

Figure 7.10 The benchmark scene with (a) 10, (b) 50, (c) 100, and (d) 165 buildings.

from the linear trend may occur, depending on how the building are distributed over the scene.

It must be noted that a reduction of the computational time of about two orders of magnitude is observed if the simulation does not include diffractions. This is particularly useful in actual, very large, scenarios. For instance, quick simulations can be performed in order to have a first look at the overall field distribution, before launching the complete simulation. In addition, there are several applications in which the user is interested only in the highest field levels, so that diffracted contributions can be neglected. For instance, recall that this is the case in applications aimed at verifying that the field intensity is compliant with the exposure limits imposed by the regulations.

Finally, it is useful to note that processing time is, with good approximation, directly proportional to the number of anxels (hence inversely proportional to the anxel width) and to the number of output grid points (hence inversely proportional to the square of grid spacing).

7.2.4 Advantages and Limits

This tool is a powerful instrument for obtaining, in a reasonable time, a quantitative prediction of the field propagated in a specific urban site. Value-added products can also be obtained by the prediction. The binary coverage map, indicating points where the electromagnetic field is higher than a threshold value (for instance, the receiving characteristic of the most common commercial mobile phones), can be quickly evaluated.

In addition, the employed ray launching makes it possible to record all of the physical properties necessary to evaluate the channel response; it is possible to evaluate the rms delay spread, as well as its frequency domain counterpart (coherence bandwidth), according to the user needs.

This tool does not limit the computation of reflections and diffractions to a given number, but it stops the calculation in accordance with a threshold value. This is a physical-based approach that guarantees good computational efficiency and the inclusion of all the contributions relevant for the propagation.

The main limits of the tool are related to the scene description. Actual building walls are not smooth due to balconies, architectonic elements, and other irregularities. In addition, the streets are usually filled with people and vehicles, which are not modeled, as well as vegetation and other random elements.

Determining how the inaccuracies of the scene description influence the final results was the objective of the measurement campaign presented in Section 7.3.

7.3 Measurement Issues

The results of a measurement campaign in an open environment depend on several uncontrollable factors (including weather conditions and the presence of various unpredictable elements that affect the field values). A meaningful comparison between data and predicted values requires the nonambiguous definition of the quantities to be compared.

The simulated field is the superposition of multipath contributions, and its intensity is evaluated as the average value in a resolution cell area. A meaningful comparison requires analogous space filtering of the measured data. The measurement setup that was employed in the considered measurement campaign is composed of instruments able to perform the required average, detailed as follows.

A continuous-wave (CW) generator working at 940 MHz powered a transmitting antenna with 8-dB gain. The antenna was placed on the top of the building marked with the red triangle in Figure 7.6.

A receiver was mounted on a car and plugged to a laptop connected with a GPS system, able to georeference the acquired field levels. The rate of receiver measurement acquisitions was 10^5 samples per second, and its sensitivity was −120 dBm. The GPS system is a portable device, with acquisition time of 1s and nominal precision of 5m. The car traveled in the environs surrounding the antenna, in areas with different building densities. Its velocity was about 10 m/s, so that a large number of received field intensity values could be averaged over each 1-m path of the car. The measurements were taken also in proximity to the San Paolo-Maradona stadium (the big oval building in Figure 7.6), whose presence strongly influences the propagation of the electromagnetic field. Note that the stadium structure is actually rather different from a prism, and it is not a closed structure, as it is instead interpreted by the solver. Therefore, this area is expected to be very challenging for the considered solver.

7.4 Comparison of Solver Predictions and Measured Data

A wide set of measures was acquired during the experimental campaign, in areas with different characteristics. Here we present a meaningful case study that makes it possible to verify the tool performance. The chosen measurement path is presented in Figure 7.11, along with the software predictions.

In order to show matches and discrepancies between prediction and measured data, in Figure 7.12, the measurements (gray line) and the software results (black line) are plotted. Considering the challenging features of the

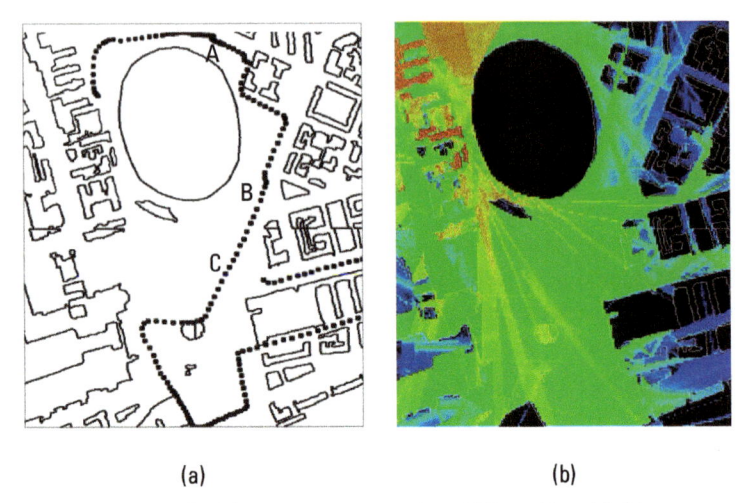

(a)   (b)

Figure 7.11   A case study: (a) measurement path and (b) predicted field map.

considered scene, the comparison shows good matching on average, with the mean difference between measurements and predictions being 2.06 dB, with a rather high standard deviation of 8.20 dB. Correspondingly, the root mean square error (RMSE) is 8.45 dB.

A deeper analysis of Figure 7.12 shows that these mean results are the combination of results coming from areas where the local trend is very different. The reasons for the observed discrepancies are illustrated as follows.

Let us first consider the data between markers A and B of Figures 7.11(a) and 7.12, where the software underestimates the field values. This data is relative to an area where the field is mainly due to the contribution that is transmitted through the stadium, via several apertures in its structure. Such a contribution is not accounted for by the software, where the stadium is described as a full building. Therefore, in the A-B path, the mean difference between measured and predicted values is 7.19 dB with a 4.79-dB standard deviation. By excluding the A-B path from the comparison, we get an RMSE of about 5 dB, with a negligible mean error, which is in agreement with results of most commercially available electromagnetic solvers based on ray-tracing.

Then, in the area between markers B and C and beyond, the simulated data is significantly more oscillating in nature than the measured data. Such an effect is due to the fact that the reflection model considers flat walls; therefore, a discontinuity is created between the area hit by the reflected field and the surrounding pixels. However, in reality, the building walls are not flat, and they scatter the field more diffusely than what is predicted by the software tool.

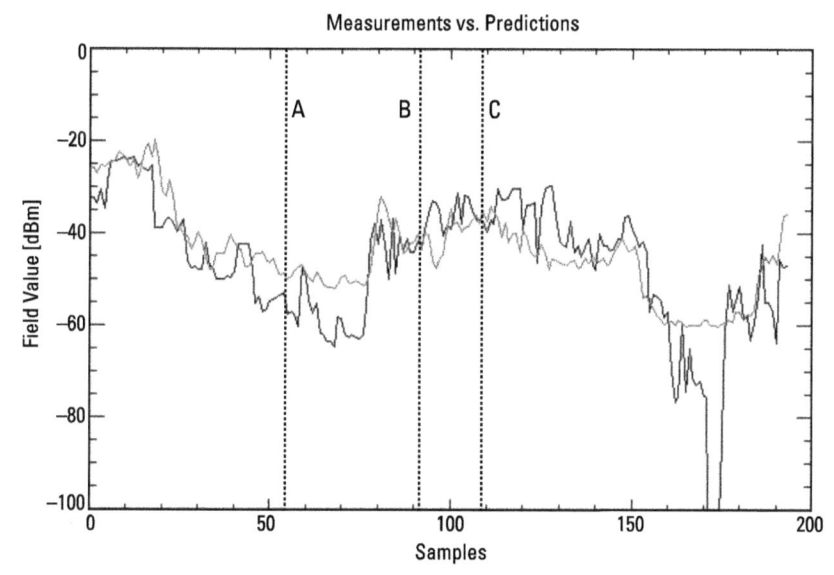

Figure 7.12 Comparison between measured (gray line) and predicted (black line) electric field values. Markers A, B, and C correspond to points A, B, and C of Figure 7.11(a).

The two discussed phenomena shed light on the main limit of the solver, that its performance is strongly influenced by the accuracy level of the scene description. Therefore, to improve software performance, it is necessary to obtain a more precise building description, or, if this is not possible, a proper way to account for diffuse scattering.

Finally, in Figure 7.12 we observe that a prediction value, the sample no. 173, is several decibels lower than the corresponding measured value. Such a phenomenon can be explained by considering that the GPS has an absolute precision on the order of some meters, and when we operate in narrow streets, the field level measured in the street can be erroneously compared with the field level predicted on the building roof. This is very likely what happened for the sample no. 173.

In conclusion, results show that the overall field behavior is well accounted for and that the observed discrepancies can be reliably interpreted in terms of the limitations of the scene description that is used. As already highlighted in Section 6.4, the main limit of the ray-launching tool is the description of the simulation environment from both geometric and electromagnetic points of view. The achieved accuracy is strictly related to the accuracy of the environment description; the more accurate the description, the more reliable the prediction and the lower the RMSE.

References

[1] Rappaport, T. S, et al., "Millimeter Wave Mobile Communications for 5G Cellular: It will Work!," *IEEE Access*, Vol. 1, 2013, pp. 335–349.

[2] Rappaport, T. S., et al., "Overview of Millimeter Wave Communications for Fifth-Generation (5G) Wireless Networks—With a Focus on Propagation Models," *IEEE Trans. Antennas Propag.*, Vol. 65, No. 12, 2017, pp. 6213–6230.

[3] Franceschetti, G., et al., "A Tool for Planning Electromagnetic Field Levels in Urban Areas," *Proc. of the IEEE Antennas and Propagation Society International Symposium*, 2004, pp. 2211–2214.

[4] Rossi, J.-P., et al., "A Ray Launching Method for Radio-Mobile Propagation in Urban Area," *Proc. Antennas Propagat. Soc. Symp.*, 1991, pp. 1540–1543.

[5] Liang, G., and H. L. Bertoni, "A New Approach to 3-D Ray Tracing for Propagation Prediction in Cities," *IEEE Trans. Antennas Propag.*, Vol. 46, No. 6, 1998, pp. 853–863.

[6] Gradshteyn, I. S., and I. M. Ryzhik, *Tables of Integrals, Series, and Products*, Orlando, FL: Academic, 1980.

8

New Propagation Scenarios in 5G Telecommunication Systems[1]

This chapter is devoted to the analysis and discussion of the new propagation features arising in 5G telecommunication systems. To date, only a limited number of 5G services have been released for commercial use. As a matter of fact, the fulfillment of all the pledges associated to 5G is still far from being accomplished and, even if the key technological aspects have been largely fixed, research is still ongoing on advanced setups and deployment scenarios. The new propagation scenarios arising in 5G telecommunication systems are strictly related to the exploitation of new higher-frequency bands for electromagnetic wave propagation. The use of frequencies higher than a few gigahertz, pertaining to the so-called millimeter-wave range, modifies the relative weights of the propagation mechanisms described in Chapters 1–7. For higher frequency, free-space loss increases, rain and atmospheric effects can never be ignored, and edge diffraction becomes increasingly negligible, thus leading to significant shadowing by even small obstacles; finally, scattering becomes more important as the wavelength decreases. In summary, the millimeter-wave range faces a seemingly harsh propagation environment but offers, however, an

---

1. This chapter was written by the authors of this book and Gerardo Di Martino, who is with the University of Naples "Federico II."

unprecedented amount of available bandwidth. Consider, for example, that the licensed frequencies in the 24–27-GHz band and the unlicensed spectrum in the 57–71-GHz band (frequently proposed for 5G use) largely exceed the whole bandwidth used until now in mobile communication networks. Moreover, the rise in frequency can be fruitfully exploited to devise new technologies able not only to solve many of the millimeter-wave propagation problems, but even to provide advanced solutions hardly deployable at lower frequencies [e.g., adaptive beamforming and massive multiple-input and multiple-output (MIMO)].

This chapter provides a synthetic description of 5G networks and their expected performance in Section 8.1, and then introduces the main electromagnetic features relevant for these kinds of networks (i.e., millimeter-wave propagation and channel modeling) in Section 8.2. Finally, Section 8.3 presents the basic concepts of beamforming.

8.1 Description of 5G Networks and Expected Performance

The growing quest for large-bandwidth, low-latency, ultra-reliable, ubiquitous communications has been the driving factor toward the rapid development of mobile telecommunication systems, leading to the succession of new generations with a frequency of about 10 years. As predicted by Cooper's law of spectral efficiency, capacity has increased exponentially over the past decades. Spectral reuse has played a crucial role in dictating this trend, and the improvements obtained in terms of network capacity, with peak throughputs higher than 3 Gbps for 4G, have been mainly achieved through the deployment of effective strategies for bandwidth and network resource management, or effective modulation schemes, rather than through an actual increase in available bandwidth. Over the last two decades, mobile communications have operated in a relatively narrow frequency range, roughly spanning from 800 to 2,600 MHz, as detailed in Table 8.1. Until now, a total bandwidth of less than 2 GHz has been devoted to mobile telecommunication systems, whereas in the millimeter-wave spectrum region, tens of gigahertz are available. The exploitation of these frequencies represents the most important novelty of 5G systems, and it is the enabling basis for the effective deployment of physical layer strategies, such as adaptive beamforming and massive MIMO, whose success is largely based on the ability to devise electrically large (i.e., large with respect to wavelength) antenna arrays.

The fundamental step toward the availability of spectrum resources adequate to meet the goals of 5G is the shift toward higher frequencies, including the millimeter-wave band. However, other lower frequencies are also of

Table 8.1
Operating Frequencies of Mobile Phone System Generations

Mobile Generation	Frequency (MHz)	Year
1G	150–900	1970/1980
2G	800–1,900	1990
3G	800–2,600	2000
4G	700–2,600	2010

interest for use in 5G. In fact, 5G bands are roughly divided into two frequency ranges (FRs): FR1, which includes frequencies lower than 10 GHz, including both bands traditionally used by previous standards and newly allocated ones (e.g., the 700-MHz band previously allocated for TV broadcast systems); and FR2, which includes frequencies higher than 10 GHz, in the range between 25 and 55 GHZ. These two ranges, FR1 and FR2, are considered in the Third-Generation Partnership Project (3GPP) standard [1], but even higher frequencies have been and continue to be under investigation. For example, the large unlicensed portion of the spectrum around 60 GHz has attracted great attention and is also being investigated via meaningful experimental studies. These two frequency ranges are, in turn, divided into three different bands, suitable for different operational scenarios. The low band (frequencies below 1 GHz) and the mid band (1 to 10 GHz) are both part of FR1, and are intended to optimize the trade-off between coverage and bandwidth in different situations. Specifically, coverage-demanding rural scenarios are of interest for low band, whereas crowded urban scenarios are relevant to mid bands. The high band (frequencies above 10 GHz), pertaining to FR2, including the millimeter-wave spectrum, is intended to be used to provide a large amount of bandwidth to meet the demanding expectations on peak data rates in the urban small-cell use case.

The key performance indicators (KPIs) of 5G networks are organized according to the following three application scenarios, as defined by ITU in [2]:

- *Enhanced Mobile Broadband (eMBB):* This usage scenario focuses on applications requiring wide bandwidth and high throughput capabilities to obtain an increasingly seamless user experience. A typical example of these kinds of applications are 3-D and ultra-high-definition real-time streaming.
- *Massive machine-type communications (mMTC):* This usage scenario focuses on the connection of a very large number of devices that

typically transmit a relatively low volume of non-delay-sensitive data. These applications are typical of the IoT, such as smart homes and smart cities.

- *Ultra-reliable and low-latency communications (URLLC):* This usage scenario focuses on applications with stringent constraints on latency and reliability, including self-driving cars and telesurgery.

To provide a graphical display of envisioned 5G applications, the so-called *triangle of 5G* (see Figure 8.1) is frequently used. The vertexes of the triangle represent the eMBB, mMTC, and URLLC scenarios, and the applications are placed in the triangle space according to their proximity to the different usage scenarios. An example of a 5G triangle, inspired by that provided by ITU in [2], is shown in Figure 8.1. Table 8.2 lists the main KPIs, noting their degree of relevance in each usage scenario [2]. The main KPIs are listed as follows:

- Peak data rate [bit/s (bps)]: The maximum achievable data rate under ideal conditions per user/device;
- Peak spectral efficiency (bps/Hz): The maximum data throughput per unit of spectrum resource and per cell;
- User-experienced data rate (bps): The achievable data rate that is available ubiquitously across the coverage area to a mobile user/device;

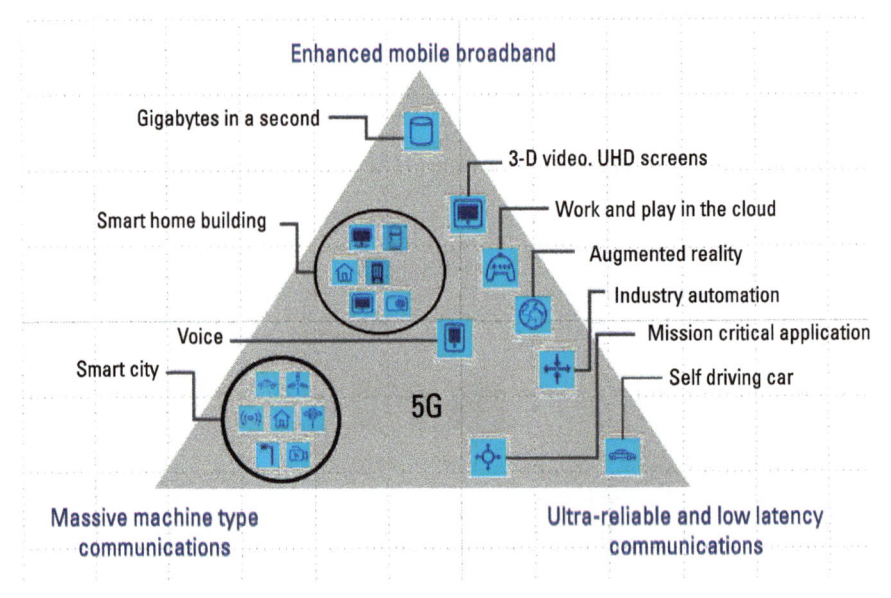

Figure 8.1   Triangle of 5G.

Table 8.2

Main 5G KPI and Usage Scenarios

KPI	Target Value	Degree of Relevance in the Considered Usage Scenario		
		eMBB	mMTC	URLLC
Peak data rate	Downlink: 20 Gbps	High	Low	Low
	Uplink: 10 Gbps			
Peak spectral efficiency	Downlink: 30 bps/Hz	High	Low	Low
	Uplink: 15 bps/Hz			
User experienced data rate	Downlink: 100 Mbps	High	Low	Low
	Uplink: 50 Mbps			
Area traffic capacity	10 Mbps/m^2	High	Low	Low
Latency (user plane)	eMBB: 4 ms	Medium	Low	High
	URLLC: 1 ms			
Connection density	10^6 devices/km^2	Medium	High	Low
Network energy efficiency	No target value specified	High	Medium	Low
Mobility	1.5 bps/Hz at 10 km/h	High	Low	High
	1.12 bps/Hz at 30 km/h			
	0.8 bps/Hz at 120 km/h			
	0.45 bps/Hz at 500 km/h			

- Area traffic capacity (bps/m^2): The total traffic throughput served per geographic area;
- Latency [seconds (s)]: The contribution by the radio network to the time from when the source sends a packet to when the destination receives it;
- Connection density (devices/km^2): The total number of connected and/or accessible devices per unit area;
- Network energy efficiency: The radio interface technologies should support a high sleep ratio and long sleep duration;

- Mobility [m/s]: The maximum speed (declined in four mobility classes) at which a defined spectral efficiency value in the uplink is obtained, and a seamless transfer between radio nodes can be achieved.

For a more complete and precise description of the KPIs, readers can refer to [3]. Moreover, for a detailed report on the scenarios and procedures that must be adopted to assess the compliance to the target values of the KPI interested readers can refer to [4].

Many novel technical paradigms are envisioned in the frame of 5G networks to support compliance with these KPIs, and the deployment of these paradigms is ongoing and nowhere close to the end. As already experimented within previous generation networks, the reduction of cell size is an effective solution for increasing spectral efficiency, since the reduced number of users can each benefit from larger spectral resources. However, in the framework of 5G, small cells are not simply a means for increasing spectral efficiency; in fact, they are a necessary technical solution to support the deployment of services in the millimeter-wave band, and, thereby, cope with the harsh propagation environment, and, more specifically, increase the probability of LOS propagation. In addition, small cells imply a decrease in the power transmitted by the antennas, potentially also increasing the energy efficiency of the overall network.

The analysis of the peculiarities of the propagation scenario in this band, discussed in Section 8.2, is, therefore, of great importance to assess 5G potentialities. The success of millimeter-wave propagation will largely be based on the adoption of appropriate antenna solutions, which can compensate for higher free-space attenuation, which is the topic of Section 8.3.

8.2 Millimeter-Wave Propagation

Propagation of millimeter waves has frequently been believed to be problematic, one of the main reasons being a dramatic increase in the free-space loss for millimeter-wave frequencies. Indeed, looking at the Friis formula in (2.167) one may conclude that the received power decreases with the square of frequency, according to the free-space loss in (2.169). However, this conclusion is based on the assumption that antenna gains are invariant with frequency. This assumption is verified for wire antennas (see Table 2.4) but not for aperture antennas (e.g., paraboloidal reflector antennas, described in the last row of Table 2.4) and antenna arrays, as it emerges from (2.164) and, more in general, from Section 2.5.4. Mobile networks are largely based on the use of antenna arrays, so a reconsideration of the Friis formula is mandatory. In particular, the effective areas of electrically large reflector antennas and planar arrays have the desirable property of being independent of frequency and approximately related to their physical areas. Therefore, by using (2.160), we can rewrite (2.167) as a function of the effective area, thus obtaining

$$P_r = P_T A_T A_R \left(\frac{1}{\lambda r} \right)^2 \tag{8.1}$$

Expressing (8.1) in decibels leads to

$$\left[P_r \right]_{dB} = \left[P_T \right]_{dB} + 10\mathrm{Log}\left(A_T \right) + 10\mathrm{Log}\left(A_R \right) + 20\mathrm{Log}\left(\frac{f}{rc} \right) \tag{8.2}$$

In its new form, the Friis formula shows that the received power increases with the square of frequency, a result that is the opposite of the previous case of wire antennas.

In the case of antenna arrays, keeping the physical size of the array unchanged, when frequency is increased, requires an increase in the number of elements of the array sufficient to meet the $\lambda/2$-spacing condition needed to avoid the appearance of grating lobes; see Section 2.5.4.

In conclusion, *electrically large* arrays employed to support millimeter-wave propagation have large gains that overcompensate for the increased free-space loss. However, since the transmitted beam is narrow, the larger antenna gains call for the development of phased arrays with appropriate electronic beam steering, or adaptive beamforming capabilities, which is the subject of Section 8.3.

If the much-feared increase of the free-space loss is a kind of myth, since its solution just requires appropriate attention in the development of the antennas, other unfavorable conditions affect the propagation of millimeter waves. In Chapter 6 it is shown that diffraction is one of the main propagation mechanisms in a conventional outdoor scenario, especially in the NLOS case (*around-the-corner propagation*). However, as shown in Sections 3.4 and 3.5 [e.g., see (3.80) to (3.82)], the amplitude of the diffraction contribution vanishes as the square root of the frequency. Therefore, the relative weight between LOS and NLOS propagation is significantly changed, with LOS propagation acquiring a dominant role, especially in the 5G high band. Furthermore, blockage effects by even small objects and human bodies present in the propagation environment are also responsible for high levels of fading.

Since millimeter wavelengths become comparable with the typical roughness of building walls, the diffuse scattering contribution cannot be neglected. Construction materials, such as concrete, bricks, and asphalt, present a surface-roughness standard deviation on the order of 1–2 mm (i.e., the roughness becomes comparable with the wavelength of millimeter waves). As discussed in Section 3.6.2, this results in a decrease of the coherent reflected component and an increase of the incoherent component scattered in a potentially wide angular sector around the specular reflection, whose width increases with increasing roughness. The presence of the diffuse component, in turn, may introduce significant signal variations over very short spatial scales, with a level of fading up to 20 dB [5]. Moreover, depolarization in the presence of rough surface scattering may also be a problem for the deployment of polarization diversity techniques [5].

Another important issue is related to atmospheric absorption, both in clear air conditions and in the presence of rain or fog. As discussed in Section 4.4, in the frequency range up to 100 GHz in clear air conditions, the specific attenuations of water and oxygen are always lower than 1 dB/km, except for the absorption peak of oxygen of about 10 dB/km at 60 GHz. However, as Figure 4.7 demonstrated, this peak is sufficiently sharp, so that the band interested by this absorption is less than 10 GHz. In conclusion, in clear air conditions the atmospheric absorption is not an issue in the design of small-cell links over short distances, apart for frequencies around 60 GHz, which are suitable only for very short-range links. Regarding the attenuation due to rain, fog, and snow, the results presented in Section 4.4.1 apply. In particular, the specific attenuation due to rain must be always considered in the network design for frequencies higher than 10 GHz. Indeed, according to (4.29) and Table 4.2, assuming the worst case of a 100-GHz frequency in horizontal polarization and a rain rate of 50 mm/h, a specific attenuation of almost 20 dB/km is obtained. Considerations on the desired fault probability of the link should drive the choice of the appropriate rain rate to be used in (4.29), as discussed in Section 4.4.1. Regarding fog and snow, their effects will be, in general, lower than those of rain, so that if the wireless link has been designed to overcome rain attenuation limits, it will certainly be able to cope with fog and snow attenuation. Obviously, these attenuation factors pose a limit on the range of transmission; however, this should not be dramatic for small-cell applications, also in view of the use of high-gain antennas.

An even more important factor to consider is the high penetration loss in building materials. ITU [6] provides the following heuristic parametric expression for the real part of the relative dielectric constant ε' and for the effective conductivity σ_{eq} of building materials as a function of frequency:

$$\varepsilon' = a f_{GHz}^{b} \tag{8.3}$$

$$\sigma_{eq} = c f_{GHz}^{d} \tag{8.4}$$

where f_{GHz} is the frequency expressed in gigahertz, and the values of the parameters a, b, c, and d for selected building materials are reported in Table 8.3, which is adapted from [6]. The effective conductivity accounts for losses due to both the Joule effect and dielectric hysteresis, and it is equal to $\sigma_{eq} = \sigma + \omega\varepsilon''$, where ε'' is the imaginary part of the dielectric constant, so that the equivalent dielectric constant defined in (2.40) can be written as $\varepsilon_{eq} = \varepsilon' + \sigma_{eq}/(j\omega)$. Equations (8.3) and (8.4) can be used whenever more precise measurement-derived values are not available in the frequency range of interest. The use of these values for building material characterization enables the estimation of the power penetration loss as a function of frequency. This can be accomplished exploiting the framework presented in Section 2.3.6 for the analytical evaluation of the transmission coefficient of a homogeneous wall; indeed, the power penetration loss (L_{PL}) expressed in decibels can be evaluated as

Table 8.3
Values of the Parameters Used in (8.4) and (8.5) for Different Building Materials

Material class	Real Part of Relative Permittivity		Conductivity S/m		Frequency Range
	a	b	c	d	GHz
Concrete	5.24	0	0.0462	0.7822	1–100
Brick	3.91	0	0.0238	0.16	1–40
Plasterboard	2.73	0	0.0085	0.9395	1–100
Wood	1.99	0	0.0047	1.0718	0.001–100
Glass	6.31	0	0.0036	1.3394	0.1–100
Ceiling board	5.79	0	0.0011	1.0750	1–100
Chipboard	1.48	0	0.0217	0.7800	1–100
Plywood	1.52	0	0.33	0	1–40
Marble	2.58	0	0.0055	0.9262	1–60
Floorboard	2.71	0	0.0044	1.3515	50–100
Metal	7.074	0	10^7	0	1–100

$$L_{PL} = -20\mathrm{Log}\big(|T_{AB}|\big) \tag{8.5}$$

where the wall transmission coefficient can be computed according to (2.104) to (2.109). As an example, Figure 8.2 shows the graph of L_{PL} as a function of frequency for a 15-cm-thick concrete wall, assuming normal incidence. The dramatic increase of power penetration loss with frequency represents a clear indication of the poor practicability of outdoor-indoor propagation in the 5G high band. Empirical models for the prediction of penetration loss are also available (e.g., see the O2I model proposed by 3GPP [7]), leading to quite similar results and conclusions.

8.2.1   Empirical Channel Models

The framework for the definition of millimeter-wave empirical propagation models is similar to that outlined for the general case in Section 6.2, and it is based on the development of a parametric analytical expression for propagation loss through measurement and fitting procedures. This section focuses on models for urban microcell and indoor office scenarios, which are especially relevant for 5G high bands. The most common modeling approach is based on the definition of separate LOS and NLOS models, which must be combined with models quantifying the probability (as a function of distance) that a given radio link is in line of sight.

For the outdoor case, when no quantitative description of the propagation environment is available, the ABC models described in Section 6.2.1 can be used, given by (6.3) with the parameters listed in Table 6.1. This model

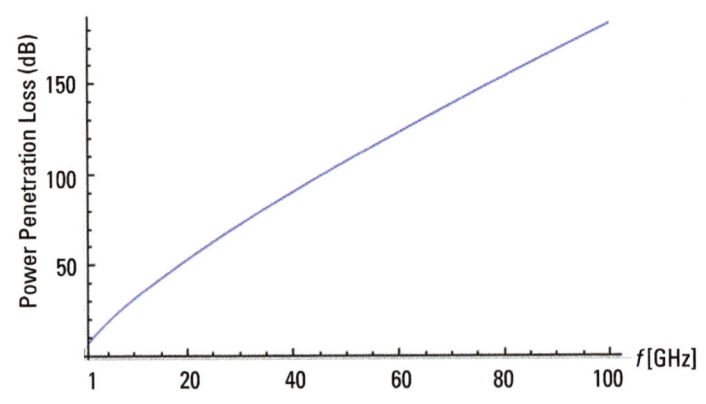

Figure 8.2   Power penetration loss as a function of frequency for a 15-cm-thick concrete wall, assuming normal incidence.

is also valid for the millimeter-wave band, at least up to 82 GHz for LOS propagation and 73 or 82 GHz according to the type of scenario for NLOS propagation (see Table 6.1).

The urban microcell scenario, when both antennas are placed below rooftop level, is of particular interest for 5G small-cell propagation. In this situation, as discussed in Section 6.2.1, in the LOS case, empirical models based on the two-ray model are frequently used. However, for frequencies above about 10 GHz, the breakpoint distance r_{bp} in (6.10) largely exceeds the typical maximum cell radius (~500m), so that the path loss is similar to that of free space. Therefore, in this frequency range, so called "close-in" (CI) path-loss models are commonly recommended for the LOS case. Their general expression is given by

$$L_{CI} = 32.4 + 10n\text{Log}(r) + 20\text{Log}(f) \qquad (8.6)$$

with f measured in gigahertz, and the distance r in meters.

In contrast to the ABC model, the CI model requires the optimization of a single coefficient [i.e., the power loss exponent, n (which may be noninteger)]. For millimeter waves it is recognized that the CI model can describe the path loss with sufficient accuracy, at least for the LOS case [5, 7, 8]. For instance, for the urban microcell street-canyon scenario, 3GPP [7] recommends a CI model with $n = 2.1$. Attenuation due to atmospheric gases and rain (see Section 4.4) must be added to (8.6) to obtain the overall millimeter-wave LOS path loss.

The provided expressions of the path loss should be intended as mean (or median) values. They usually come with indication of the associated shadow-fading standard deviation σ_{SF} in decibels, where shadow fading is modeled as a log-normal random variable. For instance, for the abovementioned 3GPP street-canyon LOS model, $\sigma_{SF} = 4$ dB.

With regard to the NLOS case, the ABC model (6.3) is often employed. In particular, 3GPP recommends use of the following coefficients: $A = 22.4$, $B = 35.3$, $C = 21.3$, and $\sigma_{SF} = 7.82$ [7]. In this model a correction term is also introduced to account for mobile terminals placed at heights greater than 1.5m. The model is applicable for frequencies between 500 MHz and 100 GHz, for distances between 10m and 5 km, for mobile terminals with heights between 1.5 and 22.5m, and for a base station placed at a height of 10m. As an alternative option, 3GPP also provides a CI model with $n = 3.19$ and $\sigma_{SF} = 8.2$. Note that the models by 3GPP have been reported and adopted also by ITU in [4].

In order to appropriately combine LOS and NLOS path-loss models, the approach recommended by 3GPP in [7] and ITU in [4] is based on the definition of a LOS probability. A widespread LOS probability (Pr_{LOS}) model is the so-called d_1/d_2 model, named after the two curve-fitting parameters used in the model, which has the following expression

$$Pr_{\text{LoS}} = \begin{cases} 1 & r \le d_1 \\ \dfrac{d_1}{r} + \exp\left(-\dfrac{r}{d_2}\right)\left(1 - \dfrac{d_1}{r}\right) & r > d_1 \end{cases} \qquad (8.7)$$

In [4, 7] the two parameters are set to $d_1 = 18\text{m}$ and $d_2 = 36\text{m}$. Similar values of the parameters are reported for other similar d_1/d_2 LOS probability models [5].

By moving to the case of indoor propagation, both ABC and CI models are available for millimeter waves. In particular, in [9] ITU recommends the ABC model expressed as (6.3) with the coefficients reported in Table 8.4. As a site-specific model, a CI model expressed as (8.6) is recommended, with the addition of a further attenuation factor used to account for transmitters and receivers located at different floors; however, for millimeter waves the penetration through the floor quickly becomes negligible. Exemplary values of the power loss exponent of the ITU CI model are reported in Table 8.5. Typical values of σ_{SF} for these models are also reported in [9].

As an alternative, 3GPP [7] and ITU [4] propose the use of an ABC model for both the LOS and the NLOS indoor office cases in the frequency range from 500 MHz to 100 GHz and for distances up to 150m: coefficients for the LOS case are $A = 32.4$, $B = 17.3$, $C = 20$, and $\sigma_{SF} = 3$, whereas for the NLOS case $A = 17.3$, $B = 38.3$, $C = 24.9$, and $\sigma_{SF} = 8.03$. An optional CI model is also provided with $n = 3.19$ and $\sigma_{SF} = 8.29$.

Table 8.4
Transmission Loss Coefficients for ITU ABC Indoor Models [9]

Environment	LOS/NLOS	Frequency Range (GHz)	Distance Range (m)	A	B	C	σ
Office	LOS	0.3–83.5	2–27	34.62	14.6	20.3	3.76
	NLOS	0.3–82.0	4–30	29.53	24.6	23.8	5.04
Corridor	LOS	0.3–83.5	2–160	28.12	16.3	22.5	4.07
	NLOS	0.625–83.5	4–94	29.27	27.7	24.8	7.63
Industrial	LOS	0.625–70.28	2–101	24.52	23.1	20.6	2.69
	NLOS	0.625–70.28	5–108	21.01	37.9	13.4	9.05

Table 8.5
Power Loss Exponent 10n for ITU CI Indoor Models [9]

Frequency (GHz)	Office	Commercial	Factory	Corridor
12.65–14.15	–	–	19.5	18.3
			39.3	44.5
25.3–28.3	–	–	19.0	19.2
			37.8	37.7
28	–	17.9	–	–
		24.8		
38	–	18.6	–	–
		25.9		
51–57	15	–	–	13
				16.3
60	–	–	–	16
67–73	19	–	18.3	18.8
			38.8	35.1

Upper and lower values are for LOS and NLOS cases, respectively.

Finally, the model for LOS probability recommended by 3GPP [7] and ITU [4] has the following expression

$$
Pr_{\text{LoS}} = \begin{cases} 1 & r \leq 5\text{m} \\ \exp\left(-\dfrac{r-5}{70.8}\right) & 5\text{m} < r \leq 49\text{m} \\ \exp\left(-\dfrac{r-49}{211.7}\right)\cdot 0.54 & r > 49\text{m} \end{cases} \tag{8.8}
$$

8.2.2 Ray Tracing

When a digital description of the propagation environment is available, ray-tracing algorithms, whose general principles are widely discussed in Section 6.4 and Chapter 7, can be devised. The use of ray tracing has gained new success in applications for 5G and beyond [10]. Indeed, both ITU [4] and 3GPP [7] recommend the use of hybrid map-based models whenever possible, ray tracing being a key enabling step in this kind of approach. Ray-tracing software suitable for the millimeter-wave scenario should be able to account for all the specific propagation aspects mentioned in this section, with some of them calling for additional prior knowledge of the propagation environment. In particular, scattering from building walls requires the characterization of the wall roughness, whereas scattering from small objects can also be evaluated under the assumption of omnidirectional scattering (e.g., Lambertian scattering). Models for atmosphere and rain attenuation should be also included in the software [11]. The introduction of random elements (e.g., random objects, people, vegetation, and cars), along with the intrinsic modeling inaccuracies due to a necessarily nonperfect characterization of the propagation scenario (note that these inaccuracies increase with decreasing wavelength) imply that the results obtained from these algorithms must be always interpreted in a statistical way, as discussed in Section 6.4.

Anyway, the possibility to include accurate and advanced characterizations of the antennas represents an important advantage of ray tracing with respect to the heuristic models described in Section 8.2.1, which are mostly developed as omnidirectional path-loss models, assuming unitary gain antennas. Omnidirectional models are hardly usable in the presence of highly directive antenna systems unless temporal and spatial multipaths are known or modeled. Conversely, ray tracing makes it possible to evaluate multipath parameters, such as delay spread (see Sections 5.2.2 and 5.4) and angular spread (see Section 5.2.1). For this reason, ray-tracing algorithms are frequently used to support the development of appropriate, site-specific, multipath statistical models, standing as the core of hybrid channel modeling approaches [10, 12].

8.2.3 Example: Downlink in a High-Band 5G Wireless System

We assume that a base-station antenna located below rooftop at 10m above the ground plane is serving an urban microcell and that its gain G_T is 19 dB. A typical 5G high-band carrier frequency $f = 28$ GHz is considered, and the transmitted power P_T is 40 dBm. We want to compute the median value of the power received by a user's mobile device placed at a 500-m distance. We assume that the receiving antenna is at 1.5m above the ground and that its gain is close to unity. Since we only know the heights of the antennas, and no other scene parameter, and considering the employed frequency, we can use the empirical models described in Section 8.2.1 to get an estimate of the received power. First, we can use (8.7) to compute the LOS probability, obtaining $P_{T_{LOS}} = 0.036$, which is quite low at this distance. In addition, we can use the CI model of (8.6) [with $n = 2.1$, according to [4, 7] (see Section 8.2.1)] to compute the path loss in LOS, obtaining $L_{LOS} = 118$ dB and $P_R =$

−59 dBm. Finally, for the evaluation of the path loss in NLOS, we can use the ABC model of (6.3) (with $A = 22.4$, $B = 35.3$, $C = 21.3$, according to [4, 7]—see Section 8.2.1), getting $L_{\text{NLOS}} = 148.5$ dB and $P_R = −89.5$ dBm. In this case the Friis formula would lead to a received power of about −56.3 dBm. Note that the typical sensitivity of current receiver mobile devices is −95 to −100 dBm.

The proposed example makes use of the empirical models introduced in Section 8.2.1, which, as already noted, are intrinsically omnidirectional and cannot account for advanced antenna configurations, such as beamforming. Conversely, the approach based on ray tracing, outlined in Section 8.2.2, allows for the inclusion of advanced antenna settings but requires a reasonably accurate knowledge of the propagation scenario. The possibility of including the effects of beamforming in ray-tracing simulations is illustrated in Figure 8.3, where a phased-array antenna, able to synthesize a narrow pencil beam both in the horizontal and vertical planes, is considered [11]. The highly spatial-dependent nature of the path loss can be easily appreciated from Figure 8.3, where the possibility to point the beam in different directions is represented.

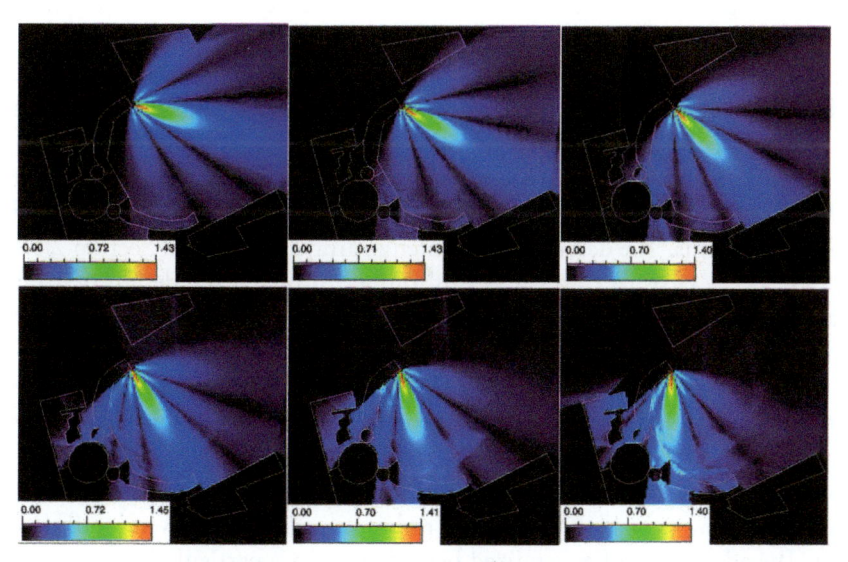

Figure 8.3 Ray-tracing simulated electromagnetic field levels at 1.5m from the ground in the area of Piazza del Plebiscito, Naples, Italy, for the six control beams of a 5G base station operating at 28 GHz and pointing in different directions. The reported values of the amplitude of the electric field are measured in V/m.

8.3 Beamforming

Equation (8.2) shows that whenever planar arrays with fixed areas are used both in transmission and reception, the received power increases as the square of frequency. However, two other meaningful consequences stem from keeping the antenna area fixed as frequency varies: 1) the number of array elements scales up with the square of frequency [assuming here that the spacing between array elements is fixed (e.g., equal to $\lambda/2$ in both dimensions)]; 2) the beamwidth decreases as the inverse of frequency, as can be inferred recalling that the beamwidth in radians is on the order of the ratio of wavelength over the array linear size (see Section 2.5.4). Therefore, the increase in antenna gains with frequency is associated with both a significant increase in the number of antenna elements (and, thus, in the complexity of the excitation network) and a significant decrease of the beamwidth. In practice, instead of keeping the array areas unchanged, it might be convenient to reduce their size to, for example, a factor that guarantees compensation of the free-space loss dependency on frequency (consider the fact that the reduction rate can also be different for the base station and the mobile terminal). In this way, we obtain a less marked reduction of the beamwidth. This also has the advantage of allowing for a certain degree of antenna miniaturization. Moreover, we incidentally note that the use of electrically too large antennas implies a significant increase in the far-field distance according to (2.140), which may impair the application of standard beamforming techniques, relying on the far-field assumption, in actual small-cell propagation scenarios.

In any case, the presence of narrow beams calls for the deployment of appropriate beamforming strategies to adequately pair transmitting and receiving beams. The description of antenna arrays provided in Section 2.5.4 stands as the basis of classical beamforming, involving the design of array element (complex) input currents in order to obtain the desired main beam shape and pointing direction. In addition, more refined, generalized beamforming strategies can be used in a MIMO framework [13] to take advantage of the properties of multipath propagation described in Section 5.2. In this latter context, beamforming consists of the synthesis of overall radiation patterns optimized for complex propagation environments, so that frequently no dominating strong lobe exists, but rather several lobes or several nulls [13, 14]. The advantages of generalized beamforming are particularly relevant in the presence of significant angular spread. Conversely, classical beamforming based on the synthesis of narrow pencil beams aims at filtering out most of the multipath contributions, thus focusing on the direct path. Therefore, pencil beams can be fruitfully used to foster LOS propagation through highly spatially selective links between the base station and the mobile terminal. Since, as discussed in Section 8.2, LOS propagation takes on a fundamental role at millimeter waves, classical beamforming is particularly important in the context of the 5G high band. For this reason, the following focuses on classical beamforming strategies.

The analysis reported in Section 5.2 regarding multipath channel characterization, and, in particular, the expression of the complex attenuation coefficients A_n reported in (5.15), highlights the role of the antennas' effective lengths, and, hence, of the array factors, in dictating the amplitudes of the multipath contributions. Therefore, a pencil-beam array factor able to null (or rather, strongly attenuate) most of the A_n related to NLOS propagation represents an effective way of reducing the number of multipath rays arriving at the receiver, thus reducing both angular spread and delay spread [see (5.32)]. On the other hand, beam-tracking/steering techniques can cope with unexpected link blockage, which can be disruptive for the reliability of the connection.

According to (2.161), the far field radiated by an array is proportional to the product of the array factor and the element factor, with the array factor depending on the normalized input currents a_n. A straightforward way to synthesize beams steered in desired directions is provided by (2.163) for a uniform linear array (ULA). This can be obtained via simple phase-only electronic control of the excitations a_n. This result can be readily extended to uniform planar arrays.

The phase-only approach can be fruitfully used to synthesize a finite set of predefined beams, each transmitting a different signal, which can be simultaneously radiated provided that appropriate orthogonality constraints are enforced between beam excitations and radiation patterns. (Usually, radiation pattern nulls are forced in the positions of the peaks of adjacent beams.) For narrowband signals, the beams can be obtained through analog processing via phase shifters, by properly setting the phase-shift values in (2.163). As illustrated in Figure 8.4, in this case one radiofrequency (RF) chain per beam is necessary. The fully connected architecture shown in Figure 8.4, where each RF chain is connected to all the antenna elements, is able to fully exploit the advantages of the complete array (i.e., high gain and narrow beamwidth). However, this is obtained at the expense of significant combining/dividing losses, so that the use of architectures based on the definition of subarrays, in which each RF chain is connected only to a subset of antenna elements, is frequently preferred, even if this implies a degradation of gain and beamwidth performance.

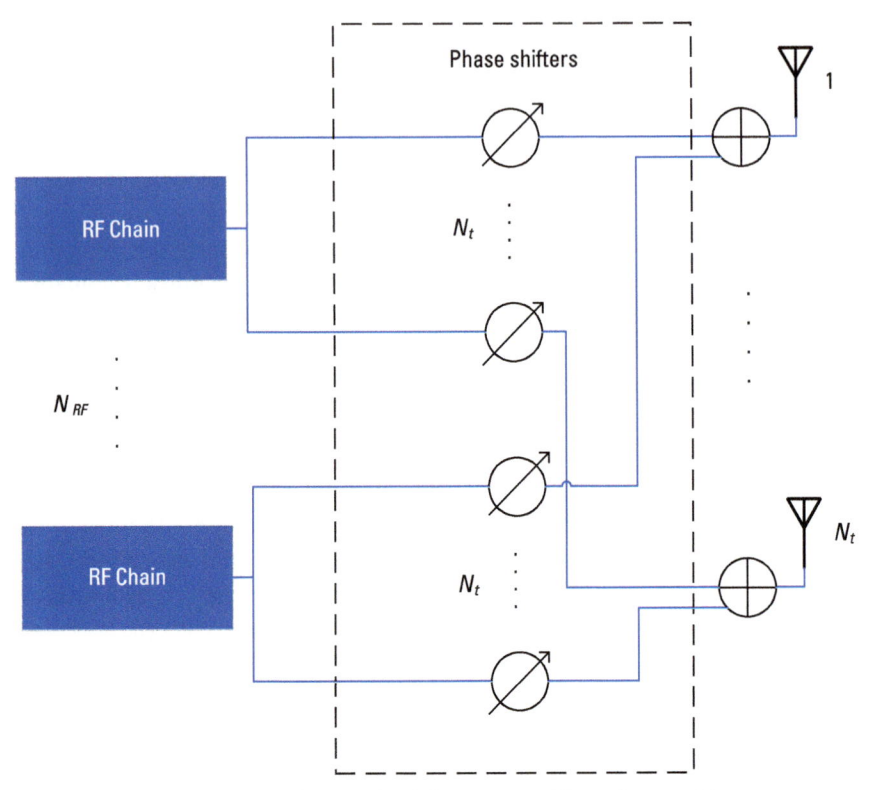

Figure 8.4   Basic structure of an analog fully connected beamformer able to synthesize N_{RF} beams with N_t antennas.

For wideband signals, the use of phase shifters should be avoided because it gives rise to undesired frequency-dependent shifts in beam orientations (*beam squint* effect). In fact, the maximum of the array factor in (2.163) is obtained for a frequency-depending angle ϑ_0 such that $(\omega d\cos\vartheta_0)/c = u_0$. This can be avoided by designing the excitation network in terms of progressive time delays $n\tau_0$, rather than progressive phases nu_0, so that the maximum of the array factor is obtained for a frequency-independent angle ϑ_0 such that $(d\cos\vartheta_0)/c = \tau_0$. This is the basic rationale of the classical *delay-and-sum* beamformer.

To gain increased beamforming flexibility, both amplitude and phase of the excitations must be controlled. Indeed, if the phase mainly controls the orientation of the beams, the amplitude can be used to set the desired trade-off between sidelobe levels and beamwidth. Although amplitude control can be obtained with analog circuitry by using attenuators, their use implies significant losses, which may be hardly manageable.

Compared to analog approaches, digital beamforming solutions provide more flexibility in the synthesis of the excitations and, hence, on the attainable patterns. However, as shown in Figure 8.5, in the digital configuration, the use of one RF chain for each antenna is necessary, implying increased complexity and power consumption. For this reason, digital beamforming is not a viable choice in situations where the number of antenna elements is much larger than the number of beams.

In the 5G high band, analog time-domain beamforming approaches have been the first choice. These kinds of approaches present a small number of radio-frequency chains, thus avoiding the excessive use of expensive and energy-consuming digital-to-analog converters. Even if the flexibility of these architectures can be low if compared to their digital counterpart, it enables the establishment of highly directional links that are suitable for millimeter-wave LOS communications. An essential part of these techniques stands in the beam management procedure, which should provide fast and efficient algorithms able to guarantee correct time-varying beam pointing and beam refinement (i.e., the selection and shaping of the best beam to serve a specific mobile terminal).

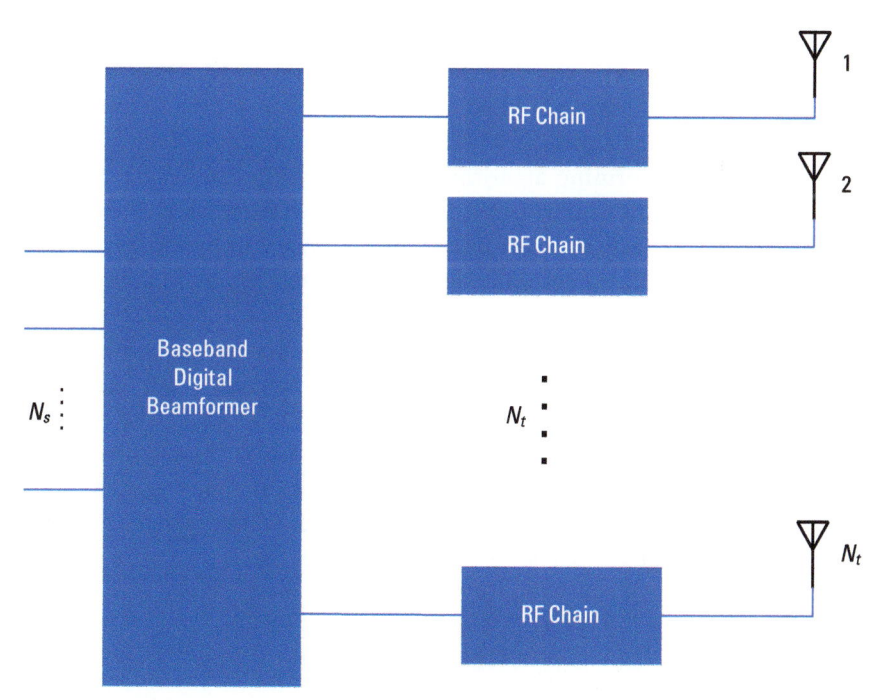

Figure 8.5 Basic structure of a digital beamformer with N_s input signals and N_t antennas.

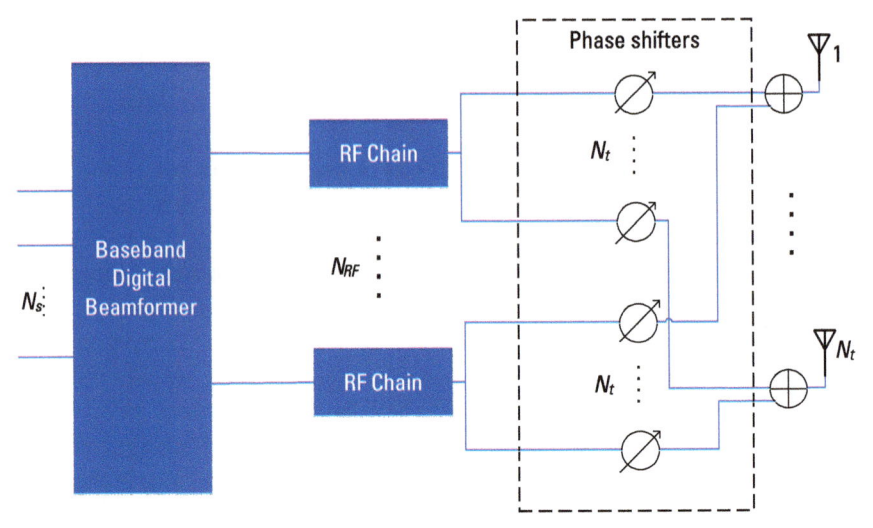

Figure 8.6 Basic structure of a fully connected hybrid beamformer with N_s input signals, N_{RF} RF chains, and N_t antennas.

This can be obtained through the use of synchronization signal blocks (SSBs) [13]. A common strategy relies on the combined use of wide *broadcast beams* and narrow *traffic beams*, whose pointing is refined according to mobile terminal position and channel information. In this context, the increasing need for adaptive beamforming algorithms, able to synthesize flexible patterns based on channel state information in real time, with continuous variations of the excitation coefficients, stimulated the proposal of hybrid beamforming architectures, as a compromise between the flexibility of digital solutions and the affordability and low power consumption of analog ones. Figure 8.6 shows this architecture, in which an analog beamformer is fed with signals that have undergone a previous baseband digital-processing step. Appropriate design stems in this case from the definition of a well-balanced trade-off among beamwidth, throughput, and latency. In particular, the latter parameter (fundamental in 5G, as mentioned in Section 8.1) cannot be neglected, since the higher the complexity of the (possibly adaptive) beamforming algorithm, the higher might be the latency experienced by the user.

References

[1]   *NR; User Equipment (UE) Radio Transmission and Reception,* Technical Specification TS 38.101 Release 17, 3rd Generation Partnership Project, Mar. 2022.

[2] *IMT Vision—Framework and overall objectives of the future development of IMT for 2020 and beyond*, Recommendation ITU-R M.2083-0, International Telecommunication Union, Mobile, radiodetermination, amateur and related satellite services M Series, Sep. 2015.

[3] *Minimum requirements related to technical performance for IMT-2020 radio interface(s)*, Report ITU-R M.2410-0, International Telecommunication Union, Mobile, radiodetermination, amateur and related satellite services M Series, Nov. 2017.

[4] *Guidelines for evaluation of radio interface technologies for IMT-2020*, Report ITU-R M.2412-0, International Telecommunication Union, Mobile, radiodetermination, amateur and related satellite services M Series, Dec. 2017.

[5] Rappaport, T. S., et al., "Overview of Millimeter Wave Communications for Fifth-Generation (5G) Wireless Networks—With a Focus on Propagation Models," *IEEE Transactions on Antennas and Propagation*, Vol. 65, No. 12, Dec. 2017, pp. 6213–6230.

[6] *Effects of building materials and structures on radiowave propagation above about 100 MHz*, Recommendation ITU-R P.2040-2, International Telecommunication Union, Mobile, radiodetermination, amateur and related satellite services M Series, Sep. 2021.

[7] *Study on channel model for frequencies from 0.5 to 100 GHz*, Technical Specification TS 38.901 Release 17, 3rd Generation Partnership Project, Mar. 2022.

[8] *Propagation data and prediction methods for the planning of short-range outdoor radiocommunication systems and radio local area networks in the frequency range 300 MHz to 100 GHz*, document ITU-R P.1411-11, International Telecommunication Union, Radiowave Propagation P Series, Sep. 2021.

[9] *Propagation data and prediction methods for the planning of indoor radiocommunication systems and radio local area networks in the frequency range 300 MHz to 450 GHz*, document ITU-R P.1238-11, International Telecommunication Union, Radiowave Propagation P Series, Sep. 2021.

[10] He, D., et al., "The Design and Applications of High-Performance Ray-Tracing Simulation Platform for 5G and Beyond Wireless Communications: A Tutorial," *IEEE Communications Surveys & Tutorials*, Vol. 21, No. 1, first quarter 2019, pp. 10–27.

[11] Gargiulo, M., et al., "Electromagnetic Propagation Software Tool for Planning and Analyzing 5G Networks," *2021 IEEE 6th International Forum on Research and Technology for Society and Industry (RTSI)*, 2021, pp. 424–428.

[12] Samimi, M. K., and T. S. Rappaport, "3-D Millimeter-Wave Statistical Channel Model for 5G Wireless System Design," *IEEE Transactions on Microwave Theory and Techniques*, Vol. 64, No. 7, July 2016, pp. 2207–2225.

[13] Asplund, H., et al., *Advanced Antenna Systems for 5G Network Deployments: Bridging the Gap Between Theory and Practice*, New York: Academic, 2020.

[14] Van Veen, B. D., and K. M. Buckley, "Beamforming: A Versatile Approach to Spatial Filtering," *IEEE ASSP Magazine*, Vol. 5, No. 2, April 1988, pp. 4–24.

9

Regulations on the Exposure of the General Public to Electromagnetic Fields

This chapter provides an overview of regulations on human exposure to electromagnetic fields. With this in mind, it is useful to recall that radio waves, microwaves, and millimeter waves (as well as infrared radiation and visible light) are nonionizing radiations (see Section 1.2). Accordingly, their interaction with biological tissues mainly produces *thermal effects*; in fact, in the presence of an electromagnetic field, biological tissues, like all lossy media (see Section 2.2), absorb electromagnetic energy, which is transformed into heat. Therefore, exposure to very high field levels can be harmful due to its ability to heat biological tissues rapidly. Other different, *nonthermal effects* may occur (e.g., changes to the permeability of cell membranes and *long-term* effects), whose mechanisms are in general not fully understood. However, at field levels lower than those that would produce significant heating, as of 2022, there is no confirmed evidence of the production of harmful health effects. After analyzing available experimental evidence, on May 31, 2011, the World Health Organization's International Agency for Research on Cancer (IARC) classified radiofrequency fields as "possibly carcinogenic to humans," meaning that "there is *limited evidence* in humans for carcinogenicity of radiofrequency radiation" and that "there is *limited evidence* in experimental animals for carcinogenicity of radiofrequency radiation" [1]. According to the IARC, scientific evidence is *limited* "when the evidence of the effect is related only to some study or when there are unanswered questions related to the adequacy of the design or to the research interpretation. When doubt cannot be removed with reasonable confidence."

A huge number of further studies have taken place since then, but no update of this classification has been considered necessary by IARC. It is useful to note that, for instance, caffeic acid, found in coffee and other common fruits and vegetables, is also classified by IARC as "possibly carcinogenic to humans."

Various international organizations have developed standards and guidelines that recommend safe levels of exposure. Based on these guidelines, many national governments have enforced regulations that limit exposure to electromagnetic fields. In particular, in many countries, including those of the European Union, regulations are based on the exposure guidelines developed by the International Commission on Non-Ionizing Radiation Protection (ICNIRP) [2], which are described in Section 9.1. In the United States, the Federal Communications Commission (FCC) has adopted exposure guidelines partially based on the standard developed by the Institute of Electrical and Electronics Engineers (IEEE) [3]; these are summarized in Section 9.2. Finally, Section 9.3 provides an overview of regulations adopted by different countries around the world.

9.1   ICNIRP Guidelines

ICNIRP guidelines were first published in 1998 [4], and the radio-frequency–related recommendations were recently revised [2] in view of both the significant technological developments and of the considerable body of science gathered since then, further addressing the relation between radiofrequency electromagnetic fields (EMFs) and adverse health outcomes.

The guidelines set different limitations for *occupationally exposed individuals* and *members of the general public*. Occupationally exposed individuals are defined as "adults who are exposed under controlled conditions associated with their occupational duties, trained to be aware of potential radio-frequency EMF risks and to employ appropriate harm-mitigation measures, and who have the sensory and behavioral capacity for such awareness and harm-mitigation response" [2]. Conversely, the definition of the general public is "individuals of all ages and of differing health statuses, which includes more vulnerable groups or individuals, and who may have no knowledge of, or control over, their exposure to EMFs" [2]. Of course, restrictions for the general public are more stringent than those for occupationally exposed individuals. It is specified that, for a pregnant woman, restrictions for the general public must be always applied [2].

These guidelines are intended to provide protection against *scientifically substantiated* short- and long-term adverse health effects of electromagnetic fields. According to the ICNIRP, scientifically substantiated means that "reported adverse effects of radiofrequency EMFs on health need to be independently verified, be of sufficient scientific quality and consistent with current scientific understanding, in order to be taken as *evidence* and used for setting exposure restrictions" [2]. It is here important to distinguish between *biological effects* and *adverse health effects*. The former are measurable changes in cells, tissues, or organs, as a result of a given stress, and they are not necessarily harmful to one's health. The human body can compensate for many types of changes and/or adapt to them, and this is part of normal life in an environment that continuously changes. A biological effect can become an adverse health effect only if the body is stressed for long periods and/or with high intensity, so that compensation is not possible.

Based on the available scientific literature, for each effect, ICNIRP identifies a threshold level, which is the lowest exposure level known to cause the adverse health effect. Proper reduction factors are then applied as a conservative measure to obtain exposure-restriction values, which are referred to as *basic restrictions*. Some of these are enforced on physical quantities inside an exposed body, and therefore they cannot be easily measured. To provide a more practical means of demonstrating compliance with the guidelines, quantities that are more easily evaluated, referred to as *reference levels*, have been derived from the basic restrictions in the worst-case condition, so that compliance to reference levels implies compliance to basic restrictions.

Although ICNIRP guidelines also consider lower frequencies [4], this section illustrates basic restrictions and reference levels for electromagnetic fields in the range from 100 kHz to 300 GHz, which are of interest in the frame of this book. They are determined on the basis of the thermal effects (i.e., they are aimed at avoiding dangerous temperature increases in tissues). In fact, at these frequencies, other effects have not been substantiated, at least at exposure levels that protect against adverse thermal effects.

In Table 9.1, $\mathbf{E}$ and $\mathbf{H}$ are the internal field vectors (i.e., the fields inside the exposed tissue), $\hat{\mathbf{n}}$ is the normal unit vector of the body surface, σ_{eq} is the equivalent conductivity (defined in Section 8.2), $-\varepsilon''_{eq}$ is the imaginary part of the equivalent dielectric constant (defined in Section 2.2.3), and ρ_m is the tissue mass density, measured in kg/m^3. $\mathbf{E}_{inc}$ and $\mathbf{H}_{inc}$ are the electric and magnetic incident fields (i.e., the fields in the absence of the exposed body, namely, the unperturbed fields).

Table 9.1
Physical Quantities Used in the ICNIRP Guidelines

Quantity	Definition	Units
Specific absorption rate	$$SAR = \frac{1}{2}\frac{\sigma_{eq}|\mathbf{E}|^2}{\rho_m} = \frac{1}{2}\frac{\omega\varepsilon''_{eq}|\mathbf{E}|^2}{\rho_m}$$	W/kg
Specific absorption	$$SA = \int_0^t SAR\,d\tau$$	J/kg
Absorbed power density	$$S_{ab} = \mathrm{Re}\left\{\frac{1}{2}\mathbf{E}\times\mathbf{H}^*\cdot\hat{n}\right\}$$	W/m²
Absorbed energy density	$$U_{ab} = \int_0^t S_{ab}\,d\tau$$	J/m²
Incident electric field strength	$$E_{inc} = \frac{1}{\sqrt{2}}|\mathbf{E}_{inc}|$$	V/m
Incident magnetic field strength	$$H_{inc} = \frac{1}{\sqrt{2}}|\mathbf{H}_{inc}|$$	A/m
Incident power density	$$S_{inc} = \frac{1}{2}|\mathbf{E}_{inc}\times\mathbf{H}^*_{inc}|$$	W/m²
Incident energy density	$$U_{inc} = \int_0^t S_{inc}\,d\tau$$	J/m²

9.1.1 Basic Restrictions

Basic restrictions are enforced on the following physical quantities (see also Table 9.1):

- Specific absorption rate (SAR) (i.e., the absorbed power per unit mass, measured in W/kg);
- Specific absorption (SA) (i.e., the absorbed energy per unit mass, measured in joule/kg (J/kg));
- Absorbed power density (S_{ab}) (i.e., the absorbed power per unit area, namely the flux of the real part of the Poynting vector per unit area of the body surface, measured in W/m²);
- Absorbed energy density (U_{ab}) (i.e., the absorbed energy per unit area, or the time integral of S_{ab}, measured in J/m²).

The following basic restrictions are considered by ICNIRP: *whole-body average SAR* (frequency range: 100 kHz–300 GHz), which protects against excessive temperature increase of the body's core temperature; *local SAR* (100 kHz–6 GHz); and *local S_{ab}* (6 GHz–300 GHz), which protects against excessive localized temperature increase; and local SA (400 MHz–6 GHz) and *local U_{ab}* (6–300 GHz), which protect against very short intense localized radiation that may produce fast temperature increase. Whole-body and local basic restrictions must be simultaneously met. They are fully presented and discussed in Sections 9.1.1.1 to 9.1.1.5 and summarized in Tables 9.2 and 9.3.

Table 9.2

Basic Restrictions for Electromagnetic Field Exposure from 100 kHz to 300 GHz, for Averaging Intervals ≥6 min

Exposure Scenario	Frequency Range	Whole-Body Average SAR	Head/Torso Local SAR	Limb Local SAR	Local S_{ab}
Occupational	100 kHz–6 GHz	0.4 W/kg	10 W/kg	20 W/kg	NA
	6 GHz–300 GHz	0.4 W/kg	NA	NA	100 W/m²
General public	100 kHz–6 GHz	0.08 W/kg	2 W/kg	4 W/kg	NA
	6 GHz–300 GHz	0.08 W/kg	NA	NA	20 W/m²

NA = not applicable.

Table 9.3

Basic Restrictions for Electromagnetic Field Exposure from 400 MHz to 300 GHz, for Integrating Intervals <6 min

Exposure Scenario	Frequency Range	Head/Torso Local SA	Limb Local SA	Local U_{ab}
Occupational	400 MHz–6 GHz	$3.6[0.05 + 0.95(t/360)^{0.5}]$ kJ/kg	$7.2[0.025 + 0.975(t/360)^{0.5}]$ kJ/kg	NA
	6 GHz–300 GHz	NA	NA	$36[0.05 + 0.95(t/360)^{0.5}]$ kJ/m²
General public	400 MHz–6 GHz	$0.72[0.05 + 0.95(t/360)^{0.5}]$ kJ/kg	$1.44[0.025 + 0.975(t/360)^{0.5}]$ kJ/kg	NA
	6 GHz–300 GHz	NA	NA	$7.2[0.05 + 0.95(t/360)^{0.5}]$ kJ/m²

NA = not applicable.

9.1.1.1   Whole-Body Average SAR

The adverse health effect threshold level for this basic restriction is assumed to be the level that is able to increase the body core temperature by 1°C. ICNIRP evaluates that this happens for a SAR of 4 W/kg, averaged over the entire body mass and a 30-minute time interval. Reduction factors of 10 and 50 have been applied for occupational exposure and the general public, respectively. Accordingly, the basic restrictions for occupational exposure and for the general public are a whole-body average SAR of 0.4 W/kg and 0.08 W/kg, respectively, both averaged over 30 minutes. These restrictions apply to the entire frequency range from 100 kHz to 300 GHz.

9.1.1.2   Local SAR

To protect against local thermal effects, different restrictions are set for the body head and torso on one side, and for limbs on the other side.

Regarding head and torso, the adverse health effect threshold level for local exposure is assumed to be the one able to increase the local temperature by 2°C or 5°C, according to the tissue type [2]. This happens for a SAR of 20 W/kg, averaged over a 10-gram cubic mass and a six-minute interval. Reduction factors of 2 and 10 have been applied for occupational exposure and the general public, respectively. Accordingly, the basic restrictions for occupational exposure and for the general public are a local (i.e., averaged over a 10-g cubic mass) SAR of 10 W/kg and 2 W/kg, respectively, both averaged over six minutes.

With regard to the limbs, the adverse health effect threshold level for local exposure is assumed to be the one able to increase the local temperature by 5°C. This happens for a SAR of 40 W/kg, averaged over a 10-gram cubic mass and a six-minute interval. Reduction factors of 2 and 10 have been again applied for occupational exposure and the general public, respectively. Accordingly, the basic restrictions for occupational exposure and for the general public are a local SAR of 20 W/kg and 4 W/kg, respectively, both averaged over six minutes.

Local SAR restrictions apply to the frequency range from 100 kHz to 6 GHz. At higher frequencies, the penetration depth of electromagnetic radiation within the human skin tissue (dermis) is limited to less than a few millimeters, so that the SAR average over a 10-g cubic mass is meaningless. In fact, assuming a tissue density on the order of the one of the water (i.e., 1,000 kg/m^3), a 10-g cubic mass corresponds to a cube whose side is about 2.15 cm, much larger than the penetration depth. Therefore, for frequencies higher than 6 GHz, ICNIRP found it more appropriate to enforce a basic restriction for local exposure on the absorbed power density S_{ab}, rather than on the SAR.

9.1.1.3 Local S_{ab}

At frequencies larger than 6 GHz, both for head and torso, and for limbs, the adverse health effect threshold level for local exposure is assumed to be the one able to increase the local temperature by 2°C or 5°C, according to the tissue type [2]. This happens for an absorbed power density S_{ab} of 200 W/m^2, averaged over a square 4-cm^2 surface area of the body and a six-minute interval. Again, reduction factors of 2 and 10 have been applied for occupational exposure and the general public, respectively. Accordingly, the basic restrictions for occupational exposure and for the general public are a local (i.e., averaged over a square 4-cm^2 surface area) S_{ab} of 100 W/m^2 and 20 W/m^2, respectively, both averaged over six minutes.

These local S_{ab} restrictions apply to the frequency range from 6 GHz to 300 GHz.

In addition, to account for focal beam exposure from 30 to 300 GHz (at these small wavelengths very narrow, focused beams can be radiated), S_{ab} averaged over a square 1-cm^2 surface area of the body must not exceed two times the S_{ab} value of the 4-cm^2 basic restrictions for workers, as well as the general public.

9.1.1.4 Local SA

A very short exposure to a radiation of very high intensity may meet the six-minute averaged local SAR restriction, and yet produce a dangerously high cumulative absorbed energy. In order to avoid such an event, ICNIRP sets an SA restriction for exposure intervals of less than six minutes, as a function of time, to keep the temperature rise below the operational adverse health effect threshold. This local (i.e., averaged over any 10-g cubic mass of the body) SA restriction is given by $7.2[0.05 + 0.95(t/360)^{0.5}]$ kJ/kg for head and torso, and $14.4[0.025 + 0.975(t/360)^{0.5}]$ kJ/kg for the limbs, where t is the exposure duration in seconds.

Similar to the local SAR case, reduction factors of 2 and 10 have been applied for occupational exposure and the general public, respectively. Accordingly, the basic restrictions for occupational exposure are a local SA of $3.6[0.05 + 0.95(t/360)^{0.5}]$ kJ/kg for head and torso, and $7.2[0.025 + 0.975(t/360)^{0.5}]$ kJ/kg for the limbs, while for the general public they are a local SA of $0.72[0.05 + 0.95(t/360)^{0.5}]$ kJ/kg for head and torso, and $1.44[0.025 + 0.975(t/360)^{0.5}]$ kJ/kg for the limbs.

Local SA restrictions apply to the frequency range from 400 MHz to 6 GHz. At higher frequencies, as already noted for the local SAR case, the penetration depth is so small that the 10-g average SA is meaningless, and local energy density U_{ab} must be employed instead.

9.1.1.5 Local U_{ab}

For frequencies higher than 6 GHz, to protect against very short intense radiation that may produce fast high-energy absorption, ICNIRP sets a U_{ab} restriction for exposure intervals of less than six minutes as a function of time to keep temperature rise below the operational adverse health effect threshold. This local (i.e., averaged over any square 4-cm^2 surface area of the body) U_{ab} restriction, accounting for the same reduction factors as local SA, is given by $36[0.05 + 0.95(t/360)^{0.5}]$ kJ/m^2 for occupational exposure and $7.2[0.05 + 0.95(t/360)^{0.5}]$ kJ/m^2 for the general public, where t is the exposure duration in seconds, for both head/torso and limbs. Above 30 GHz, to account for focal beam exposure, an additional constraint is imposed, such that U_{ab}, averaged over a square 1-cm^2 surface area of the body, is restricted to $72[0.025 + 0.975(t/360)^{0.5}]$ kJ/m^2 for occupational exposure and $14.4[0.025 + 0.975(t/360)^{0.5}]$ kJ/m^2 for the general public.

9.1.2 Reference Levels

Reference levels are enforced on quantities that are more easily measured than those involved in basic restrictions. These quantities (see Table 9.1) are listed as follows:

- Incident electric field strength E_{inc}; that is, the rms value of the unperturbed (i.e., in the absence of the exposed body) electric field vector, measured in V/m;
- Incident magnetic field strength H_{inc} (i.e., the rms value of the unperturbed magnetic field vector), measured in A/m;
- Incident power density S_{inc} (i.e., the modulus of the unperturbed Poynting vector), measured in W/m^2;
- Incident energy density U_{inc} (i.e., the time integral of S_{inc}), measured in J/m^2.

A combination of numerical simulations and measurements led ICNIRP to conclude that compliance with reference levels ensures that basic restrictions are not exceeded. In most cases, reference levels are more conservative than basic restrictions, to account for the uncertainty of numerical simulations and measurements.

The values of the ICNIRP reference levels are reported in Tables 9.4–9.6. Restrictions are different for 30-minute-average whole-body exposure (Table 9.4), six-minute-average local exposure (Table 9.5), and for short, high-intensity exposure (Table 9.6). Here, *local exposure* means that the peak value of

Table 9.4

Reference Levels for Exposure, Averaged over 30 Minutes and the Whole Body, to Electromagnetic Fields from 100 kHz to 300 GHz

Exposure Scenario	Frequency Range	E_{inc}	H_{inc}	S_{inc}
Occupational	100 kHz–30 MHz	$660/f_M^{0.7}$ V/m	$4.9/f_M$ A/m	NA
	30–400 MHz	61 V/m	0.16 A/m	10 W/m²
	400 MHz–2 GHz	$3\,f_M^{0.5}$ V/m	$0.008\,f_M^{0.5}$ A/m	$f_M/40$ W/m²
	2–300 GHz	NA	NA	50 W/m²
General public	100 kHz–30 MHz	$300/f_M^{0.7}$ V/m	$2.2/f_M$ A/m	NA
	30–400 MHz	27.7 V/m	0.073 A/m	2 W/m²
	400 MHz–2 GHz	$1.375\,f_M^{0.5}$ V/m	$0.0037\,f_M^{0.5}$ A/m	$f_M/200$ W/m²
	2–300 GHz	NA	NA	10 W/m²

f_M is the frequency in megahertz; NA = not applicable.

Table 9.5

Reference Levels for Local Exposure, Averaged over Six Minutes, to Electromagnetic Fields from 100 kHz to 300 GHz

Exposure Scenario	Frequency Range	E_{inc}	H_{inc}	S_{inc}
Occupational	100 kHz–30 MHz	$1504/f_M^{0.7}$ V/m	$10.8/f_M$ A/m	NA
	30–400 MHz	139 V/m	0.36 A/m	50 W/m²
	400 MHz–2 GHz	$10.58\,f_M^{0.43}$ V/m	$0.0274\,f_M^{0.43}$ A/m	$0.29\,f_M^{0.86}$ W/m²
	2–6 GHz	NA	NA	200 W/m²
	6–300 GHz	NA	NA	$275/f_M^{0.177}$ W/m²
General public	100 kHz–30 MHz	$671/f_M^{0.7}$ V/m	$4.9/f_M$ A/m	NA
	30–400 MHz	62 V/m	0.163 A/m	10 W/m²
	400 MHz–2 GHz	$4.72\,f_M^{0.43}$ V/m	$0.0123\,f_M^{0.5}$ A/m	$0.058\,f_M^{0.86}$ W/m²
	2–6 GHz	NA	NA	40 W/m²
	6–300 GHz	NA	NA	$55/f_G^{0.177}$ W/m²

NA = not applicable.

*f_M is the frequency in megahertz; f_G is the frequency in gigahertz.

Table 9.6
Reference Levels for Local Exposure, Integrated over Intervals of Between Zero and Six Minutes, to Electromagnetic Fields from 100 kHz to 300 GHz

Exposure Scenario	Frequency Range	U_{inc}
Occupational	100 kHz–400 MHz	NA
	400 MHz–2 GHz	$0.29 f_M^{0.86} \times 36[0.05 + 0.95(t/360)^{0.5}]$ kJ/m²
	2–6 GHz	$200 \times 36[0.05 + 0.95(t/360)^{0.5}]$ kJ/m²
	6–300 GHz	$275/f_G^{0.177} \times 36[0.05 + 0.95(t/360)^{0.5}]$ kJ/m²
General public	100 kHz–400 MHz	NA
	400 MHz–2 GHz	$0.058 f_M^{0.86} \times 36[0.05 + 0.95(t/360)^{0.5}]$ kJ/m²
	2–6 GHz	$40 \times 36[0.05 + 0.95(t/360)^{0.5}]$ kJ/m²
	6–300 GHz	$55/f_G^{0.177} \times 36[0.05 + 0.95(t/360)^{0.5}]$ kJ/m²

NA = not applicable.

*f_M is the frequency in megahertz; f_G is the frequency in gigahertz.

the relevant quantity over the exposed body must be considered for frequencies smaller than 6 GHz, and its average value over a 4-cm² body surface for higher frequency. The time and space averages of the incident field strength are intended as root mean square averages, specifically,

$$\langle E_{inc} \rangle_t = \sqrt{\frac{1}{t}\int_0^t \frac{1}{2}\left|E_{inc}\right|^2 d\tau}, \quad \langle H_{inc} \rangle_t = \sqrt{\frac{1}{t}\int_0^t \frac{1}{2}\left|H_{inc}\right|^2 d\tau} \quad (9.1)$$

$$\langle E_{inc} \rangle_A = \sqrt{\frac{1}{A}\int_A \frac{1}{2}\left|E_{inc}\right|^2 dS}, \quad \langle H_{inc} \rangle_A = \sqrt{\frac{1}{A}\int_A \frac{1}{2}\left|H_{inc}\right|^2 dS} \quad (9.2)$$

where $<X>_t$ and $<X>_A$ stand for average of X on a time interval of duration t and over a surface of area A, respectively. Conditions for the compliance with reference levels change based on the frequency and on the distance of the exposed body from the radiation source. For frequencies smaller than 30 MHz, ICNIRP requires compliance for both E_{inc} and H_{inc} values.

For frequencies larger than 30 MHz but smaller than 2 GHz, if the exposed body is in the far-field zone [see (2.140)], then ICNIRP requires compliance to be demonstrated for only one of the E_{inc}, H_{inc}, or S_{inc} quantities, since in this case they are strictly related one to each other (see Chapter 2). Conversely, if the exposed body is in the near field, and its distance from the source is larger than one wavelength, ICNIRP requires compliance to be demonstrated for both E_{inc} and H_{inc}, or for S_{inc}. For smaller distances, reference levels are not sufficient to guarantee compliance with basic restrictions, so that directly basic restrictions must be enforced; this is the case for checking compliance for mobile phones.

Finally, for frequencies larger than 2 GHz, in the far field, compliance must be demonstrated only for S_{inc}, whereas in the near field the reference levels are not sufficient to guarantee compliance with basic restrictions, so that directly basic restrictions must be enforced.

General public whole-body reference levels are also shown, in graphical form, in Figure 9.1.

It is interesting to note that reference levels are lower in the frequency range from 30 to 400 MHz; as a matter of fact, in this range of frequency for a given value of the incident field strength a higher SAR is obtained, which can be physically motivated as follows.

A human being penetrated by an electromagnetic wave approximately behaves like a receiving wire antenna. As illustrated in Section 2.5.3, the most efficient wire antenna is the halfwave dipole, whose length is one half of the wavelength. For a 1.80-m-tall person, this *resonance* condition is obtained for a wavelength of 3.60m (i.e., for a frequency of 83.3 MHz). Considering a range of possible heights (including children) from 1 to 2m, we get a range of frequencies from 75 to 150 MHz. For a person standing on a conducting ground, by using the image theorem (see Section 2.4.4), we get that the lower

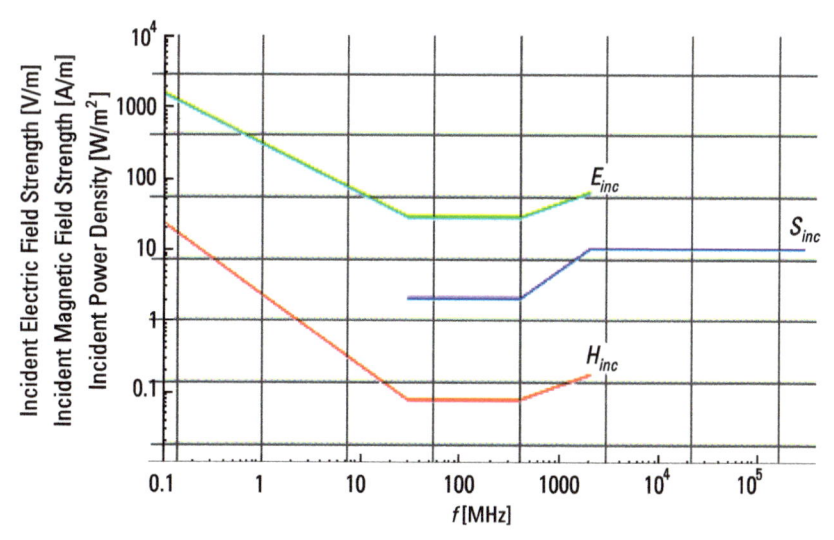

Figure 9.1   ICNIRP general public reference levels for exposure, averaged over 30 minutes and the whole body, to electromagnetic fields from 100 kHz to 300 GHz.

frequency must be halved, thus obtaining 37.5 MHz. In addition, individual limbs may act as receiving wire antennas as well, and, considering limbs as short as about 35 cm, we get the approximately 400-MHz upper limit of the frequency range.

We finally note that if sources at different frequencies are present in the same area, for which, according to Tables 9.4 and 9.5, different reference levels apply, then compliance of reference levels is achieved if

$$\sum_{n=1}^{N}\left(\frac{E_{\text{inc},n}}{E_{L,n}}\right)^2 + \sum_{n=N+1}^{N+M}\left(\frac{S_{\text{inc},n}}{S_{L,n}}\right) \le 1 \qquad (9.3)$$

where N is the number of sources at frequencies smaller than 2 GHz; M is the number of sources at frequencies larger than or equal to 2 GHz; $E_{\text{inc},n}$ and $S_{\text{inc},n}$ are the incident electric field strength and the incident power density of the nth source; and $E_{L,n}$ and $S_{L,n}$ are the reference levels for the incident electric field strength and for the incident power density, at the frequency of the nth source.

9.2 IEEE Standard

The IEEE Standard on human exposure to electromagnetic fields was first issued in 1992 [5], then modified in 2005 [6], and finally revised in 2019 [3], mainly to update the exposure limits for frequencies larger than 6 GHz based on the latest numerical and experimental studies.

The rationale of the IEEE standard is very similar to the one of ICNIRP guidelines, and the recent revised versions of both IEEE and ICNIRP recommendations are more similar to each other than their original issues. However, subtle but important differences are still present.

The IEEE standard sets different limitations for *restricted environments* and *unrestricted environments*. The former are areas where only individuals who follow the applicable safety program guidance and procedures are admitted, while the latter are areas opened to the general public. Note that, at variance with the ICNIRP guidelines, the IEEE standard specifically avoids the declaration that only individuals who are exposed because of their occupation may enter restricted environments [3]. Of course, exposure limits in unrestricted environments are more stringent than those for restricted ones.

The IEEE standard is intended to provide protection against any *established adverse health effect* of electromagnetic fields; that is, in [3] any "effect detrimental to the health of an individual due to exposure to an electric, magnetic, or electromagnetic field, or to induced or contact currents, with the following characteristics:

1. It is supported by the weight of the evidence of that effect in studies published in the scientific literature.
2. The effect has been demonstrated by independent laboratories.
3. There is consensus in the scientific community that the effect occurs for the specified exposure conditions."

Similar to the ICNIRP case, the IEEE identifies, for each effect and on the basis of the available scientific literature, a *lowest observed adverse effect level*, which is the lowest exposure level known to cause the adverse health effect. Proper reduction factors are then applied as a conservative measure, so to obtain exposure restriction values referred to as *dosimetric reference limits* (DRLs), which are the IEEE counterparts of the ICNIRP basic restrictions. DRLs are applied on physical quantities inside the tissue, so that, in order to provide more practical means of demonstrating compliance with the standard, quantities that are more easily evaluated, referred to as *exposure reference levels* (ERLs), have been derived from the DRLs in worst-case conditions. ERLs are the IEEE counterparts of the ICNIRP reference levels.

Although the IEEE standard also considers lower frequencies [3, 5], Sections 9.2.1 and 9.2.2 illustrate DRLs and ERLs for electromagnetic fields in the range from 100 kHz to 300 GHz, which are of interest for this book.

9.2.1 DRLs

DRLs are enforced on SAR and SA, as well as on the *epithelial power density*, which is defined in the same way as the ICNIRP absorbed power density S_{ab}. The following DRLs are considered by IEEE (see Table 9.7): *whole-body average SAR* (frequency range: 100 kHz–6 GHz), which protects against excessive temperature increase of the body core temperature; *local SAR* (100 kHz–6 GHz), and *local* S_{ab} (6 GHz–300 GHz), which protect against excessive localized temperature increase; and local SA (100 kHz–6 GHz), which protects against very short (peaked) intense localized radiation that may produce fast temperature increase. Note that, at variance with the ICNIRP basic restrictions, no DRL is enforced on whole-body average SAR for frequencies higher than 6 GHz, and no DRLs are enforced on absorbed energy density. Similar to the ICNIRP case, SAR is averaged over 30 minutes for whole-body exposure, and over six minutes and a 10-g cube for local exposure. The epithelial power density is averaged over six minutes and over any 4 cm^2 of the body

Table 9.7
DRLs for Electromagnetic Field Exposure from 100 kHz to 300 GHz

Exposure Scenario	Frequency Range	Whole-Body Average SAR	Head/ Torso Local SAR	Limb Local SAR	Local Epithelial Power Density S_{ab}
Restricted environment	100 kHz–6 GHz	0.4 W/kg	10 W/kg	20 W/kg	NA
	6–300 GHz	NA	NA	NA	100 W/m²
Unrestricted environment	100 kHz–6 GHz	0.08 W/kg	2 W/kg	4 W/kg	NA
	6–300 GHz	NA	NA	NA	20 W/m²

*NA stands for not applicable.

surface. In addition, to account for exposures of very small areas of the body in the frequency range from 30 to 300 GHz, it is required that S_{ab} averaged over a square 1-cm² surface area of the body does not exceed two times the 4-cm² DRL.

IEEE standard DRLs for SAR and S_{ab} are reported in Table 9.7. They can be compared with the ICNIRP guidelines' basic restrictions of Table 9.2. With regard to local SA, its value in joules per kilogram, computed over any 100-millisecond time interval, must not exceed 1/5 of the DRL on local SAR multiplied by the average time of six minutes (i.e., 360 seconds). This DRL is applicable to very short, intense pulses (or pulse trains).

9.2.2 ERLs

ERLs are enforced on quantities that are more easily measured than those involved in DRLs (i.e., incident electric and magnetic field strengths E_{inc} and H_{inc}, incident power density S_{inc}, and incident energy density U_{inc}).

The IEEE standard ERLs are reported in Tables 9.8 and 9.9. Restrictions are different for 30-minute-average whole-body exposure (Table 9.8) and six-minute-average local exposure (Table 9.9). The quoted IEEE ERLs of Tables 9.8 and 9.9 can be compared to the ICNIRP reference levels of Tables 9.4 and 9.5.

For frequencies smaller than 100 MHz, for which the ERL on power density is not enforced, electric and magnetic field plane-wave-equivalent power densities S_E and S_H may be used in the place of incident fields, as a convenient comparison with ERLs at higher frequencies:

Table 9.8
ERLs for Exposure, Averaged over 30 Minutes and the Whole Body, to Electromagnetic Fields from 100 kHz to 300 GHz

Exposure Scenario	Frequency Range	E_{inc}	H_{inc}	S_{inc}
Restricted environment	100 kHz–1 MHz	1842 V/m	$16.3/f_M$ A/m	NA
	1 MHz–30 MHz	$1842/f_M$ V/m	$16.3/f_M$ A/m	NA
	30–100 MHz	61.4 V/m	$16.3/f_M$ A/m	NA
	100–400 MHz	61.4 V/m	0.163 A/m	10 W/m²
	400 MHz–2 GHz	NA	NA	$f_M/40$ W/m²
	2–300 GHz	NA	NA	50 W/m²
Unrestricted environment	100 kHz–1.34 MHz	614 V/m	$16.3/f_M$ A/m	NA
	1.34–30 MHz	$823.8/f_M$ V/m	$16.3/f_M$ A/m	NA
	30–100 MHz	27.5 V/m	$158.3/f_M^{1.668}$ A/m	NA
	100–400 MHz	27.5 V/m	0.0719 A/m	2 W/m²
	400 MHz–2 GHz	NA	NA	$f_M/200$ W/m²
	2–300 GHz	NA	NA	10 W/m²

*NA stands for not applicable.
*f_M is the frequency in megahertz.

$$S_E = \frac{E_{inc}^2}{\zeta_0}, \quad S_H = \zeta_0 H_{inc}^2 \qquad (9.4)$$

where ζ_0 is the intrinsic impedance of vacuum, equal to about 377Ω; see Section 2.3.1.

With regard to local incident energy density, its value in joules per square meter, computed over any 100-ms time interval, must not exceed 1/5 of the ERL on local power density multiplied by the average time of six minutes (i.e., 360 seconds). This ERL is applicable to very short, intense pulses (or pulse trains).

For frequencies smaller than 30 MHz or if the exposed body is in near field, compliance to the ERLs of both electric and magnetic fields is required. Conversely, if the exposed body is in the far field and frequency is higher than 30 MHz, it is sufficient to verify the compliance to the ERL of either the electric or the magnetic field.

Table 9.9
ERLs for Local Exposure,* Averaged over Six Minutes, to Electromagnetic Fields from 100 kHz to 300 GHz

Exposure Scenario	Frequency Range	E_{inc}	H_{inc}	S_{inc}
Restricted environment	100 kHz–1 MHz	4119 V/m	$36.4/f_M$ A/m	NA
	1–30 MHz	$4119/f_M$ V/m	$36.4/f_M$ A/m	NA
	30–100 MHz	137.3 V/m	$36.4/f_M$ A/m	NA
	100–400 MHz	$47.3\,f_M^{0.232}$ V/m	$0.125\,f_M^{0.232}$ A/m	$5.93\,f_M^{0.463}$ W/m²
	400 MHz–2 GHz	NA	NA	$5.93\,f_M^{0.463}$ W/m²
	2–6 GHz	NA	NA	200 W/m²
	6–300 GHz	NA	NA	$274.8\,f_G^{-0.177}$ W/m²
Unrestricted environment	100 kHz–1.34 MHz	1373 V/m	$36.4/f_M$ A/m	NA
	1.34–30 MHz	$1842/f_M$ V/m	$36.4/f_M$ A/m	NA
	30–100 MHz	61.4 V/m	$353/f_M^{1.668}$ A/m	NA
	100–400 MHz	$21.2\,f_M^{0.232}$ V/m	$0.0562\,f_M^{0.232}$ A/m	$1.19\,f_M^{0.463}$ W/m²
	400 MHz–2 GHz	NA	NA	$1.19\,f_M^{0.463}$ W/m²
	2–6 GHz	NA	NA	40 W/m²
	6–300 GHz	NA	NA	$55\,f_G^{-0.177}$ W/m²

*The peak value of the relevant quantity over the exposed body for frequencies smaller than 6 GHz and its average value over a 4-cm² body surface for higher frequency.

†NA stands for not applicable.

‡f_M is the frequency in megahertz, and f_G is the frequency in gigahertz.

9.3 Exposure Limits in Countries Across the World

National legislation of most countries in the world include regulations on the exposure to radio-frequency electromagnetic fields. An overview of exposure limits around the world can be found in [9].

With regard to the general public's exposure, on which we focus here, the legislations of countries of most part of Europe, Africa, South America, and the Middle East are mainly based on the ICNIRP guidelines, while the legislation of the United States, Canada, and some countries in Asia are mainly based on the IEEE standard.

The countries of the European Union (EU) follow the EU Council recommendation 1999/519/EC [7], which is based on the ICNIRP guidelines in their original version [4], and at the time of writing is under revision to account for the last version [2]. In particular, recommendation 1999/519/EC adopts the general public basic restrictions of Table 9.2, with the separation frequency set at 10 GHz instead of 6 GHz, and general public reference levels that for frequencies higher than 30 MHz are those of Table 9.4. In their legislation, the member states of the EU adopt these exposure limitations or may set even stricter limits. This is the case in Poland and Greece, which set slightly lower exposure limits, and in Italy, which adopted a regulation that is significantly stricter than the EU recommendation.

In the United States, proposed or existing transmitting facilities, operations, or devices must comply with limits for human exposure to radio-frequency fields adopted by the FCC [8]. The latter are partly based on the IEEE standard [5], and at time of writing they are under revision to account for the last version of the IEEE standard [3].

Due to their differences with respect to ICNIRP guidelines and to the IEEE standard, Sections 9.3.1 and 9.3.2 summarize the exposure limit regulations in Italy and in the United States.

9.3.1 Exposure Limits in Italy

Italian legislation on the exposure of the general public to electromagnetic fields is based on the Law No. 36 of 22-2-2001 "Framework Act 36 on protection against exposure to electric, magnetic and electromagnetic fields" (Official Gazette no. 55 of 7-3-2001), promulgated on February 22, 2001, together with the Actuation Decree, DPCM 8-7-2003 (Official Gazette no. 199 of 28-8-2003). With regard to radio-frequency fields, this law substantially confirms regulations previously enforced by the Ministerial Decree 381 of 10-9-98 (MD 381/98) issued by the Ministry of the Environment, in agreement with the Ministries of Health and Communications, on the "regulations establishing rules for the determination of maximum radiofrequency levels compatible with human health" (Official Gazette no. 257 of 3-11-1998). DPCM 8-7-2003 sets maximum levels of exposure "for protection from short-term effects and possible long-term effects in the population due to exposure to electromagnetic fields generated by fixed sources with frequency between 100 kHz and 300 GHz" (mobile phone base station transmitters, radio-television broadcasting antennas, and radars), while for mobile sources (mobile phone devices and other transmitting mobile devices) ICNIRP basic restrictions are adopted, see Section 9.1.1. Exposure limits enforced by the abovementioned laws are reported in Table 9.10, and they are different for areas where people are expected to stay for less than four hours per day (e.g., streets, gardens, parks, and open rural

Table 9.10
Limits for Exposure to Electromagnetic Fields from 100 kHz to 300 GHz According to
Italian Regulations

Exposure Scenario	Frequency Range	E_{inc}	H_{inc}	S_{inc}
Streets	100 kHz–3 MHz	60 V/m	20 A/m	NA
	3 MHz–3 GHz	20 V/m	0.05 A/m	1 W/m²
	3–300 GHz	40 V/m	0.1 A/m	4 W/m²
Buildings	100 kHz–3 MHz	6 V/m	0.016 A/m	NA
	3 MHz–300 GHz	6 V/m	0.016 A/m	0.1 W/m²

NA = not applicable.

areas) and for all other areas (houses, schools, hospitals, factories, and, more in general, inhabited buildings). The considered quantities must be averaged over the whole body and, according to the MD 381/98, over a time interval of six minutes. Subsequent modifications of the rules set the average time interval to 24 hours.

A comparison of Tables 9.4 and 9.10 shows that exposure limits in inhabited buildings are at least two orders of magnitude lower than ICNIRP reference levels, when expressed in terms of incident power density (and, hence, at least one order of magnitude lower when expressed in terms of incident electric or magnetic field). Note finally that only whole-body exposure is considered, whereas no restriction on local exposure is enforced at the moment (year 2022).

9.3.2 Exposure Limits in the United States

Legislation on the exposure to electromagnetic fields in the United States is based on the FCC guidelines, summarized in [8]. These guidelines set limitations on the EIRP (see Section 2.5.1) of transmitting antennas, according to their intended use. In addition, limits for maximum possible exposure (MPE), partly based on IEEE ERLs, are enforced; they are different for occupational/controlled exposure (equivalent to IEEE restricted area exposure) and the general population/uncontrolled exposure (equivalent to IEEE unrestricted area exposure). FCC MPEs are reported in Table 9.11. The considered quantities must be averaged over the whole body and over a time interval of six minutes for occupational/controlled exposure, and of 30 minutes for general population/uncontrolled exposure.

Table 9.11
FCC MPE to Electromagnetic Fields from 300 kHz to 100 GHz

Exposure Scenario	Frequency Range	E_{inc}	H_{inc}	S_{inc}
Occupational/ controlled exposure	300 kHz–3 MHz	614 V/m	1.63 A/m	NA
	3–30 MHz	$1842/f_M$ V/m	$4.89/f_M$ A/m	NA
	30–300 MHz	61.4 V/m	0.163 A/m	10 W/m²
	300 MHz–1.5 GHz	NA	NA	$f_M/30$ W/m²
	1.5–100 GHz	NA	NA	50 W/m²
General population/ uncontrolled exposure	300 kHz–1.34 MHz	614 V/m	1.63 A/m	NA
	1.34–30 MHz	$824/f_M$ V/m	$2.19/f_M$ A/m	NA
	30–300 MHz	27.5 V/m	0.073 A/m	2 W/m²
	300 MHz–1.5 GHz	NA	NA	$f_M/150$ W/m²
	1.5 GHz–100 GHz	NA	NA	10 W/m²

f_M = frequency in megahertz; NA = not applicable.

FCC guidelines, too, like the Italian legislation, only consider whole-body exposure, whereas no restriction on local exposure is enforced at the moment (year 2022). This will probably change with the spread of millimeter-wave technology, which will allow the use of very narrow beams (see Chapter 8) and make the occurrence of local exposure more likely.

References

[1] World Health Organization's International Agency for Research on Cancer (IARC), *Non-Ionizing Radiation, Part 2: Radiofrequency Electromagnetic Fields*, IARC Monographs on the Evaluation of Carcinogenic Risks to Humans, 2013.

[2] International Commission on Non-Ionizing Radiation Protection (ICNIRP), *ICNIRP Guidelines for Limiting Exposure to Electromagnetic Fields (100 kHz to 300 GHz)*, ICNIRP Publication, 2020.

[3] Institute of Electrical and Electronics Engineers (IEEE), *IEEE Standard for Safety Levels with Respect to Human Exposure to Electric, Magnetic, and Electromagnetic Fields, 0 Hz to 300 GHz*, IEEE Std C95.1-2019.

[4] ICNIRP, "Guidelines for limiting exposure to time-varying electric, magnetic, and electromagnetic fields (up to 300 GHz)," *Health Phys*, Vol. 74, 1998, pp. 494–521.

[5]   IEEE, *IEEE Standard for Safety Levels with Respect to Human Exposure to Radio Frequency Electromagnetic Fields, 3 kHz to 300 GHz*, ANSI/IEEE C95.1-1992.

[6]   IEEE, *IEEE Standard for Safety Levels with Respect to Human Exposure to Electric, Magnetic, and Electromagnetic Fields, 0 Hz to 300 GHz*, IEEE Std C95.1-2005.

[7]   The Council of the European Union, "1999/519/EC: Council Recommendation of 12 July 1999 on the limitation of exposure of the general public to electromagnetic fields (0 Hz to 300 GHz)," *Official Journal of the European Communities*, July 30, 1999, pp. 59–70.

[8]   Federal Communications Commission, Office of Engineering & Technology, "Evaluating Compliance with FCC Guidelines for Human Exposure to Radiofrequency Electromagnetic Fields," *OET Bulletin* 65, Aug. 1997.

[9]   https://www.who.int/data/gho/data/themes/topics/topic-details/GHO/electromagnetic-fields (consulted on November 17, 2022).

10

Conclusion and Future Perspectives

This final chapter is devoted, first of all, to summarizing the main concepts and tools presented in this book.

Chapter 2 detailed the mathematical foundations of electromagnetic wave propagation, presenting basic concepts of electromagnetic fields, namely propagation and radiation. The main practical tools introduced in Chapter 2 were Snell's reflection and refraction laws, Fresnel reflection and transmission coefficients, antennas gain and radiation pattern, the Friis formula, and the concept of receiver noise temperature. In particular, the two last tools allow designing free-space links (i.e., links between very directive antennas placed in LOS).

Chapter 3 discussed asymptotic methods, namely GO, GTD, and UTD; they are the mathematical tools that make it possible to account for the presence of obstacles of various shapes. This is achieved by describing the propagation of electromagnetic waves in terms of rays, which is an accurate approximation when obstacles are much larger than the electromagnetic wavelength. The application of these methods to different scenarios was the subject of Chapters 4–8.

In particular, the two-ray model, presented in Chapter 4, can be used to design the link between two nondirective antennas over a flat or spherical Earth, at the frequencies employed in modern communication systems. Chapter 4 also provided practical tools to compute attenuation by atmospheric gases and by rain; in addition, the chapter covered propagation through the ionosphere, although this is of no direct interest for propagation in the urban scenario.

Chapter 5 described the main features of electromagnetic propagation in complex environments like urban areas. In particular, the concepts of multipath, fading, and delay spread were considered. A statistical approach to deal with fading (i.e., field amplitude random fluctuations) was provided, and the characterization of the delay spread in terms of rms delay spread or, equivalently, coherence bandwidth, was presented.

Chapter 6 served a sort of handbook of available methods to predict field levels in urban areas and, hence, to plan radio coverage in cellular networks. Both empirical and raytracing methods were presented, with their advantages and disadvantages highlighted. Subsequently, Chapter 7 presented an illustrative example of a ray-tracing solver based on the so-called vertical plane launching approach.

Chapter 8 described new scenarios arising from the deployment of last-generation 5G wireless networks. In particular, from the electromagnetic propagation viewpoint, the main novelties introduced by 5G are the use of higher frequencies, up to the millimeter-wave band, and of (analog, digital, or hybrid) beamforming.

Finally, Chapter 9 detailed regulations on human exposure to electromagnetic fields. Both ICNIRP guidelines and the IEEE standard were considered, as well as the regulations in specific countries, namely the United States and Italy. In particular, Chapter 9 also detailed the limits on measurable electromagnetic quantities (incident field, or incident power density) in tables and graphs.

In conclusion, this book has provided concepts and tools useful for engineers and technicians involved in wireless network planning, including information regarding the last-generation 5G wireless networks.

5G networks have not been fully deployed yet but, in spite of that, researchers from all over the world are already trying to envision the main features of the next generation of wireless networks (*beyond-5G*, or *6G*); see for example [1, 2]. These next-generation networks will embody artificial intelligence and, continuing a trend already started with 5G, will be able to fully support *virtual reality*, *augmented reality*, and *extended reality* [1, 2]. From the electromagnetic propagation viewpoint, the main novelties will be the use of higher and higher frequencies, and of *reconfigurable intelligent surfaces* (RISs), also known as *intelligent reflecting surfaces* (IRSs) [2–4]. With regard to frequencies, the whole millimeter-wave band is being explored for potential use, and short-distance links in the terahertz and infrared bands are envisioned [2]. RISs (i.e., reconfigurable material sheets that are able to adaptively modify their radio reflecting characteristics) can be attached to environmental surfaces (e.g., walls, glass, and ceilings) in order to convert parts of the environment into smart reconfigurable reflectors. In other words, engineers will design not only transmitting and receiving devices, but also some objects in the surrounding environment, which will then become a smart radio environment [2].

We believe that, by building on the basis of the concepts presented in this book, readers will be able to face current and future challenges posed by this forthcoming technology.

References

[1] Uusitalo, M. I., et al., "6G Vision, Value, Use Cases and Technologies from European 6G Flagship Project Hexa-X," *IEEE Access*, Vol. 9, pp. 160004–160020, 2021.

[2] Jiang, W., et al., "The Road Towards 6G: A Comprehensive Survey," *IEEE Open Journal of Communications Society*, Vol. 2, 2021, pp. 334–366.

[3] Pan, C., et al., "Reconfigurable Intelligent Surfaces for 6G Systems: Principles, Applications, and Research Directions," *IEEE Communications Magazine*, June 2021, pp. 14–20.

[4] Wu, Q., et al., "Intelligent Reflecting Surface-Aided Wireless Communications: A Tutorial," *IEEE Transactions on Communications*, Vol. 69, No. 5, May 2021, pp. 3313–3351.

Appendix A

Vector Analysis

In the following, $\mathbf{A}$, $\mathbf{B}$, and $\mathbf{C}$ are vector functions, whereas ϕ and ψ are scalar functions.

A.1 Vector Multiplications

$$\mathbf{A} \times \mathbf{B} \cdot \mathbf{C} = \mathbf{B} \times \mathbf{C} \cdot \mathbf{A} = \mathbf{C} \times \mathbf{A} \cdot \mathbf{B} \qquad (A.1)$$

$$\mathbf{A} \times (\mathbf{B} \times \mathbf{C}) = \mathbf{B}(\mathbf{A} \cdot \mathbf{C}) - \mathbf{C}(\mathbf{A} \cdot \mathbf{B}) \qquad (A.2)$$

A.2 Differential Relationships

$$\nabla \cdot (\mathbf{A} \times \mathbf{B}) = \mathbf{B} \cdot (\nabla \times \mathbf{A}) - \mathbf{A} \cdot (\nabla \times \mathbf{B}) \qquad (A.3)$$

$$\nabla \cdot (\phi \mathbf{A}) = \phi \nabla \cdot \mathbf{A} + \mathbf{A} \cdot \nabla \phi \qquad (A.4)$$

$$\nabla \times (\phi \mathbf{A}) = \nabla \phi \times \mathbf{A} + \phi \nabla \times \mathbf{A} \qquad (A.5)$$

$$\nabla \times \nabla \phi = 0 \qquad (A.6)$$

$$\nabla \cdot \nabla \times \mathbf{A} = 0 \qquad (A.7)$$

$$\nabla \cdot \nabla \phi = \nabla^2 \phi \qquad (A.8)$$

$$\nabla \nabla \cdot \mathbf{A} - \nabla \times \nabla \times \mathbf{A} = \nabla^2 \mathbf{A} \qquad (A.9)$$

A.3 Integral Relationships

$$\iiint_V \nabla \cdot A \, dV = \oiint_S A \cdot \hat{n} \, dS \quad \text{(Divergence, or Gauss, theorem, see Figure 2.1(a))}$$

$$(A.10)$$

$$\iint_S \nabla \times A \cdot \hat{n} \, dS = \oint_C A \cdot \hat{c} \, dc \quad \text{(Stokes theorem, see Figure 2.1(b))} \qquad (A.11)$$

A.4 Cartesian Coordinates

Cartesian coordinates are depicted in Figure A.1.

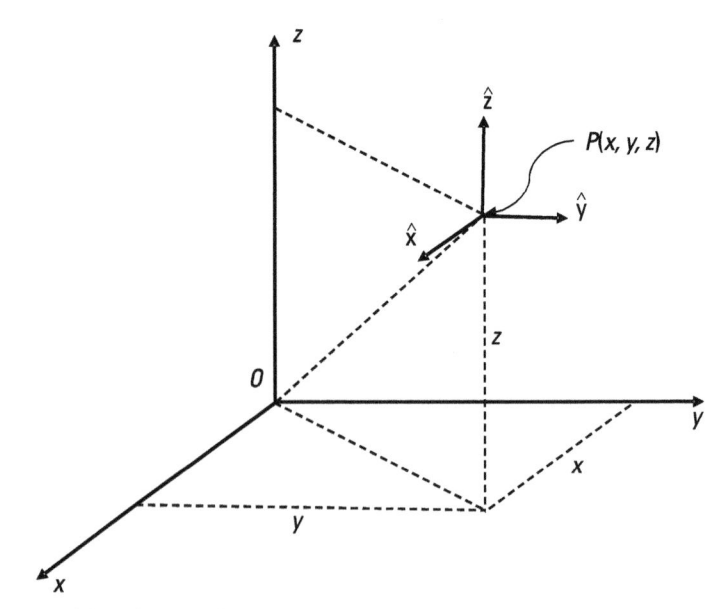

Figure A.1 Cartesian coordinates.

$$\nabla\phi = \frac{\partial\phi}{\partial x}\hat{\mathbf{x}} + \frac{\partial\phi}{\partial y}\hat{\mathbf{y}} + \frac{\partial\phi}{\partial z}\hat{\mathbf{z}} \quad \text{(gradient)} \tag{A.12}$$

$$\nabla\cdot\mathbf{A} = \frac{\partial A_x}{\partial x} + \frac{\partial A_y}{\partial y} + \frac{\partial A_z}{\partial z} \quad \text{(divergence)} \tag{A.13}$$

$$\nabla\times\mathbf{A} = \left(\frac{\partial A_z}{\partial y} - \frac{\partial A_y}{\partial z}\right)\hat{\mathbf{x}} + \left(\frac{\partial A_x}{\partial z} - \frac{\partial A_z}{\partial x}\right)\hat{\mathbf{y}} + \left(\frac{\partial A_y}{\partial x} - \frac{\partial A_x}{\partial y}\right)\hat{\mathbf{z}} \quad \text{(curl)} \tag{A.14}$$

$$\nabla^2\phi = \frac{\partial^2\phi}{\partial x^2} + \frac{\partial^2\phi}{\partial y^2} + \frac{\partial^2\phi}{\partial z^2} \quad \text{(Laplacian)} \tag{A.15}$$

$$\nabla^2\mathbf{A} = \frac{\partial^2\mathbf{A}}{\partial x^2} + \frac{\partial^2\mathbf{A}}{\partial y^2} + \frac{\partial^2\mathbf{A}}{\partial z^2} = \left(\nabla^2 A_x\right)\hat{\mathbf{x}} + \left(\nabla^2 A_y\right)\hat{\mathbf{y}} + \left(\nabla^2 A_z\right)\hat{\mathbf{z}} \quad \text{(vector Laplacian)}$$

$$\tag{A.16}$$

A.5 Cylindrical Coordinates

Cylindrical coordinates are depicted in Figure A.2.

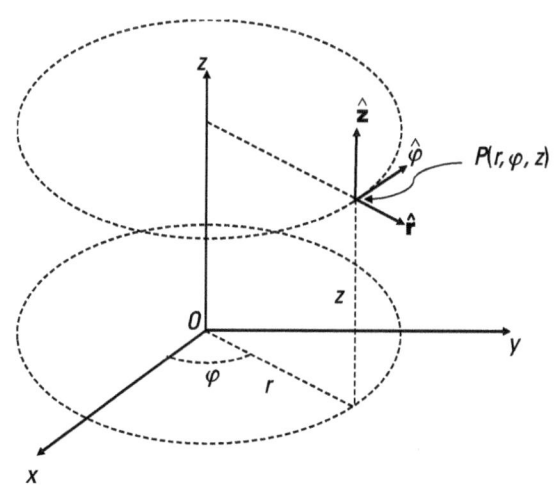

Figure A.2 Cylindrical coordinates.

$$\begin{cases} x = r\cos\varphi \\ y = r\sin\varphi \\ z = z \end{cases} \quad \begin{cases} \hat{\mathbf{r}} = \hat{\mathbf{x}}\cos\varphi + \hat{\mathbf{y}}\sin\varphi \\ \hat{\boldsymbol{\varphi}} = \hat{\mathbf{x}}\sin\varphi + \hat{\mathbf{y}}\cos\varphi \\ \hat{\mathbf{z}} = \hat{\mathbf{z}} \end{cases} \quad \begin{cases} \hat{\mathbf{x}} = \hat{\mathbf{r}}\cos\varphi - \hat{\boldsymbol{\varphi}}\sin\varphi \\ \hat{\mathbf{y}} = \hat{\mathbf{r}}\sin\varphi + \hat{\boldsymbol{\varphi}}\cos\varphi \\ \hat{\mathbf{z}} = \hat{\mathbf{z}} \end{cases} \quad \text{(A.17)}$$

$$\nabla\phi = \frac{\partial\phi}{\partial r}\hat{\mathbf{r}} + \frac{1}{r}\frac{\partial\phi}{\partial\varphi}\hat{\boldsymbol{\varphi}} + \frac{\partial\phi}{\partial z}\hat{\mathbf{z}} \tag{A.18}$$

$$\nabla\cdot\mathbf{A} = \frac{1}{r}\frac{\partial(rA_r)}{\partial r} + \frac{1}{r}\frac{\partial A_\varphi}{\partial\varphi} + \frac{\partial A_z}{\partial z} \tag{A.19}$$

$$\nabla\times\mathbf{A} = \left(\frac{1}{r}\frac{\partial A_z}{\partial\varphi} - \frac{\partial A_\varphi}{\partial z}\right)\hat{\mathbf{r}} + \left(\frac{\partial A_r}{\partial z} - \frac{\partial A_z}{\partial r}\right)\hat{\boldsymbol{\varphi}} + \frac{1}{r}\left[\frac{\partial(rA_\varphi)}{\partial r} - \frac{\partial A_r}{\partial\varphi}\right]\hat{\mathbf{z}} \tag{A.20}$$

$$\nabla^2\phi = \frac{1}{r}\frac{\partial}{\partial r}\left(r\frac{\partial\phi}{\partial r}\right) + \frac{1}{r^2}\frac{\partial^2\phi}{\partial\phi^2} + \frac{\partial^2\phi}{\partial z^2} \tag{A.21}$$

A.6   Spherical Coordinates

Spherical coordinates are depicted in Figure A.3.

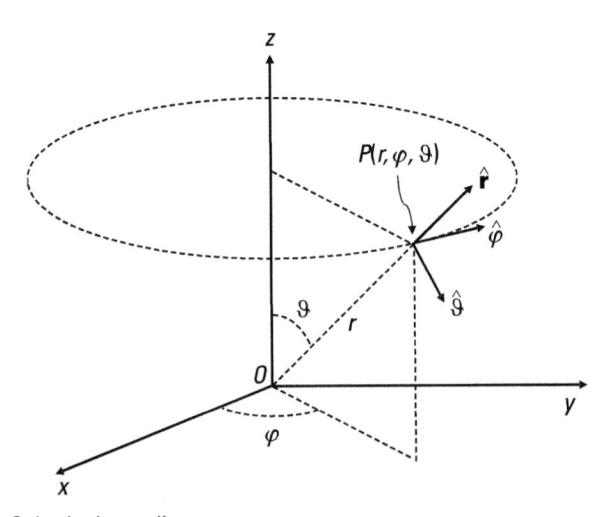

Figure A.3   Spherical coordinates.

$$\begin{cases} x = r\sin\vartheta\cos\varphi \\ y = r\sin\vartheta\sin\varphi \\ z = r\cos\vartheta \end{cases} \tag{A.22}$$

$$\begin{cases} \hat{\mathbf{r}} = \hat{\mathbf{x}}\sin\vartheta\cos\varphi + \hat{\mathbf{y}}\sin\vartheta\sin\varphi + \hat{\mathbf{z}}\sin\vartheta \\ \hat{\boldsymbol{\vartheta}} = \hat{\mathbf{x}}\cos\vartheta\sin\varphi + \hat{\mathbf{y}}\cos\vartheta\cos\varphi - \hat{\mathbf{z}}\sin\vartheta \\ \hat{\boldsymbol{\varphi}} = -\hat{\mathbf{x}}\sin\varphi + \hat{\mathbf{y}}\cos\varphi \end{cases} \quad \begin{cases} \hat{\mathbf{x}} = \hat{\mathbf{r}}\sin\vartheta\cos\varphi + \hat{\boldsymbol{\vartheta}}\cos\vartheta\cos\varphi - \hat{\boldsymbol{\varphi}}\sin\varphi \\ \hat{\mathbf{y}} = \hat{\mathbf{r}}\sin\vartheta\sin\varphi + \hat{\boldsymbol{\vartheta}}\cos\vartheta\sin\varphi + \hat{\boldsymbol{\varphi}}\cos\varphi \\ \hat{\mathbf{z}} = \hat{\mathbf{r}}\cos\vartheta - \hat{\boldsymbol{\vartheta}}\sin\vartheta \end{cases}$$

$$\tag{A.23}$$

$$\nabla\phi = \frac{\partial\phi}{\partial r}\hat{\mathbf{r}} + \frac{1}{r}\frac{\partial\phi}{\partial\vartheta}\hat{\boldsymbol{\vartheta}} + \frac{1}{r\sin\vartheta}\frac{\partial\phi}{\partial\phi}\hat{\boldsymbol{\phi}} \tag{A.24}$$

$$\nabla\cdot\mathbf{A} = \frac{1}{r^2}\frac{\partial\left(r^2 A_r\right)}{\partial r} + \frac{1}{r\sin\vartheta}\frac{\partial\left(\sin\vartheta A_\vartheta\right)}{\partial\vartheta} + \frac{1}{r\sin\vartheta}\frac{\partial A_\varphi}{\partial\varphi} \tag{A.25}$$

$$\nabla\times\mathbf{A} = \frac{1}{r\sin\vartheta}\left[\frac{\partial\left(\sin\vartheta A_\varphi\right)}{\partial\vartheta} - \frac{\partial A_\vartheta}{\partial\varphi}\right]\hat{\mathbf{r}} + \frac{1}{r}\left[\frac{1}{\sin\vartheta}\frac{\partial A_r}{\partial\varphi} - \frac{\partial\left(r A_\varphi\right)}{\partial r}\right]\hat{\boldsymbol{\vartheta}} + \frac{1}{r}\left[\frac{\partial\left(r A_\vartheta\right)}{\partial r} - \frac{\partial A_r}{\partial\vartheta}\right]\hat{\boldsymbol{\varphi}}$$

$$\tag{A.26}$$

$$\nabla^2\phi = \frac{1}{r^2}\frac{\partial}{\partial r}\left(r^2\frac{\partial\phi}{\partial r}\right) + \frac{1}{r^2\sin\vartheta}\frac{\partial}{\partial\vartheta}\left(\sin\vartheta\frac{\partial\phi}{\partial\vartheta}\right) + \frac{1}{r^2\sin^2\vartheta}\frac{\partial^2\phi}{\partial\varphi^2} \tag{A.27}$$

A.7 Matrices

In the following, D, E, and F are matrix functions, whereas **A**, **B**, and **C** are vector functions.

$$D = \begin{pmatrix} D_{xx} & D_{xy} & D_{xz} \\ D_{yx} & D_{yy} & D_{yz} \\ D_{zx} & D_{zy} & D_{zz} \end{pmatrix} \tag{A.28}$$

$$D\mathbf{A} = \mathbf{C}, \text{ with } C_p = D_{px}A_x + D_{py}A_y + D_{pz}A_z \tag{A.29}$$

$$DE = F, \text{ with } F_{pq} = D_{px}E_{xq} + D_{py}E_{yq} + D_{pz}E_{zq} \tag{A.30}$$

Equations (A.29) and (A.30) hold for p and q equal to x, y, and z.

$$I = \begin{pmatrix} 1 & 0 & 0 \\ 0 & 1 & 0 \\ 0 & 0 & 1 \end{pmatrix} \quad \text{(identity matrix)} \tag{A.31}$$

$$DI = ID = D, \quad IA = A \tag{A.32}$$

$$D^{-1}D = I \tag{A.33}$$

$$\text{Det}(D) = D_{xx}\left(D_{yy}D_{zz} - D_{yz}D_{zy}\right) + D_{xy}\left(D_{yz}D_{zx} - D_{yx}D_{zz}\right) + D_{xz}\left(D_{yx}D_{zy} - D_{yy}D_{zx}\right) \tag{A.34}$$

$$\mathbf{AB \cdot C} = \mathbf{DC} \text{ where } D = \begin{pmatrix} A_x B_x & A_x B_y & A_x B_z \\ A_y B_x & A_y B_y & A_y B_z \\ A_z B_x & A_z B_y & A_z B_z \end{pmatrix} \tag{A.35}$$

Appendix B

Dirac Pulse

The *Dirac pulse* (or *Dirac delta*, or *delta function*) $\delta(t)$ can be defined as the limit of a succession of functions that become increasingly peaked in the immediate vicinity of the origin, their integral over the real axis remaining constant and equal to unity (see Figure B.1). In symbols,

$$\delta(t) = \lim_{n \to \infty} g_n(t) \text{ with } g_n(t) \text{ such that } \lim_{n \to \infty} g_n(0) = +\infty \text{ and } \int_{-\infty}^{+\infty} g_n(t)\,dt = 1$$

$$(\text{B.1})$$

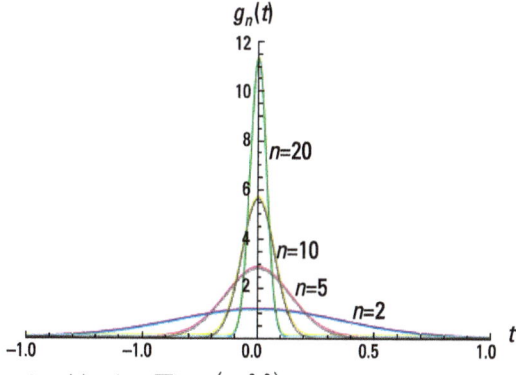

Figure B.1 Plots of $g_n(t) = (n/\sqrt{\pi})\exp(-n^2 t^2)$.

Possible choices for the succession of functions are, for instance,

$$g_n(t) = \frac{n}{\sqrt{\pi}}\exp\left(-n^2t^2\right) \text{ or } g_n(t) = \frac{\sin(nt)}{\pi t} \text{ or } g_n(t) = n\text{rect}(nt) \quad \text{(B.2)}$$

where rect(t) is equal to unity if $|t| < 0.5$ and is zero otherwise.

Strictly speaking, the Dirac pulse is not an ordinary function, but a *generalized function* or *distribution*. Therefore, its rigorous treatment would require the use of the *theory of distributions*. However, in the following, we derive some properties of the Dirac pulse by simply assuming that the limit of the succession of functions in (B.1) exists in some sense.

If $f(t)$ is a continuous function, we have

$$f(t)\delta(t) = \lim_{n\to\infty} f(t)g_n(t) = f(0)\delta(t) \quad \text{(B.3)}$$

and

$$\int_{-\infty}^{+\infty} \delta(t)f(t)\,dt = \lim_{n\to\infty} \int_{-\infty}^{+\infty} g_n(t)f(t)\,dt = f(0)\int_{-\infty}^{+\infty} g_n(t)\,dt = f(0) \quad \text{(B.4)}$$

More in general,

$$f(t)\delta(t-t_0) = \lim_{n\to\infty} f(t)g_n(t-t_0) = f(t_0)\delta(t-t_0) \quad \text{(B.5)}$$

and

$$\int_{-\infty}^{+\infty} \delta(t-t_0)f(t)\,dt = \lim_{n\to\infty} \int_{-\infty}^{+\infty} g_n(t-t_0)f(t)\,dt = f(t_0)\int_{-\infty}^{+\infty} g_n(t-t_0)\,dt = f(t_0)$$

$$\text{(B.6)}$$

which is the *sampling property* (or *sifting property*) of the Dirac pulse. Note that this property also holds if the integral is not extended over the entire real axis, but only over a finite interval including t_0.

From (B.4) with $f(t) = \exp(-j\omega t)$ we get

$$\int_{-\infty}^{+\infty} \delta(t)\exp\left(-j\omega t\right)\,dt = 1 \quad \text{(B.7)}$$

In addition,

$$\delta(t) = \lim_{n \to \infty} \frac{\sin(nt)}{\pi t} = \lim_{n \to \infty} \frac{1}{2\pi} \int_{-n}^{+n} \exp(j\omega t)\, d\omega = \frac{1}{2\pi} \int_{-\infty}^{+\infty} \exp(j\omega t)\, d\omega \qquad \text{(B.8)}$$

Therefore, the (generalized) FT of the Dirac pulse $\delta(t)$ is the constant function $F(\omega) = 1$. Similarly,

$$2\pi\delta(\omega) = \int_{-\infty}^{+\infty} \exp(-j\omega t)\, dt \quad \text{and} \quad \frac{1}{2\pi} \int_{-\infty}^{+\infty} 2\pi\delta(\omega)\exp(j\omega t)\, d\omega = 1 \qquad \text{(B.9)}$$

so that the (generalized) FT of the constant function $f(t) = 1$ is $F(\omega) = 2\pi\delta(\omega)$. In three dimensions,

$$\delta(\mathbf{r}) = \lim_{n \to \infty} g_n(\mathbf{r}) \text{ with } g_n(\mathbf{r}) \text{ such that } \lim_{n \to \infty} g_n(0) = +\infty \text{ and } \iiint_{\mathfrak{R}^3} g_n(\mathbf{r})\, dV = 1$$

$$\text{(B.10)}$$

where $\mathfrak{R}^3$ is the whole tridimensional space. In addition,

$$\iiint_V \delta(\mathbf{r} - \mathbf{r}_0) f(\mathbf{r})\, dV = f(\mathbf{r}_0) \qquad \text{(B.11)}$$

where V is any volume including the point $\mathbf{r}_0$.

Appendix C

Useful Integrals

$$\int_{-\infty}^{\infty} \exp(-ax^2)\,dx = 2\int_{0}^{\infty} \exp(-ax^2)\,dx = \sqrt{\frac{\pi}{a}} \qquad \text{with } a > 0 \qquad (C.1)$$

$$\int_{-\infty}^{\infty} \exp(jax^2)\,dx = 2\int_{0}^{\infty} \exp(jax^2)\,dx = \sqrt{\frac{\pi j}{a}} \qquad \text{with } a \neq 0 \qquad (C.2)$$

$$\int_{-\infty}^{\infty} \exp\left(-\frac{t^2}{2\sigma^2}\right)\exp(-j\omega t)\,dt = \sqrt{2\pi}\,\sigma \exp\left(-\frac{\sigma^2 \omega^2}{2}\right) \qquad (C.3)$$

Appendix D

Derivation of the Transition Function for the Uniform Geometrical Theory of Diffraction

This appendix is devoted to demonstrating the results presented in Section 3.3.2, or in other words, to obtaining an accurate approximate evaluation of the integral

$$I_a(\Omega) = \int_a^\infty f(x)\exp\left[j\Omega q(x)\right]dx \tag{D.1}$$

when the stationary phase point (see Section 3.3) is close to the limit of integration, $x_s \cong a$, so that the results of Section 3.3.1 cannot be used.

Considering that the main contribution to the integral comes from the region around the stationary point and hence, in this case, close to the limit of integration, it is possible to write

$$I_a(\Omega) \cong f(a)\exp\left[j\Omega q(a)\right]\int_a^\infty \exp\left\{j\Omega\left[q'(a)(x-a)+\frac{q''(a)}{2}(x-a)^2\right]\right\}dx$$

$$\tag{D.2}$$

so that $x_s \cong a - q'(a)/q''(a)$, and x_s belongs to the integration domain if $q'(a)/q''(a) < 0$, while it does not belong to the integration domain if $q'(a)/q''(a) > 0$. In the former case we have

$$
\begin{aligned}
I_a(\Omega) &= \int_{-\infty}^{+\infty} f(x)\exp\left[j\Omega q(x)\right]dx - \int_{-\infty}^{a} f(x)\exp\left[j\Omega q(x)\right]dx \\
&\cong f(x_s)\exp\left[j\Omega q(x_s)\right]\sqrt{\frac{2\pi j}{\Omega q''(x_s)}} \\
&\quad - f(a)\exp\left[j\Omega q(a)\right]\int_{-\infty}^{a} \exp\left\{j\Omega\left[q'(a)(x-a)+\frac{q''(a)}{2}(x-a)^2\right]\right\}dx \\
&= f(x_s)\exp\left[j\Omega q(x_s)\right]\sqrt{\frac{2\pi j}{\Omega q''(x_s)}} \\
&\quad - f(a)\exp\left[j\Omega q(a)\right]\int_{0}^{+\infty} \exp\left\{j\Omega\left[-q'(a)x'+\frac{q''(a)}{2}x'^2\right]\right\}dx'
\end{aligned}
$$

(D.3)

in which we have separated the end-point contribution from the stationary phase point one and have set $x' = -(x - a)$; while in the latter case [i.e., for $q'(a)/q''(a) > 0$] we have

$$
\begin{aligned}
I_a(\Omega) &= \int_{a}^{+\infty} f(x)\exp\left[j\Omega q(x)\right]dx \\
&\cong f(a)\exp\left[j\Omega q(a)\right]\int_{a}^{+\infty} \exp\left\{j\Omega\left[q'(a)(x-a)+\frac{q''(a)}{2}(x-a)^2\right]\right\}dx \\
&= f(a)\exp\left[j\Omega q(a)\right]\int_{0}^{+\infty} \exp\left\{j\Omega\left[q'(a)x'+\frac{q''(a)}{2}x'^2\right]\right\}dx'
\end{aligned}
$$

(D.4)

in which the substitution $x' = x - a$ has been made. The end-point contribution can be then expressed as

$$I_a^{\pm}(\Omega) = \pm f(a)\exp\left[j\Omega q(a)\right]\int_0^{+\infty}\exp\left\{j\Omega\frac{q''(a)}{2}\left[2\left|\frac{q'(a)}{q''(a)}\right|x' + x'^2\right]\right\}dx'$$

$$= \pm f(a)\exp\left[j\Omega q(a)\right]\int_0^{+\infty}\exp\left\{j\Omega\frac{q''(a)}{2}\left[\frac{q'(a)^2}{q''(a)^2} + 2\left|\frac{q'(a)}{q''(a)}\right|x' + x'^2 - \frac{q'(a)^2}{q''(a)^2}\right]\right\}dx'$$

$$= \pm f(a)\exp\left[j\Omega q(a)\right]\exp\left\{-j\Omega\frac{q'(a)^2}{2q''(a)}\right\}\int_0^{+\infty}\exp\left\{j\Omega\frac{q''(a)}{2}\left[\left|\frac{q'(a)}{q''(a)}\right| + x'\right]^2\right\}dx'$$

$$\tag{D.5}$$

where the plus or minus sign is selected according to the signum of $q'(a)/q''(a)$, so that (D.5) is equal to (D.4) if $q'(a)/q''(a) > 0$ or to the second term of (D.3) if $q'(a)/q''(a) < 0$. By making the following change of integration variable

$$t = \sqrt{\frac{\Omega|q''(a)|}{2}}\left(\left|\frac{q'(a)}{q''(a)}\right| + x'\right) \quad \text{so that} \quad dx' = \sqrt{\frac{2}{\Omega|q''(a)|}}dt \tag{D.6}$$

and then letting

$$X = \Omega\frac{|q'(a)|^2}{2|q''(a)|} \tag{D.7}$$

and finally multiplying and dividing by $j\Omega q'(a)$, we get

$$I_a^{\pm}(\Omega) = \pm\frac{f(a)}{j\Omega q'(a)}\exp\left[j\Omega q(a)\right]2j\sqrt{X}\exp\{\mp jX\}\int_{\sqrt{X}}^{+\infty}\exp\{\pm jt^2\}dt \tag{D.8}$$

where this time the upper or lower sign must be selected according to the signum of $q''(a)$. We now define the *transition function* $F(X)$ as follows:

$$F(X) = 2j\sqrt{X}\exp\{jX\}\int_{\sqrt{X}}^{+\infty}\exp\{-jt^2\}dt \tag{D.9}$$

Accordingly,

$$I_a^{\pm}(\Omega) = \begin{cases} -\dfrac{f(a)}{j\Omega q'(a)}\exp\left[j\Omega q(a)\right]F(X) & \text{for } q''(a) < 0 \\[3mm] -\dfrac{f(a)}{j\Omega q'(a)}\exp\left[j\Omega q(a)\right]F^*(X) & \text{for } q''(a) > 0 \end{cases} \tag{D.10}$$

The result is then equal to the end-point asymptotic evaluation of (3.44) or (3.47), multiplied by the transition function (or its complex conjugate). The latter can be evaluated numerically and is usually available in the form of look-up table; see Section 3.3.2 and Figure 3.13.

By iteratively integrating by part the integral in (D.9) according to the same procedure as in (3.43), we obtain that $F(X) \to 1$ for $X \to \infty$, so that for x_s far from a and $\Omega \to \infty$ we recover the end-point asymptotic evaluation of (3.44) or (3.46).

Conversely, for $X \to 0$, using (C.2) and (D.7), it turns out that

$$F(X) \sim j\sqrt{j\pi X} = j\sqrt{\frac{j\pi\Omega}{2q''(a)}}|q'(a)| \tag{D.11}$$

so that for $x_s \to a$ we get $I_a^{\pm}(\Omega) = -I_{x_s}(\Omega)$ if $x_s > a$ and $I_a^{\pm}(\Omega) = I_{x_s}(\Omega)$ if $x_s < a$. The overall $I_a(\Omega)$ integral value is then in agreement with (3.47).

About the Authors

Giorgio Franceschetti is a professor emeritus at the University of Naples Federico II, Italy; an honorary professor at the University of Trento, Italy; and a distinguished visiting scientist at the Jet Propulsion Laboratory, NASA. In addition, he holds an honorary PhD in physics from the University of Santiago de Compostela, Spain. He was an adjunct professor at UCLA (1992–2008) and a lecturer at the Top-Tech Master program, Delft University from 1998 to 2010. In addition, he has worked as a visiting professor in Europe, the United States, and Somalia and a lecturer in China and India. Franceschetti is the author of about 200 peer-reviewed journal papers and 14 books and the recipient of several national and international awards, culminating with the 2016 IEEE Electromagnetics Award for leadership in the academic world, teaching, research, and scientific activities in advanced electromagnetics. He is a Life Fellow of the IEEE.

Antonio Iodice is a full professor of electromagnetic fields at the University of Naples Federico II, Italy. In addition, he is the coordinator of the BS and MS degree programs in telecommunications and digital media engineering at the University of Naples Federico II. He has been a principal investigator for research projects on remote sensing and on wireless propagation, and he has authored one book and more than 100 peer-reviewed scientific journal papers. He was the recipient of the 2009 Sergei A. Schelkunoff Transactions Prize Paper Award from the IEEE Antennas and Propagation Society. Iodice is a senior member of the IEEE and the chair of the IEEE Geoscience and Remote Sensing South Italy Chapter.

Daniele Riccio is a full professor of electromagnetic fields at the University of Naples Federico II, Italy. He taught abroad as a lecturer to the PhD program at the Universitat Politecnica de Catalunya, Barcelona, Spain (2006) and at the Czech Technical University, Prague, Czech Republic (2012). He is the coordinator of the PhD Schools in Information and Communication Technology for Health and in Information Technology and Electrical Engineering at the University of Naples Federico II. Riccio has authored three books and more than 100 peer-reviewed scientific journal papers, and he was the recipient of the 2009 Sergei A. Schelkunoff Transactions Prize Paper Award from the IEEE Antennas and Propagation Society. He is a Fellow of the IEEE.

Index

For further information on these and other Artech House titles, including previously considered out-of-print books now available through our In-Print-Forever® (IPF®) program, contact:

Artech House
685 Canton Street
Norwood, MA 02062
Phone: 781-769-9750
Fax: 781-769-6334
e-mail: artech@artechhouse.com

Artech House
16 Sussex Street
London SW1V HRW UK
Phone: +44 (0)20 7596-8750
Fax: +44 (0)20 7630 0166
e-mail: artech-uk@artechhouse.com

Find us on the World Wide Web at: www.artechhouse.com